Rethinking Innateness

A connectionist perspective on development

Neural Network Modeling and Connectionism
Jeffrey L. Elman, Editor

Connectionist Modeling and Brain Function: The Developing Interface
Stephen José Hanson and Carl R. Olson, editors

Neural Network Design and the Complexity of Learning
J. Stephen Judd

Neural Networks for Control
W. Thomas Miller, Richard S. Sutton, and Paul J. Werbos, editors

The Perception of Multiple Objects: A Connectionist Approach
Michael C. Mozer

Neural Computation of Pattern Motion: Modeling Stages of Motion
Analysis in the Primate Visual Cortex
Margaret Euphrasia Sereno

Subsymbolic Natural Language Processing: An Integrated Model of
Scripts, Lexicon, and Memory
Risto Miikkulainen

Analogy-Making as Perception: A Computer Model
Melanie Mitchell

Mechanisms of Implicit Learning: Connectionist Models of Sequence
Processing
Axel Cleeremans

The Human Semantic Potential: Spatial Language and Constrained
Connectionism
Terry Regier

Rethinking Innateness: A Connectionist Perspective on Development
Jeffrey L. Elman, Elizabeth A. Bates, Mark H. Johnson,
Annette Karmiloff-Smith, Domenico Parisi, and Kim Plunkett

Rethinking Innateness

A connectionist perspective on development

Jeffrey L. Elman, Elizabeth A. Bates,
Mark H. Johnson, Annette Karmiloff-Smith,
Domenico Parisi, Kim Plunkett

A Bradford Book
The MIT Press
Cambridge, Massachusetts
London, England

© 1996 Massachusetts Institute of Technology

Printed and bound in the United States of America.

Library of Congress Cataloging-in-Publication Data

Rethinking innateness: a connectionist perspective on development/
    Jeffrey L. Elman . . . [et al.]
        p. cm.—(Neural network modeling and connectionism : X)
    "A Bradford book."
    Includes bibliographical references and index.
    ISBN 0-262-05052-8 (hb : alk. paper)
        1. Nature and nurture. 2. Connectionism. 3. Nativism
(Psychology) I. Elman, Jeffrey L. II. Series: Neural network modeling
and connectionism : 10.
BF341.R35 1996
155.7—dc20   96-15522
                                                       CIP

"This is dedicated to the ones we love."

Contents

Series foreword

The goal of this series, Neural Network Modeling and Connectionism, is to identify and bring to the public the best work in the exciting field of neural network and connectionist modeling. The series includes monographs based on dissertations, extended reports of work by leaders in the field, edited volumes and collections on topics of special interest, major reference works, and undergraduate and graduate-level texts. The field is highly interdisciplinary, and works published in the series will touch on a wide variety of topics ranging from low-level vision to the philosophical foundations of theories of representation.

Jeffrey L. Elman, Editor

Associate Editors:

James Anderson, Brown University
Andrew Barto, University of Massachusetts, Amherst
Gary Dell, University of Illinois
Jerome Feldman, University of California, Berkeley
Stephen Grossberg, Boston University
Stephen Hanson, Princeton University
Geoffrey Hinton, University of Toronto
Michael Jordan, MIT
James McClelland, Carnegie-Mellon University
Domenico Parisi, Instituto di Psicologia del CNR
David Rumelhart, Stanford University
Terrence Sejnowski, The Salk Institute
Paul Smolensky, Johns Hopkins University
Stephen P. Stich, Rutgers University
David Touretzky, Carnegie-Mellon University
David Zipser, University of California, San Diego

Preface

Where does knowledge come from?

That is the central question we pose in this book.

Human children and adults know many things. They know how to speak a language (many people know several). They know how to read maps, and how to go from San Diego to London and Rome. Some of us know how to build cars, and others know how to solve partial differential equations.

Most people feel that knowledge comes from two kinds of sources: What is given us by virtue of our nature, and what we know as a consequence of our nurture. In reality, though, it is far from clear what is meant by either nature or nurture. Nature is usually understood to mean "present in the genotype," and nurture usually means "learned by experience." The difficulty is that when we look at the genome, we don't really see arms or legs (as the preformationists thought we might) and we certainly don't see complex behaviors. As we learn more about the basic mechanisms of genetics, we have come to understand that the distal effects of gene products are highly indirect, complicated, and most often dependent on interactions not only with other gene products but also with external events.

Learning is similarly problematic. We know that learning probably involves changes in synaptic connections, and it is now believed that these changes are effected by the products of specific genes which are expressed only under the conditions which give rise to learning.

The obvious conclusion is that the real answer to the question, *Where does knowledge come from*, is that it comes from the *interaction* between nature and nurture, or what has been called "epigenesis."

Genetic constraints interact with internal and external environmental influences, and they jointly give rise to the phenotype. Unfortunately, as compelling and sensible as this claim seems, it is less a conclusion than a starting point. The problem does not go away, it is simply rephrased. In fact, epigenetic interactions must, if anything, be more complicated than the simpler more static view that x% of behavior comes from genes and y% comes from the environment. For this reason, the interactionist (or constructivist) approach has engendered a certain amount of skepticism on the part of developmentalists. To paraphrase David Klahr (whose complaint was about Piaget's concepts of assimilation and accommodation), nature and nurture are like the Batman and Robin of developmental theory: They hang around waiting in the wings, swoop in and solve a problem, and then disappear before they can be unmasked.

In fact, we believe that the interactionist view is not only the correct one, but that the field is now in a position where we can flesh this approach out in some detail. Our optimism springs from two sources. First, there has been extraordinary progress made in recent years in genetics, embryology, and developmental neuroscience. We are beginning to have an idea of how genes do their job. In addition, much has been discovered about the cortical basis for complex behavior. We also know more now than we did two decades ago about brain development; current evidence suggests a far higher degree of cortical plasticity than was anticipated, and this has obviously far-reaching consequences for theories of development. An impressive array of tools for studying brain processes has been developed, which permit non-invasive access to events in the brain and a spatial and temporal granularity that is quite remarkable.

Second, we have seen in recent years dramatic advances in a framework for computation which is particularly appropriate for understanding neural processing. This framework has been variously called parallel distributed processing, neural network modeling, or connectionism (a term introduced by Donald Hebb in the 1940's, and the name we adopt here). This approach has demonstrated that a great deal of information is latent in the environment and can be extracted using simple but powerful learning rules. But importantly, connectionist models also suggest new ways in which things can be innate. Furthermore, by using connectionist models together with genetic algorithms and artificial life models, it is possible to study within one and the same simulation evolutionary change at the level of populations of neural networks, maturation and learning in individual neural networks, and the interactions between the two.

This book offers our perspective on development, the nature/ nurture controversy, and on the issue of innateness. The definition of innateness itself is controversial. We take the question to be essentially how to account for those behaviors which, given the normal experiences encountered during development, are universal across a species. This is a much broader perspective than many might adopt, but it lets us then ask what are the sources of constraint which lead to these universal outcomes.

We take a connectionist perspective, but we are very aware that ours is a specific and significantly enlarged conception of what connectionism is. In some ways, it is our view of what connectionism *should* (and hopefully, will) be. We are convinced that connectionism has a great deal to offer for understanding development. We also think that connectionists can only profit from the encounter with development. In a very deep sense, we believe that development is not just an accidental path on the way from being small to getting big. Rather, with Piaget, we are convinced that only by understanding the secrets of the process of development will we ever understand complex behaviors and adult forms; but our solution will be somewhat different from Piaget's.

It seems appropriate to say something about how we came to write this book and the process by which it was written. The reader should know that this book is a truly collaborative effort; thus, chapters are unsigned and each reflects our joint efforts. The collaboration began in the late 1980's. At that time, the John D. and Catherine T. MacArthur Foundation awarded a training grant to the Center for Research in Language at UCSD; this was part of the larger MacArthur Network in Transitions from Infancy to Early Childhood, directed by Bob Emde. The basic goal was to introduce developmentalists to the tools and methodology of connectionist modeling. But the training program was unusual in several respects, and reflects the open and innovative approach encouraged by Emde. There was a high degree of flexibility in the program. We were able to bring senior as well as junior fellows to UCSD, and for varying degrees of time, depending on the schedules and goals of the fellows. In some cases, the goal was simple literacy in connectionist modeling. In other cases, fellows developed computer simulations of data they had brought with them. An important component of the program was the set of simulation exercises which were created to illustrate properties of connectionist models that are especially relevant to developmental theory. These simulations have been extended and amplified and form the core of the companion volume to this book.

After several years of the program, a workshop was held at UCSD in 1991. Alumni of the training program returned for a four-day reunion and presented the work that they had done as a result of their participation in the program. This was an extremely exciting event, because it impressed upon us the extent to which connectionism not only provides a very natural computational framework for modeling many developmental phenomena, but also gives us concepts for rethinking some of the old chestnuts. Furthermore, we realized that a critical mass was building; as a group, we felt we had a great deal to say. Thus was born the idea of summarizing this work in book form.

About the same time, the organizers of the Society for Research in Child Development invited several of us to organize a special colloquium on connectionism and development for their 1992 meeting. Betty Stanton of Bradford Books/MIT Press was present at that symposium. Afterwards, she enthusiastically suggested that we turn the symposium contents into a book. Having just decided ourselves that the time was ripe to do this, we were pleased with Betty's excitement and support.

Primarily for logistical reasons, a subset of the original group proceeded to work on the book, with Jeff Elman being chiefly responsible for coordinating the joint efforts. We realized that we had several goals which would require more than one volume.

First, we wanted to make a theoretical statement; this warranted a volume of its own. Second, we very much wanted to make the methodology accessible to as broad an audience as possible; so we designed a second volume with this pedagogical goal in mind. The second volume contains software and simulation exercises which allow the interested reader to replicate many of the simulations we describe in the first volume. The software is general purpose and can also be used by readers to carry out their own simulations. Finally, there is now a considerable body of literature in using connectionism to model development. Although we summarize and refer to much of this literature in the first two volumes, we felt it would be useful to collect some of the best work into a reader, the third volume. The first two volumes will be published almost simultaneously; our goal is for the third volume to appear within 18 months.

Our conception of the first volume has changed dramatically in the course of writing. Our original view was that this volume would bring together chapters written by us as individuals. Our discussions as a group proved so stimulating, however, that we soon moved to a very different model: A truly coauthored volume reflecting our joint ideas. Of course, this meant having to develop these joint ideas! As congenial as our viewpoints were and as great the overlap in our attitudes, we discovered that there were many areas around which we held different opinions, and very many more about which we held no opinions at all. The process of writing the book was thus highly constructive. Our meetings became seminars; the planning of chapter contents turned into lively discussions of theory. We all have found the process of working on this project to be enormously stimulating and we have learned much from each other. If we may be permitted a bit of self-appreciation, we are very grateful to each other for the forbearance, patience, and graciousness which have made it possible to forge a synthesis out of our different perspectives. This book is more than could have been produced by any one of us, and all of us feel that we ourselves have gained from the experience.

We owe a great deal to the MacArthur Foundation for their support. The far-sighted approach of Bob Emde, Mark Appelbaum,

Kathryn Barnard, Marshall Haith, Jerry Kagan, Marion Radke-Yar-
row, and Arnold Sameroff was critical in this effort. Without their
support—both tangible as well as intellectual—we could not have
written this book.

We also wish to make clear that although the cover of this volume bears the names of only six of us, the ideas within the book
reflect an amalgam of insights and findings garnered from a much
larger group. This includes the trainees in the program as well as
other participants in the 1991 workshop: Dick Aslin, Alain Content,
Judith Goodman, Marshall Haith, Roy Higginson, Claes von
Hofsten, Jean Mandler, Michael Maratsos, Brian MacWhinney,
Bruce Pennington, Elena Pizzuto, Rob Roberts, Jim Russell, Richard
Schwartz, Joan Stiles, David Swinney, and Richard Wagner. The
trainers in the program, Virginia Marchman, Mary Hare, Arshavir
Blackwell, and Cathy Harris, were more than trainers. They were
colleagues and collaborators, and they played a pivotal role in the
program and in our thinking. We are also grateful to a number of
colleagues with whom we have interacted over the years. Some
may still not agree with our arguments, while others will undoubtedly recognize some of their own ideas in the pages that follow. We
owe to Jay McClelland the opening sentence of this Preface. We are
grateful to these friends and hope that our translation of their ideas
will not displease them.

In addition, we wish to thank those who read and commented
on various sections of this book. Dorothy Bishop, Gergely Cisbra,
Terry Deacon, Lucy Hadden, Francesca Happé, Henry Kennedy,
Herb Killackey, Jean Mandler, Jay Moody, Yuko Munakata, Andrew
Oliver, Adolfo Perinat, Paul Rodriguez, Marty Sereno, Jeff Shrager,
Tassos Stevens, Joan Stiles, Faraneh Vargha-Khadem, and members
of the UCSD DevLab have provided us with important and valu-
able feedback.

In December of 1994, the University of Higher Studies in the
Republic of San Marino sponsored a two-day workshop entitled
"Rethinking Innateness," where three of our authors were able to
air some of the ideas in this book and to listen and learn from some
of the best cognitive neuroscientists in Italy. Herb Killackey joined
us at the San Marino workshop, and we are immensely grateful to

him for his input, and for extensive discussions about the topics covered in Chapter 5.

George Carnevale played a particularly important role in Chapter 4. Much of that chapter draws on his own work (Bates & Carnevale, 1993), and George helped track down a number of errors in earlier drafts (of course, we reserve for ourselves the credit for those errors that remain). We also thank Jacqueline Johnson for making available the data from Johnson & Newport (1989), which we reanalyze in Chapter 4.

Meiti Opie not only read, commented, and proofed numerous drafts of Chapters 1, 5, and 7, but also served as general editorial assistant. She also tracked down and prepared the references. Meiti's persistence and attention to detail were extraordinary, and very much appreciated.

Betty and Harry Stanton of Bradford Books/MIT Press have been enthusiastic and eager in their support for this book from its beginning. We appreciate their faith in us, and their willingness to believe, as do we, that we have something important and exciting to say. Teri Mendelsohn, and later, Amy Pierce, of MIT Press have been an enormous help in the actual production of the first two volumes. A great deal of coordination was required, given the six co-authors, two books, and packaging and production of software. Teri and Amy made the job much easier, and their patience and encouragement is much appreciated.

When we began this effort, we did not fully appreciate the difficulties of producing a coauthored book with six authors who were located in San Diego, Pittsburgh, London, Oxford, and Rome. The logistics of travel, hotel accommodations, and arranging periodic meetings were formidable. Bob Buffington, Jan Corte, Miriam Eduvala, Larry Juarez, John Staight, and Meiti Opie of the Center for Research in Language at UCSD, and Leslie Tucker at the Cognitive Development Unit, London, all played a critical role in arranging our meetings and making the time together as productive as possible. We are very much indebted to them for their help.

We acknowledge the financial support which has been provided to the authors and made the research described here possible. In addition to the funds from the MacArthur Foundation already mentioned, this includes support from the Office of Naval Research (contract N00014-93-1-0194) and National Science Foundation (grant DBS 92-09432) to Jeff Elman; the National Institutes of Health (grants NIH/NIDCD 2-R01-DC00216, NIH/NINDS P50 NS22343, NIH/NIDCD Program Project P50 DC01289-0351) to Elizabeth Bates; Carnegie Mellon University, the National Science Foundation (grant DBS 91-20433), and the Medical Research Council of the United Kingdom to Mark Johnson; a McDonnell/Pew Visiting Fellowship and the Medical Research Council of the United Kingdom to Annette Karmiloff-Smith; and the Science and Engineering Research Council, the Biological and Biotechnical Research Council, and the Economic and Social Sciences Research Council to Kim Plunkett.

Finally, we wish to thank Marta Kutas, who is a valued colleague and a treasured friend. She wrote a number of poems on the theme of innateness, and we are pleased and flattered that she allowed us to choose one to open the book.

While the central arguments and concepts of this book represent our collaborative efforts, in any enterprise involving several authors with very different backgrounds there are bound to be areas of disagreement that cannot be resolved. One of these concerned the title, with which MJ wishes to put on record his disagreement. In MJ's view the term "innate" is better dispensed with entirely, as opposed to being re-thought.

nature   nurture
sure to be both   .
sum of each   .
interaction   .
learn teach   .
gene action   .
physical reaction   .
  mentality built of   .
   genes and   .
    experience hence   .
     the impossibility of
      assigning credit   .
       just watch .
        . the genetic
         .   debit accrue
          .   ensue as the old
           .   must combine
            .   with the new
             .   be refined
              .   according to
               .   prescribable laws
                .   cause and effect
                 .   direct indirect
                  from now and how
                         long ago?
                  who's to know?
                  predictable only
                  to some degree   .
                 cognition free only   .
               within the confines   .
             of some initial condition   .
              set free to experience   .
              to express its physicality   .
               to confess its mentality   .
              to redress a certainly   .
               uncertain reality   .

          marta kutas

CHAPTER 1

New perspectives on development

The problem of change

Things change. When things change in a positive direction (i.e., more differentiation, more organization, and usually ensuring better outcomes), we call that change "development." This is Heinz Werner's orthogenic principle (Werner, 1948).

Ironically, in the past several decades of developmental research there has been relatively little interest in the actual mechanisms responsible for change. The evidence of surprising abilities in the newborn, coupled with results from learning theory which suggest that many important things which we do as adults are not learnable, have led many researchers to conclude that development is largely a matter of working out predetermined behaviors. Change, in this view, reduces to the mere triggering of innate knowledge.

Counterposed to this is the other extreme: Change as inductive learning. Learning, in this view, involves a copying or internalizing of behaviors which are present in the environment. "Knowledge acquisition" is understood in the literal sense. Yet this extreme view is favored by few. Not only does it fail to explain the precocious abilities of infants and their final mature states, but it also fails to provide any account of how knowledge is deposited in the environment in the first place.

The third possibility, which has been the position advocated by classic developmentalists such as Waddington and Piaget, is that change arises through the *interaction* of maturational factors, under genetic control, and the environment. The interaction at issue here is not the banal kind where black and white yield gray, but a much more challenging and interesting kind where the pathways from genotype to phenotype may be highly indirect and nonobvious. The problem with this view in the past has been that, lacking a formal and precise theory of how such interactions might occur, talk of "emergent form" was at best vague. At worst, it reduces to hopeless mysticism.

Two recent developments, however, suggest that the view of development as an interactive process is indeed the correct one, and that a formal theory of emergent form may be within our grasp. The first development is the extraordinary progress that has been made in the neurosciences. The second has been the renascence of a computational framework which is particularly well suited to exploring these new biological discoveries via modeling.

Advances in neuroscience

The pace of research in molecular biology, genetics, embryology, brain development, and cognitive neuroscience has been breathtaking. Consider:

- Earlier theories of genes as static blueprints for body plans have given way to a radically different picture, in which genes move around, recombine with other genes at different points in development, give rise to products which bind directly to other genes (and so regulate their expression), and may even promote beneficial mutation (such that the rate of mutation may be increased under stressful conditions where change is desirable).

- Scientists have discovered how to create "designer genes." Human insulin can be produced in vats of bacteria, and caterpillar-resistant tomatoes can be grown. And plants have been created which produce biodegradable plastic!

- We now possess a complete and detailed picture of the embryology of at least one relatively complex organism (the nematode, C. Elegans). Scientists know, on a cell-by-cell basis, how the adult worm develops from the fertilized egg.

- Scientists have carried out ingenious plasticity experiments in which plugs of brain tissue from visual cortex (in late fetal rodents) are transplanted to sensorimotor cortex. This has led to the discovery that the old visual cortex neurons start to act like sensorimotor neurons. In other cases, researchers have shown that if information from the eyes is routed to auditory cortex early enough, regions of auditory cortex will set up retinotopic maps, and the organism will start to respond to visual stimuli based on messages going to the "borrowed" cortex. The conclusion many neuroscientists are coming to is that neocortex is basically an "organ of plasticity." Its subsequent specification and modularization appear to be an outcome of development— a result, rather than a cause.

- Although the degree of plasticity observed in the developing brain is surprising, the discovery of plasticity in adult mammals has come as an even greater surprise for those who believed in fixed and predetermined forms of neural organization. Studies have shown that somatosensory cortex will reorganize in the adult primate to reflect changes in the body surface (whether resulting from amputation or from temporary paralysis of a single digit on the hand). At first, this kind of reorganization seemed to be restricted to a very small spatial scale (a few microns at most) which suggested that a more transient local phenomenon could be responsible for the change. More recent evidence from adult animals that underwent amputation more than a decade prior to testing shows that this reorganization can extend across several centimeters of cortex. There are only two possible explanations for a finding of this kind: New wiring can be manufactured and established in the adult brain, or old patterns of connectivity can be converted (i.e., reprogrammed) to serve functions that they never served before.

- Sophisticated techniques have been developed for "eavesdropping" on brain activity with extraordinary spatial and temporal detail. Structural Magnetic Resonance Imaging (MRI), for example, provides enough spatial resolution to reveal a flea dancing on the corpus callosum (assuming there were such a flea). Evoked response potentials (ERP) gives us a temporal localization of brain processes to within thousandths of a second. Positron emission tomography (PET), magneto-encephalography (MEG), and new functional MRI techniques provide a bridge between the precise spatial resolution of structural MRI and the fine temporal resolution of EEG, showing us which parts of the brain are most active during various cognitive tasks. Taken together, these techniques provide us with potentially powerful tools both for examining the structure and functioning of the living brain, and its development over time.

These techniques make available a range of data which were simply not accessible even a decade ago. But although some might like to believe that theory follows inevitably from data, in fact it is usually the case that data may be interpreted in more than one way. What are needed are additional constraints. These come from a second development, which is a computational framework for understanding neural systems (real or artificial).

Neural computation: the connectionist revolution

Coinciding (but not coincidentally) with the dramatic advances in neuroscience, a second dramatic event has unfolded in the realm of computational modeling. This is the re-emergence of a biologically oriented framework for understanding complex behavior: Connectionism. The connectionist paradigm has provided vivid illustrations of ways in which global behaviors may emerge out of systems which operate on the basis of purely local information. A number of simple but powerful learning algorithms have been developed which allow these networks to learn by example. What can be learned (without being prespecified) has been surprising, and has demonstrated that a great deal more information and structure is latent in the environment than has been realized. Consider:

- Visual cortex in mammals is well known to include neurons which are selectively sensitive to highly specific visual inputs. These neurons include edge detectors, center-surround cells, and motion detectors. Biologically plausible network models have been constructed which demonstrate that such specialized response properties do not have to be prespecified. They emerge naturally and inevitably from cells which are initially uncommitted, simply as a function of a simple learning rule and exposure to stimulation (Linsker, 1986, 1990; Miller, Keller, & Stryker, 1989; Sereno & Sereno, 1991). These artificial networks even develop the characteristic zebra-like striped patterns seen in ocular dominance columns in real cortex (Miller, Keller, & Stryker, 1989).

- When artificial networks are trained to compute the 2-D location of an object, given as inputs the position of the stimulus on the retina and the position of the eyeballs, the networks not only learn the task but develop internal units whose response properties closely resemble those of units recorded from the parietal cortex of macaques while engaged in a similar task (Zipser & Andersen, 1988).

- Networks which are trained on tasks such as reading or verb morphology demonstrate, when "lesioned," symptoms and patterns of recovery which closely resemble the patterns of human aphasics (Farah & McClelland, 1991; Hinton & Shallice, 1991; Marchman, 1993; Martin et al., 1994; Plaut & Shallice, 1993; Seidenberg & McClelland, 1989).

- The rules of English pronunciation are complex and highly variable, and have been difficult to model with traditional Artificial Intelligence techniques. But neural networks can be taught to read out loud simply by being exposed to very large amounts of data (Sejnowski & Rosenberg, 1987).

- In learning a number of tasks, children frequently exhibit various "U-shaped" patterns of behavior; good early performance is succeeded by poorer performance, which eventually again improves. Networks which are trained on similar tasks exhibit the same patterns of behavior (MacWhinney et al., 1989; Plunkett & Marchman, 1991, 1993; Rumelhart & McClelland, 1986).

- Children are known to go through phases in which behavior changes slowly and is resistant to new learning. At other points in time children show heightened sensitivity to examples and rapid changes in behavior. Networks exhibit similar "readiness" phenomena (McClelland, 1989).

- Networks which are trained to process encrypted text (i.e., the words are not known to the network) will spontaneously discover grammatical categories such as noun, verb, as well as semantic distinctions such as animacy, human vs. animal, edible, and breakable (Elman, 1990). A curious fact lurks here which points to the importance of a developing system: Such networks can be taught complex grammar, but only if they undergo "maturational" changes in working memory or changes over time in the input (Elman, 1993).

Our perspective

Taken together, these advances—in developmental and cognitive neuroscience on the one hand, and neural computation on the other—make it possible for us to reconsider a number of basic questions which have challenged developmentalists, from a new and different perspective:

What does it mean for something to be innate? What is the nature of the "knowledge" contained in the genome?

Why does development occur in the first place?

What are the mechanisms which drive change?

What are the shapes of change? What can we infer from the shape of change about the mechanisms of change?

Can we talk meaningfully about "partial knowledge?"

How does the environment affect development, and how do genetic constraints interact with experience?

Our purpose in writing this book is to develop a theoretical framework for exploring the above questions and understanding how and why development occurs. We will cover a number of different specific topics in this book, but there are some central themes which recur throughout. We would like to identify these issues explicitly from the outset and foreshadow, briefly, what we will have to say about each one.

We begin with a discussion of genes. Although we are primarily concerned with behavior, and behavior is a very long way from gene expression, genes obviously play a central role in constraining outcomes. When we contemplate the issue of innateness, it is genes that we first think of. This discussion of genes will also help us to set the stage for what will be a recurring theme throughout this book: The developmental process is—from the most basic level up—essentially dependent at all times on interactions with *multiple* factors.

From genes to behavior

There is no getting around it: Human embryos are destined to end up as humans, and chimpanzee embryos as chimpanzees. Rearing one of the two in the environment of the other has only minimal effects on cross-species differences. Clearly, the constraints on developmental outcomes are enormously powerful, and they operate from the moment of conception. Furthermore, although there is a great deal of variability in brain organization between individuals, the assignment of various functions (vision, olfaction, audition, etc.) is not random. There are predictable and consistent localizations across the majority of individuals.

It is easy to state the obvious conclusion, which is that genes play the central role in determining both interspecies differences and intraspecies commonalities. This is true, but the real question is how, and what the genes are doing. Most developmentalists agree that a preformationist version of an answer (that these outcomes are contained in an explicit way in the genome) is unlikely to be correct (although some version of preformation may come close to capturing the nature of development in certain organisms, e.g., nematodes). There is simply too much plasticity in the development of higher organisms (as we shall discuss in Chapter 5) to ignore the critical effect of experience. We know too that there aren't enough genes to encode the final form directly, and that genes don't need to code everything. So how do genes accomplish their task?

How genes do their work

Asked what genes do, most people will report the basic facts known since Mendel (although he did not use the term gene), namely, that genes are the basic units of inheritance and that genes are the critters that determine things like hair color, gender, height. Such a view of genes is not incorrect, but it is woefully incomplete, and lurking beneath this view are a number of commonly held myths about genes which are very much at odds with recent findings in molecular genetics.

For instance, according to conventional wisdom, genes are discrete in both their effects and their location. Thus, one might imagine a gene for eye color which in one form (allele) specifies blue eyes, and in another specifies brown eyes. Genes are also thought of as being discrete with regard to location. As with the memory of a computer, under this view one should be able to point to some region of a chromosome and identify the starting point and ending point of a gene (which is itself made up of a sequence of base pairs).

In fact, the reality of genes and how they function is far more complex and interesting. Consider the following.

Genes are often physically distributed in space. In eukaryotes (e.g., humans, fruitflies, and corn are eukaryotes), DNA has been found to be made up of stretches of base pairs called exons, which code for the production of proteins, but which are interrupted by sequences of noncoding base pairs called introns. In some cases, the quantity of noncoding DNA may be more than 100 times greater than the coding DNA. What happens during protein synthesis, which is how most genes actually accomplish their work, is that the RNA copy of the gene-to-be—which includes introns—has to be cut up and respliced by specialized molecular machinery (see Figure 1.1). The

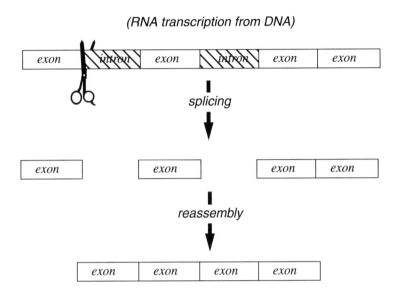

FIGURE 1.1 DNA often includes nonfunctional base pairs (*introns*) as well as sequences which code for proteins and other products (*exons*). During synthesis, the RNA transcript (but not the original DNA) is cut and spliced so that only the exons remain; the revised RNA is then used for actual synthesis.

result is a temporary "cleaned up" version of the gene transcript which can then be used for protein synthesis.

Moreover, the same portion of DNA can be spliced in different ways. For some purposes, a sequence of base pairs may be treated as an intron (noncoding), but for other purposes, the same region may be spliced to yield a different gene transcript and end up as an exon (coding). Finally, although the structure of DNA base pairs is basically stable, some sequences move around. This movement turns out to play a much more important role in genetic expression than was thought when "jumping genes" were first discovered.

Genes are not binary in their effects. What about the view of genes as discrete in their effects? This too turns out to be a misleading idea. To be sure, there are base pair sequences which code directly for specific and well-defined traits. But in many cases the encoding is continuously valued. A subsequence of base pairs may be repeated or there may be multiple copies of the gene; this causes more of the protein product to be produced and may result in a more strongly expressed trait.

Genes do their work with other genes. Sometimes, but rarely, it is possible to tie the effects of a single gene's products to some clearly defined trait. However, such "single action" genes either tend to be associated with evolutionarily primitive mechanisms or they work as switches to turn on and off some other function which is coded by a group of genes.

For example, the fruitfly, *Drosophila melanogaster*, has a gene called Antennapaedia (Antp). If the Antp gene undergoes a certain mutation, then instead of antennae the fruitfly will develop an extra pair of feet growing out of its head where the antennae would be. Notice that this bizarre effect relies on the fact that what the Antp gene does is to regulate the expression of other gene complexes which actually produce the feet (or antennae). Even simple traits such as eye color in the fruitfly may depend on joint action of 13 or more genes. Thus, while there are single-action genes, more typical are cases where multiple genes are involved in producing any given trait, with some genes playing the role of switches which control and regulate the expression of other genes.

Genes are often reused for different purposes. A very large number of genes in an animal's genome are what one might call "housekeeping genes." They code for the production of basic proteins which function as enzymes, form cellular organelles, carry out cellular metabolic activities, etc.

But Nature is stingy with her solutions. Things which work in one species frequently turn up in very distantly related species. All together, probably something like 5,000 genes are needed by cells in all eukaryotes for housekeeping purposes. Essentially the same genes, modified here and there, are shared by all species and cell types. The lesson here is that there is remarkable conservatism and reusing of old solutions. By rearranging and slightly modifying only a few thousand interacting gene complexes, enormous diversity of structure is possible.

This conservatism does not rule out the abrupt appearance of what seem to be radically new structures, be they language or flippers or wings. There is a great deal of genetic redundancy in eukaryotes. The same gene may appear many times in the genome, and often slightly different genes produce similar or identical products. This redundancy accommodates many small changes in the genome before there is a dramatic shift in phenotype. Thus the appearance of abrupt changes in phenotypic outcomes may be misleading, and result from much tinier changes at the genetic level. This brings us to the next point.

The relationship between genome and phenotype is highly nonlinear. Although a linear increase in genome size (measured as the number of DNA base pairs) which correlates with phenotypic size can be observed for simple species (e.g., worms), this does not hold for so-called higher species (see Table 1.1). In the latter case, the relationship is highly nonlinear. In Chapter 4 we will discuss nonlinear phenomena in some detail. For the moment suffice it to note that one of the most dramatic nonlinear relationships in nature is that which exists between the genome and the phenotype.

Compare, for example, the genome of the chimpanzee, the Old World monkey, and the human. To the layman's (admittedly biased) eye, the Old World monkey and the chimp resemble each other much more closely than either species resembles us. Yet genetically the chimp and the human are almost indistinguishable: We have 98.4% of our genetic material in common, compared with only approximately 93% shared by the chimp and Old World monkey. Humans are also closer to chimps, genetically, than chimps are to gorillas. Whatever differences there are between us and the chimp therefore come down to the effects of the 1.6% difference.

In Chapter 7, we will discuss the implications of the above facts for what it might mean for a trait or a behavior to be innate. For the moment, the above—which reveals only the most modest glimpse of the complexity which underlies genetic functioning—is enough to help us make a simple point. Even the simplest questions, *what* a

TABLE 1.1

Organism	Amount of DNA (base pairs)					
	10^6	10^7	10^8	10^9	10^{10}	10^{11}
Flowering plants				▓▓▓	▓▓▓	▓
Birds				▓		
Mammals				▓		
Reptiles				▓		
Amphibians				▓▓▓	▓▓▓	▓
Bony fish				▓▓		
Cartilaginous fish					▓	
Echinoderms				▓		
Crustaceans				▓		
Insects			▓▓	▓▓		
Mollusks				▓▓		
Worms			▓			
Molds			▓			
Algae			▓			
Fungi		▓				
Gram-pos. bacteria	▓▓					
Gram-neg. bacteria	▓					
Mycoplasma	▓					

Adapted from Edelman (1988). For simpler species (e.g., mycoplasma through worms), there is an approximately linear increase in DNA with increasing organism size. For more complex species (the upper portion of the table), there is no obvious relationship between amount of DNA and the size or complexity of the organism.

gene is, and *where* it resides, have answers which are very much at variance with the commonplace notion of the "digital gene." Genes are fluid and graded, which gives them a distinctly analog character, and they rarely work in isolation. Genes work in concert with large numbers of other genes, and tracing a particular gene's contribution to the emerging phenotype is very indirect and rarely possible without considering the whole network of interactions in which that gene participates.

Let us turn now to a slightly higher stage of organization from genes, namely, cells and tissues. We shall see that the same sorts of interactions which are observed in genes occur in cell differentiation and tissue formation.

How cells come to be

The human body contains roughly 100 trillion cells; these are made up of about 200 different cell types, which are more or less the same types as are found in snakes, birds, and other mammals. The major difference between species lies in changes in cell number and topology. Some species have more cells than others, and those cells are arranged in different ways.

However, there is a mystery here. All cells (with one or two minor exceptions) contain the same genetic information. So the question we must ask is how cells come to be different. This process of differentiation occurs early in development. While the process is far from fully understood, enough is known for us to give examples which illustrate the importance of interactions.

Mosaics and regulators. Not all species rely on interactions to the same degree. Embryologists distinguish between two very different styles of cellular development: Mosaic development and regulatory development.

In mosaic development, cells develop more or less independently. They tend to be largely unaffected by each other and by the environment. The fate of each cell or group of cells is determined early on by their location. When and what a cell becomes is under relatively tight genetic control.

A mosaic organism that has been particularly well-studied is the nematode, *C. Elegans.* This worm has been the subject of a long-term research project, resulting in a largely complete picture of its embryological development from zygote to adult. *C. Elegans* makes a good subject for study for several reasons. It develops very quickly (about two days to reproductive maturity). It is also reasonably complex. Further, its body is transparent; this makes it possible to view internal development with a light microscope. Finally, each member of the species is virtually identical. Every C. *Elegans* contains exactly 959 somatic cells, and cell connectivity is very regular across different individuals of the species.

The path by which each C. *Elegans* arrives at adulthood is both highly circuitous but also highly invariant. Cells differentiate in ways that seem neither tidy nor logical. Organs and tissues are formed from what seem like random groupings, and cell lineage is quite eccentric. But as byzantine as the process may seem, it also seems to be largely predetermined. Genetically identical individuals have essentially the same cell morphology, cell position, and cell connectivity (by comparison, this is not at all the case for human monozygotic twins). By and large, cell lineage is not influenced by the fate of other cells; molecular level interactions are sufficient to determine its fate.

It is easy to see the advantages of the mosaic style of development. By largely prespecifying cell fate, nature ensures that evolutionarily tested solutions are well preserved. Mosaic development is independent and fast. Each cell has its own timetable and need not wait for other cells to develop. Growth can occur just as quickly as cells are able to consolidate the materials they need to divide and differentiate. Each cell does this on its own, confident that when its job is done, its fellows will be ready and waiting in place to join up. If humans developed as mosaics we could be up and running in just a short time after conception because the approximately 100 trillion cells could be produced in just under 47 binary divisions.

In fact, the human species—and most other higher organisms— have not opted for the mosaic style of development, but instead rely on what is called regulatory development. Why should this be? What are the limitations of mosaic development?

The problems with mosaic development are the flip side to their advantages. First, lack of interaction at the cellular level permits speedy development under good conditions, but it also means lack of flexibility in the face of abnormal conditions. If some cells are damaged, others cannot compensate for their absence and the organism may no longer be viable. In environments which are unstable or likely to change, there may be no single developmental solution which can be hardwired into the genome.

There is another, and probably more serious, price which is paid by mosaic organisms. The burden on the mosaic genome is considerable. The genome comes close to being a blueprint for the body; it must specify everything. For relatively small and simple organisms—up to, say, the worm—this may not be much of a problem, and may be compensated by the other advantages of mosaic organization. However, the blueprint approach to development puts an upper bound on the complexity which can be achieved. Such direct specification of the human brain alone, for example, could plausibly require something on the order of 10 trillion base pairs of DNA, which is far in excess of what is structurally feasible. Indeed, in many organisms the relationship between the amount of a species' genetic material and its morphological and behavioral complexity is highly nonlinear. Indeed, many plants have more genetic material than do humans (recall Table 1.1).

The alternative to mosaic development is regulatory development. Regulatory systems rely heavily on cellular-level interactions. The orchestration of cell differentiation and the final outcome are under broad genetic control, but the precise pathway to adulthood reflects numerous interactions at the cellular level that occur *during development*.

While most species show some elements of both types of development (see Edelman, 1988), higher vertebrates generally show more regulatory development. The effects of regulatory development may be quite dramatic. Earlier we pointed out that the average human and chimpanzee DNA differ only by 1.6%, which is less than the difference between two species of gibbons. Yet the morphologies and behaviors of humans and chimps differ considerably. It seems reasonable to believe that these differences depend far more on the evolution of regulatory mechanisms than on the evolution of new structural genes.

One advantage of regulatory development is that it allows for greater flexibility. Damage to a group of cells can often be compensated for by their neighbors. More significantly, regulatory systems probably permit far greater complexity of phenotypes than can be achieved in mosaic developmental systems. The cellular level inter-actions provide an extra source of constraint which makes it possible to develop more complex phenotypes.

However, such interactions impose their own cost. Regulatory systems require a period of interdependent development, and this may slow down the process of development since some events will require the completion of other events before they can start. The organism is likely to have a prolonged period of immaturity during which time it is vulnerable. Mosaics are simple but fast to develop; regulators are complex and slow to develop. When the phenotype is relatively simple, genetic and molecular level information are sufficient to allow parallel development of a large number of cells. This is the mosaic approach. But there are constraints on the amount of genetic material which can be safely housed in a cell and reliably replicated across generations. In order to attain a more complex outcome (phenotype) with a roughly similar number of genes, it is necessary to create hierarchical intermediate stages, which, in the case of regulatory systems, occur at the cellular level of interaction.

It is for this reason that developmental timing becomes more crucial as the hierarchical complexity of an ontogenetic system increases. And genes are algorithms which operate sequentially in time. In Chapter 6, we discuss just how timing can be controlled, and how it can be exploited in the service of building complexity. Thus the action of genes—in particular, those associated with the development of more complex behaviors—may be very indirect. The genome is, first, algorithmic rather than descriptive. The algorithms often rely on predictable regularities in the input (so that the algorithms do not need to encode information which can be counted on to be made available through experience). Two of the lessons we have learned from connectionist research are, first, that considerably more information may be latent in the environment—and extractable, using simple learning algorithms—than was previously thought; and second, that useful self-organization may occur in the absence of explicit guidance from the environment. Certain problems have a natural solution; all that may be required are a few gentle nudges in the form of prewired biases and constraints.

The problem of interaction

We have taken some time talking about genes and cells for two reasons. First, as we already pointed out, genes are what most people think of when they think of innateness. And second, even a cursory discussion reveals how much genes and cells depend on interactions.

This is not a new insight. Developmentalists have long acknowledged the role of interaction. The problem has been that these interactions are either so trivial as to be of little interest, or so complex as to resist analysis. So the interactionist position—while in principle agreed upon by virtually every developmentalist— remains a difficult one to pursue in practice.

Several decades ago, Waddington tried to illustrate this conception of development with his picture of the "epigenetic landscape" (see Figure 1.2, left panel). Embryologists knew that phenotypically very similar individuals might have wildly different genotypes; and an organism with a single (apparent) phenotype might emerge from a genome that contains a much larger array of possibilities than are ever realized. How could this be? Waddington offered the following account:

> *I envisage [development] as a set of branching valleys in a multidimensional space that includes a time dimension, along which the values extend. The development of the phenotype of an individual proceeds along a valley bottom; variations of genotype, or of epigenetic environment, may push the course of development away from the valley floor, up the neighboring hillside, but there will be a tendency for the process to find its way back.* (Waddington, 1975; p. 258)

Waddington's image of the epigenetic landscape and his account of how development proceeds bear, at one level of description, an eery resemblance to another familiar image—the way neural networks change over time. In the case of the network, the "environment" is usually a training set; the "phenotype" is the evolving set of connection strengths between artificial synapses (Figure 1.2, right panel). Over time, the network seeks the "low ground" of error in its weight space, but there are often many paths to equivalent solutions.

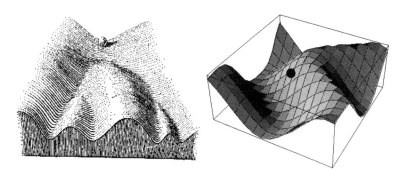

FIGURE 1.2 Waddington's epigenetic landscape on the left (Waddington, 1975); hypothetical error surface from a neural network, right.

We suspect that developmentalists would have little difficulty recognizing the image on the right as just another epigenetic landscape; while connectionists would look at Waddington's drawing and immediately think of neural network error surfaces. Is this mere coincidence?

We believe not. In fact, we shall argue in this book that the two images are more than coincidentally similar. We believe that these two images capture what is fundamentally the same process.

Taking a biological perspective

Our perspective is connectionist, but it is equally a biological perspective. Since our views probably differ from those of some connectionists, as well as of some biologists, we should explain just what our brand of "biologically oriented connectionism" entails.

(1) *We think it is crucial to pay attention to what is known about the genetic basis for behavior and about developmental neuroscience.*

Remarkable strides have been made recently in understanding the relationship between gene expression and development. If we want to know in what senses behavior might be innate, we should strive for an account which is broadly consistent with what is known about what genes do and how they work in other domains.

As we shall argue in the final chapter, the way genes work probably precludes anything like "genes for language."

Through imaging, animal models, and focal lesion data, we now know a great deal more about how the brain develops. There is good evidence that neocortex is initially equipotent (or at least, multipotent, since some outcomes are more likely than others but many outcomes are possible). In Chapter 5 we review the dramatic results of experiments involving transplanting and rewiring bits of cortex to other regions. The results of these experiments argue strongly against solutions which depend upon innately specified populations of neurons prewired for complex cognitive functions such as language.

(2) *At the level of computation and modeling, we believe it is important to understand the sorts of computations that can plausibly be carried out in neural systems.* While there is an inevitable (and desirable) tension between the abstract models and the specific systems they attempt to model, there are basic principles which should be observed.

Computation is distributed; but the information available to the computing elements is mostly local. (There are, in fact, global regulators and chemical gradients which extend over large spatial regions; but these are typically low-dimensional sources of information, and the same regulators, e.g., morphogens which trigger cell differentiation, are frequently used across systems and even across species.)

Information also is distributed; representations are graded, continuous, and spread over wide areas. Moreover, representations are often superposed, such that the same tissue participates in representing multiple pieces of information.

Finally, neural computation is often highly nonlinear. This means that under certain circumstances, small differences in input may have little effect; whereas under other circumstances, small differences in input produce qualitatively different behaviors. This is discussed in Chapter 4.

(3) *We take a broad view of biology which includes concern for the evolutionary basis of behavior.* This approach can be traced back to classical traditions in development, as articulated by pioneers like

Baldwin, Piaget, and Vygotsky. Just as it is hard to imagine studying behavior from snapshots frozen in time, it makes no sense to try to understand development without taking into account the environment within which development unfolds or the evolutionary history which gives rise to the behavior.

The study of evolutionary change is important in itself because it can give insight into the mechanisms which underlie complex behaviors. If one is interested in innate constraints, then it is important to understand possible origins for those constraints. Furthermore, if one is interested in development, evolutionary change is important because it interacts with individual development. In fact, it can be plausibly argued that individual development cannot be fully understood without understanding its evolutionary basis. An evolutionary perspective examines (through reconstruction or simulation) the process by which forms and behaviors become constrained, either by virtue of initial starting point or channeled development.

(4) *Finally, a broader biological perspective emphasizes the adaptive aspects of behaviors, and recognizes that to understand adaptation requires attention to the environment.* An important goal for simulation studies, therefore, is taking an ecological view in which behaviors are situated in context and carried out in concert. We see this as an important counterweight to the reductionist approach, which has tended to fractionate the study of behavior. Studying cells (to paraphrase David Marr) may be useful for understanding life; but understanding how a cell works will not tell us what it means to be alive.

What does it mean to be innate?

This is a question we pose now, and consider again in the final chapter. The term "innate" has a very checkered history in science. In some fields, such as ethology, the term has been virtually banned from use for the past 20 years. One of the main reasons why this has happened is that ethological studies revealed that most of the examples of behavior originally claimed to be innate (by Lorenz and his collaborators) in fact turn out to be dependent on prior interactions with the pre- or postnatal environment. For similar reasons the term has dropped from use in genetics. Since it has become evident that genes interact with their environment at all levels, including the molecular, there is virtually no interesting aspect of development that is strictly "genetic," at least in the sense that it is exclusively a product of information contained within the genes.

Nonetheless, many cognitive scientists and developmentalists have continued to use the term innate and to speak of such things as the "language instinct." We feel this reflects the entirely justifiable desire to understand how behaviors come to be; and in cases where outcomes seem inevitable, it is tempting to call these outcomes "innate." So what is wrong with this?

The problems, from our perspective, are two-fold and have to do with *mechanism* on the one hand, and *content* on the other.

First, calling a behavior innate does very little to explain the mechanisms by which that behavior comes to be inevitable. So there is little explanatory power to the term. If all that is meant by saying a behavior is innate is that it is (under normal circumstances) inevitable, then we have gained little.

What is often meant by innate is somewhat stronger: "that which is specified within the genome," under the assumption that genes code for innate behaviors. From this perspective the challenge is to elucidate what aspects of cognition or behavior, if any, are the direct result of genetic information. But as we have already argued and will repeatedly stress throughout this book, there are a multitude of molecular, cellular and system interactions that occur between the gene and its developmental products. While aware of a great many interactions at the molecular and cellular level, some developmental psychologists choose to think of these as biological details of minimal relevance to those interested in behavior and cognition. It is a reasonable approximation, they argue, to state that the component of cognition in question is coded for in the genes. As we hope to demonstrate in this book, however, this is not a position which leads to insightful accounts of development.

The second way in which in which claims regarding innateness can be problematic has to do with the *content* of what it is that is presumed to be innate. Does the fact that the vast majority of humans end up speaking (or signing) some language mean that language is innate? Possibly, but universal outcomes are not sufficient diagnostics of innate mechanisms (since the vast majority of humans living in the British Isles end up speaking English—yet their genomes obviously do not encode for one particular language). Does the fact that some individuals have difficulty learning a language at all mean that the deficit lies with an innate faculty for language learning? Or does the deficit arise more indirectly from some other problem which has a deleterious consequence for language learning?

These two problems are not easily dealt with, but lacking a more precise specification of the mechanisms which constrain development, and of the content domains over which they operate, any use of the term innate is bound to be muddled and counterproductive. To this end, we propose in the next two sections first, a way of thinking about possible mechanisms by which developmental outcomes might be constrained; and second, a way of thinking about the domain specificity of those mechanisms.

An alternative definition is to reserve the term innate to refer to developmental outcomes that are more or less inevitable for a given species. That is, given the normal environment for a species, outcomes which are more or less invariant between individuals are innate.

In considering these issues, Johnson and Morton (1991) suggest that it is useful to distinguish between the various levels of interaction between genes and their environment. Some of these are shown in Table 1.2. Here, the term innate refers to changes that arise as a result of interactions that occur within the organism itself during ontogeny. That is, interactions between the genes and their molecular and cellular environments without recourse to information from outside the organism. We adopt this working definition of the term in this book. Interactions between the organism and aspects of the external environment that are common to all members of the species, the species-typical environment, (such as patterned light, gravity, etc.) were referred to as "primal" by Johnson and Morton. Clearly, the boundary between innate and primal is often difficult to draw, and there are many instances from ethological studies where behaviors thought to be innate were found to be primal (i.e., requiring interaction at the level of the species-typical environment) on closer study.

TABLE 1.2

Level of interaction	Environment	Outcome
molecular	internal environment	INNATE
cellular	internal environment	
organism-external	species-typical environment	PRIMAL
organism-external	individual environment	LEARNING

In this book we use the term innate in the same sense as Johnson and Morton (1991) to refer to putative aspects of brain structure, cognition or behavior that are the product of interactions internal to the organism. We note that this usage of the term does not correspond, even in an approximate sense, to genetic or coded in the genes.

Ways to be innate: A framework

This brings us to a crucial question. If development truly is an interactive process, and if emergent form is the rule rather than the exception, how do we keep from lapsing into the hopeless mysticism which is too often invoked in place of real explanation? Can we be more specific about the nature of the interactions and about the things which interact? Put more generally, what are the ways in which behaviors can be innate?

We suggest that it is useful to think about development as a process which occurs on multiple levels (we use the word levels here in a heterarchical rather than hierarchical sense). For an outcome to be innate, in our terms, means that development is constrained at one or more of these levels. Interactions may occur within and also across levels. And outcomes which are observed at one level may be produced by constraints which occur at another.

Given our perspective, we have also found it useful to consider how these constraints might be implemented both in natural systems (brains) as well as in the artificial systems (networks) we use to model those natural systems. Sometimes the correspondence is ready and obvious, but there are also cases where we do not yet have a clear understanding of how a constraint might be implemented. This tension is useful, we feel.

Two words of warning. First, we recognize that Nature has no particular obligation to honor any taxonomy, so we expect to find cases where our distinctions are blurred. The primary value of this taxonomy is as a conceptual framework for formulating hypotheses about where the major determinants of cognitive behavior may lie.

Second, much of what follows necessarily makes reference to terms and concepts having to do with brains and networks. Some of these terms may not be clear to the reader, but will be defined later in the book. (Chapter 2 describes network terms, and Chapter 5 deals with brain organization; the terms themselves are also listed in the Index.) We hope this will not cause the reader undue confusion. Because the taxonomy we propose is basic to our perspective, we felt it important to present it up front. We also hope that in so doing, we will have provided a framework which will assist the reader in understanding all that follows.

So, what are the ways in which things can be innate?

We propose that constraints may operate at three different levels: **representations**, **architectures**, and **timing**. Again, we emphasize that we take these to be heterarchical rather than hierarchical levels. Typically, behaviors reflect constraints which operate at multiple interacting levels. The levels themselves may be distinguished by the degree of specificity and directness of consequence for behavior and knowledge. Representational constraints have the most specific and direct relationship to knowledge; architectural constraints operate at a more general level with less directly obvious relationship to resulting knowledge; and timing constraints are typically the most opaque with regard to outcome. Let us consider each of these possibilities. (The reader may wish to refer to Table 1.3 on page 35 for a summary of these three types of constraints.)

(1) *Representational constraints*

One of the strongest hypotheses one might advance is that knowledge and behaviors are innate by virtue of the representations which subserve them being "hard-wired" in advance. For example, it has been argued that children are born with innate knowledge about basic principles of (for example) grammar (Crain, 1992; Lightfoot, 1989; Pinker, 1994a,b; Pinker & Bloom, 1990), physics (Spelke, 1994) or mathematics (Wynn, 1992). (We discuss these claims in greater detail in Chapter 3, and again in Chapter 7.) To be sure, these authors do not argue for a simple one-to-one relationship between genes and behavior; the knowledge is taken to be shaped by experience to some extent (perhaps in the form of "triggering" or "selecting" among predetermined options, e.g., Piatelli-Palmarini, 1989); and some maturation may have to take place before the innate knowledge can be implemented in the service of some behavior (e.g., Borer & Wexler, 1987; Spelke et al., 1992). However, most of these investigators have been clear in their belief that children are born with domain-specific representations laid out somewhere in the brain.

What might this mean, in network terms and in brain terms?

In a connectionist network, representations are patterns of activations across a pool of neuron-like processing units. The form of these activation patterns is determined by the nature of the connections between the units. Thus, innate representational knowledge—by which we mean the potential to produce representations of specific sorts—would take the form of prespecified weights on the inter-unit connections.

In the brain, the most likely neural implementation for such innate knowledge would have to be in the form of fine-grained patterns of synaptic connectivity at the cortical level, i.e., cortical micro-circuitry. To the best of our knowledge at the present time, this is how the brain stores its representations, whether they are innate or acquired. In this regard, Pinker (1994b) suggests that the "language instinct" is indeed based upon specific microcircuitry, and that the same is probably true for many other cognitive processes:

> *It is a certain wiring of the microcircuitry that is essential.... If language, the quintessential higher cognitive process, is an instinct, maybe the rest of cognition is a bunch of instincts too—complex circuits designed by natural selection, each dedicated to solving a particular family of computational problems posed by the ways of life we adopted millions of years ago.* (Pinker, 1994b; pp. 93, 97)

Assuming that representations are defined in terms of cortical microcircuitry, what might it mean to say that knowledge and/or representations are innate? Although it is theoretically possible to set the weights of a network (natural or artificial) in advance through natural selection, we will argue that representational innateness (so defined) is relatively rare in higher organisms, at least at the cortical level (for some possibilities at the subcortical level, see Chapter 6). Indeed, there are many reasons to think that the cortex in higher vertebrates (especially humans) has evolved as an "organ of plasticity" which is capable of encoding a vast array of representational types.

In fact, as we shall see in some detail in Chapter 5, evidence has been mounting against the notion of innate domain-specific microcircuitry as a viable account of cortical development, i.e., against what we call "representational nativism."

In a number of recent studies with vertebrate animals, investigators have changed the nature of the input received by a specific area of cortex, either by transplanting plugs of fetal cortex from one area to another (e.g., somatosensory to visual, or vice-versa, O'Leary, 1993; O'Leary & Stanfield, 1989), by radically altering the nature of the input by deforming the sensory surface (Friedlander, Martin & Wassenhove-McCarthy, 1991; Killackey et al., 1994), or by redirecting inputs from their intended target to an unexpected area (e.g., redirecting visual inputs to auditory cortex (Frost, 1982, 1990; Pallas & Sur, 1993; Roe et al., 1990; Sur, Garraghty & Roe, 1988; Sur, Pallas & Roe, 1990; see also Molnar & Blakemore, 1991).

Surprisingly, under these aberrant conditions, the fetal cortex takes on neuroanatomical and physiological properties that are appropriate for the information it receives ("When in Rome, do as the Romans do..."), and quite different from the properties that would have emerged if the default inputs for that region had occurred. This suggests that cortex has far more representational plasticity than previously believed. Indeed, recent studies have shown that cortex retains representational plasticity into adulthood (e.g., radical remapping of somatosensory cortex after amputation, in humans and in infrahuman primates (Merzenich et al., 1988; Pons et al., 1991; Ramachandran, 1993; see also Greenough, Black, & Wallace, 1993).

One cannot entirely rule out the possibility that neurons are born "knowing" what kinds of representations they are destined to take on, but right now the case for innate representations does not look very good. However, this does not mean that there is no case for innate constraints on higher cognitive processes. Instead, it means that we have to search for other ways that genes might operate to ensure species-specific forms of brain organization, and the thoughts and behaviors mediated by that form of brain organization—which brings us to the next two sources of constraint on development.

(2) *Architectural constraints*

Although there is no good evidence that we know of that knowledge and behavior are constrained at the level of representations, it is much more likely that such constraints operate at the level of architectures. The architecture of a system encompasses a number of different characteristics. Thus, the term architecture is potentially ambiguous, meaning different things to different people. What is true of all the features we consider architectural, however, is that they operate at a higher level of granularity than representational constraints, which take the form of prespecified connections between individual neurons or nodes.

Architectural constraints can vary along a number of dimensions and degrees of granularity (from cell types to overall brain structure), but in general fall into three broad categories. We call these **unit-based architectural constraints, local architectural constraints**, and **global architectural constraints**. We recognize that these subdivisions do not perfectly classify all types of architectural features. This is simply a provisional attempt to impose some order on what is in reality a very complex and somewhat disordered system.

(a) *Unit level architectural constraints.* The lowest level of architecture deals with the specific properties of neurons (in brains) or nodes (in connectionist networks). This is the level that many neuroscientists would consider cytoarchitectural (but note that cytoarchitecture is often used in the neuroscience literature to refer to higher-level, areal organization as well).

In the brain, unit level constraints include the specification of neuron types which are found in different regions of the brain; response characteristics of neurons, including firing threshold, refractory period, etc.; type of transmitter produced (and whether it is excitatory or inhibitory); nature of pre- and postsynaptic changes (i.e., learning), etc. In network terms, unit level constraints might be realized through node activation functions, learning rules, temperature, momentum, etc.

It is clear that unit level constraints operate in brain development. There are a relatively small number of neuron types, for instance, and they are neither randomly nor homogeneously distributed throughout the brain. The unit level constraints are fundamental to brain organization, since they concern the lowest level of computation in the brain.

(b) *Local architectural constraints.* At the next higher level of granularity we have local architectural constraints. In brains, such constraints might take of the form of differences in the number of layers (e.g., the six-layered organization of cortex), packing density of cells, types of neurons, degree of interconnectivity ("fan in" and "fan out"), and nature of interconnectivity (inhibitory vs. excitatory). In network terms, local architectural differences would include feedforward vs. recurrent networks, or the layering of networks. (However, one should resist the obvious temptation to associate layers of cortex with layers in a network. We feel this would be a mistake, since the layers have traditionally served different purposes in the two domains.) In reality, current connectionist models have tended not to exploit differences in local architecture in a manner which comes anywhere near to the diversity observed in real brains.

Interestingly, the cortex itself initially appears to display relatively little in the way of local architectural differences. That is, the local architecture of the cortical mantle does differ significantly from other regions of the brain, but there is good reason to believe that initially this architectural plan is mostly the same through the cortex.

At the same time, it is also true that the adult cortex displays much greater differences in local architecture. Thus, and not surprisingly, experience and learning can significantly alter the local cortical architecture over time. What is at present unclear and controversial is the extent to which such differences are predetermined or emerge as a consequence of postnatal development. One likely candidate for possible intrinsic differentiation within the cortex is the thickening of cortex in the primary visual area (arising primarily from an enlarged layer 4, which is known to contain an unusually large number of neurons). More recently, work with mouse embryos suggests that while certain aspects of areal identity of cortical tissue can be changed by early transplant (e.g., connectivity and some cytoarchitectural features; see Chapter 5), there are other subtle characteristics of cells which may not be alterable and thus be intrinsic to that region of cortex (Cohen-Tannoudji, Babinet, & Wassef, 1994). The computational consequences of such regional differences are as yet unknown, but it has been suggested that (among other things) this form of innate constraint gives rise to the left/right-hemisphere differences in language processing that emerge in human beings under default conditions (i.e., in the absence of focal brain lesions; see Chapter 5 for details).

(c) *Global architectural constraints.* Finally, a major source of constraint arises in the way in which the various pieces of a system—be it brain or network—are connected together. Local architecture deals with the ways in which the low-level circuitry is laid out; global architecture deals with the connections at the macro level between areas and regions, and especially with the inputs and outputs to subsystems. If one thinks of the brain as a network of networks, global architectural constraints concern the manner in which these networks are interconnected.

In this form of nativism, knowledge is not innate, but the overall structure of the network (or subparts of that network) constrains or determines the kinds of information that can be received, and hence the kinds of problems that can be solved and the kinds of representations that can subsequently be stored. In other words, the macrocircuitry—meaning principally the areal patterns of input/output mappings—may be prespecified even if the microcircuitry is not.

In brain terms, such constraints could be expressed in terms of (e.g., thalamo-cortical) pathways which control where sensory afferents project to, and where efferents originate. Very few network models employ architectures for which this sort of constraint is relevant (since it presupposes a level of architectural complexity which goes beyond most current modeling). One might imagine, however, networks which are loosely connected, such that they function somewhat modularly but communicate via input/output channels. If the pattern of inter-network connections were prespecified, this would constitute an example of a global architectural constraint.

As we noted in the previous section, the representations developed by specific cortical regions are strongly determined by the input they receive. On this assumption, one good way to ensure that a region of cortex will be specialized for (say) vision, audition or language would be to guarantee that it receives a particular kind of information (e.g., that visual cortex receives its information from the retina, and auditory cortex receives its information from the ear). In addition, one might guarantee further specialization by placing more constraints on the input that a particular region receives. For instance, the differential projection of dopaminergic fibers to the frontal cortex from the substantia nigra and ventral temgental nuclei may provide constraints on what types of representations emerge in this part of cortex, since dopamine levels are thought to influence the firing thresholds of neurons.

Before continuing with the third and final source of constraint, we pause to note that although it is rarely acknowledged, architectural constraints of one kind or another are necessarily found in all connectionist networks. Specific choices are made regarding computational properties of individual units (unit level architectural constraints), and although local architectural assumptions are usually very simple and uniform throughout a network (for example, a three-layer, feedforward network in which all input units report to all units in the hidden layer, and each unit in the hidden layer is connected with all the units in the output) assumptions of one kind or another are made. These initial architectural characteristics strongly constrain the behavior of the network and are almost always critical to the success (or failure) of the model in question. (When we consider all the variations of input/output relations and cytoarchitectures that are employed in real brains and compare those with the limited set of options used in current modeling efforts, we may wonder why such simulations ever succeed at all.) The point we stress, therefore, is that almost all connectionist models assume innate architectural constraints, and very few assume innate representations.

(3) *Chronotopic constraints*

One of the crucial ways in which behavioral outcomes can be constrained is through the timing of events in the developmental process. Indeed, as Gould (and many other evolutionary biologists) has argued eloquently, changes in the developmental schedule play a critical role in evolutionary change (Gould 1977; see also McKinney & McNamara, 1991).

In networks, timing can be manipulated through exogenous means, such as control of when certain inputs are presented. Or timing can arise endogenously, as in the Marchman simulation (discussed below). In Marchman's simulation, the gradual loss of plasticity in a network comes about as a result of learning itself. In brains, very occasionally, timing is under direct genetic control. More commonly, the control of timing is highly indirect and the result of multiple interactions. Hence the onset and sequencing of events in development represents a schedule that is the joint product of genetic and environmental effects. Another example in which developmental timing plays a role can be found in the "growing networks" of Cangelosi, Parisi, and Nolfi (1994), in which nodes divide according to a genetically determined schedule.

At the level of the brain, variations in timing can play an important role in the division of labor described above, determining the specialization of cortical regions for particular cognitive functions. For example, a region of cortex may be recruited into a particular task (and develop subsequent specializations for that task) simply because it was ready at the right time. Conversely, other areas of the brain may lose their ability to perform that task because they developed too late (i.e., after the job was filled).

To offer one example, differential rates of maturation have been invoked to explain the left-hemisphere bias for language under default conditions (Annett, 1985; Corballis & Morgan, 1978; Courchesne, Townsend & Chase, 1995; Kinsbourne & Hiscock, 1983; Parmelee & Sigman, 1983; Simonds & Scheibel, 1989). Because the maturational facts are still very shaky, arguments have been offered in two opposing directions! (See Best, 1988, for a detailed discussion and an alternative view.) Some have suggested that the left hemisphere matures more quickly than the right, which leads in turn to a situation in which the left hemisphere takes on harder jobs (i.e., cognitive functions that require more computation, at greater speeds). Other have made the opposite claim, that the right hemisphere matures more quickly than the left during the first year of life, and as a result, the right hemisphere takes over visual-spatial functions that begin to develop at birth, leaving the left hemisphere to specialize in linguistic functions that do not get underway until many weeks or months after birth. Variants have been proposed that incorporate both these views, e.g., that the right hemisphere starts out at a higher level of maturity (defined in terms of synaptic density and speed of processing), but the left hemisphere grows more quickly after the first year of life, leading to a division of labor in which critical aspects of visual-spatial functioning are handled on the right while linguistic functions are mediated to a greater degree on the left.

Genetic timing has also been invoked to explain critical-period effects in language learning (Johnson & Newport, 1989; Krashen, 1973; Lenneberg, 1967; Locke, 1993). However, there are at least two versions of the critical-period hypothesis that need to be considered here, one that requires an extrinsic genetic signal and another that does not (Marchman, 1993; see also Oyama, 1992). On the "hard" maturational account, plasticity comes to an end because of some explicit and genetically determined change in learning capacity (such as a reduction in neurotrophic factors). In this case, the genetically timed stop signal is independent of the state of the system when the critical period comes to an end (see also Locke, 1993, 1994). On the "soft" maturational account, no extrinsic stop signal is required. Instead, reductions in plasticity are an end product of learning itself, due to the process of progressive cortical specialization described above. In essence, the system uses up its learning capacity by dedicating circuits to particular kinds of tasks, until it reaches a point at which there are serious limitations on the degree to which the system can respond to insult.

An example of soft maturation, mentioned above, comes from Marchman (1993), who simulated aspects of grammatical development in neural networks that were subjected to lesions (the random elimination of 2% to 44% of all connections) at different points across the course of learning. Although there were always decrements in performance immediately following the lesion, networks with small and/or early lesions were able to recover to normal levels. However, late lesions (if they were large enough) resulted in a permanent impairment of language learning. Furthermore, this impairment was more severe for some aspects of the task than it was for others (e.g., regular verb inflections were more impaired than irregular verbs). Notice that these findings mimic classical critical-period effects described for human language learning (e.g., Johnson & Newport, 1989), but without any extrinsic ("hard") changes in the state of the system. Instead, the network responds to the demands of learning through specializaion, changing its structure until it reaches a point of no return, a point at which the system can no longer start all over again to relearn the task without prejudice.

As Marchman points out, the respective hard and soft accounts of critical-period effects are not mutually exclusive. Both could contribute to the reductions in plasticity that are responsible for differences between children and adults in recovery from unilateral brain injury (see also Oyama, 1992). However, if the soft account is at least partially correct, it would help to explain why the end of the critical period for language in humans has proven so difficult to find, with estimates ranging from one year of age to adolescence (Krashen, 1973; Johnson & Newport, 1989; for a discussion, see Bialystok & Hakuta, 1994).

One major goal of Chapter 6 will be to illustrate how the brain can solve difficult problems by "arranging" the timing of input. The idea we develop is that many complex problems have good solutions which can be best found by decomposing the problem temporally. In this way a solution may be innate not by virtue of being encoded from the start, but by guaranteeing the brain will develop in such a way that the solution is inevitable. We will refer to this kind of constraint as "chronotopic nativism." This is a powerful form of innateness which plays a central role in the evolution of complex behaviors, and is similar at an abstract level to well-known examples of timing effects in the evolution and ontogeny of physical organs (e.g., Gould, 1977; McKinney & McNamara, 1991).

Finally, we note that there is one more very important source of constraint on development, namely the constraints which arise from the problem space itself. Because this source is entirely external to the developing organism, we have not included it in our taxonomy of "ways to be innate." But this fourth source may interact crucially with the other three endogenous constraints.

We refer here to the fact that certain problems may have intrinsically good (or sometimes, unique) solutions. For example, the logical function Exclusive OR (see Chapters 2, 3, and 6) readily decomposes into two simpler functions, OR and AND. Given a range of possible architectures, networks will typically "choose" this solution without being explicitly instructed to find it. The hexagonal shape of the beehive (see Chapter 3) is another example. The hexagonal cell shape is a natural consequence of rules of geometry having to do with maximizing packing density of spheres (which then deform under pressure into hexagons). Thus, in neither the network nor the bee do the solutions (AND and OR units; hexagonally shaped cells) need to be internally specified. These outcomes are immanent in the problems themselves.

TABLE 1.3

Source of constraint		Examples in brains	Examples in networks
Representations		synapses; specific microcircuitry	weights on connections
Architectures	*unit*	cytoarchitecture (neuron types); firing thresholds; transmitter types; heterosynaptic depression; learning rules (e.g., LTP)	activation function; learning algorithm; temperature; momentum; learning rate
	local	number of layers; packing density; recurrence; basic (recurring) cortical circuitry	network type (e.g., recurrent, feedforward); number of layers; number of units in layers
	global	connections between brain regions; location of sensory and motor afferents/efferents	expert networks; separate input/output channels
Timing		number of cell divisions during neurogenesis; spatiotemporal waves of synaptic growth and pruning/decay; temporal development of sensory systems	incremental presentation of data; cell division in growing networks; intrinsic changes resulting from node saturation; adaptive learning rates

Most specific/direct → *Least specific/indirect*

On domain specificity

Armed with these distinctions between forms of innateness, we can turn to the final and perhaps most hotly disputed aspect of the nature-nurture issue, which is the *content* of what is presumed to be innate. To what extent have we evolved mental/neural systems that serve only one master, i.e., are uniquely suited to and configured for a particular species-specific task, and no other task? This is the issue that is usually addressed with the twin terms "modularity" and "domain specificity." These are important but slippery issues. Indeed, the authors of this book have spent a good portion of their careers dealing with these vexing problems (e.g., Bates, Bretherton, & Snyder, 1988; Karmiloff-Smith, 1986, 1992a). And these issues are directly relevant to the question of innateness. We will deal with these issues again in Chapters 3, 5, and 7. For present purposes, let us struggle with a definition of terms.

It is vitally important to note that the word "module" is used in markedly different ways by neuroscientists and behavioral scientists. This has led to considerable confusion and unfortunate misunderstandings in the course of interdisciplinary discussions of brain and cognition. When a neuroscientist uses the word "module," s/he is usually referring to the fact that brains are structured, with cells, columns, layers, and regions which divide up the labor of information processing in various ways. There are few neuroscientists or behavioral scientists who would quibble with this use of the word module. Indeed, Karl Lashley himself probably had something similar in mind, despite his well-known claims about equipotentiality and mass action (Lashley, 1950).

In cognitive science and linguistics, on the other hand, the term module refers to a very different notion. Here, the term embodies a far stronger and more controversial claim about brain organization, and one that deserves some clarification before we proceed.

The strongest and clearest definition of modularity in cognitive science comes from Jerry Fodor's influential book, *The Modularity of Mind* (Fodor, 1983; see also Fodor, 1985). Fodor begins his book with an acknowledgment to the psycholinguist Merrill Garrett, thanking him for the inspiring line, "parsing is a reflex." This is, in fact, the central theme in Fodor's book, and it represents the version of modularity that most behavioral scientists have in mind when they use this term. *A module is a specialized, encapsulated mental organ that has evolved to handle specific information types of particular relevance to the species.* Elaborating on this definition, Fodor defines modules as cognitive systems (especially perceptual systems) that meet nine specific criteria.

Most of these criteria describe the way modules process information. These include *encapsulation* (it is impossible to interfere with the inner workings of a module), *unconsciousness* (it is difficult or impossible to think about or reflect upon the operations of a module), *speed* (modules are very fast), *shallow outputs* (modules provide limited output, without information about the intervening steps that led to that output), and *obligatory firing* (modules operate reflexively, providing predetermined outputs for predetermined inputs regardless of the context). As Fodor himself acknowledges (Fodor, 1985), these five characteristics can also be found in acquired skills that have been learned and practiced to the point of automaticity (Norman & Shallice, 1983; Schneider & Shiffrin, 1977).

Therefore, another three criteria pertain to the biological status of modules, to distinguish these behavioral systems from learned habits. These include *ontogenetic universals* (i.e., modules develop in a characteristic sequence), *localization* (i.e., modules are mediated by dedicated neural systems), and *pathological universals* (i.e., modules break down in a characteristic fashion following insult to the system). It is assumed that learned systems do not display these additional three regularities.

The ninth and most important criterion is *domain specificity*, i.e., the requirement that modules deal exclusively with a single information type, albeit one of particular relevance to the species. Aside from language, other examples might include face recognition in humans and certain other primates, echo location in bats, or fly detection in the frog. Of course, learned systems can also be domain specific (e.g., typing, driving, or baseball), but according to Fodor they lack the instinctual base that characterizes a "true" module. In the same vein, innate systems may exist that operate across domains (see below for examples). However, in Fodor's judgment such domain-general or "horizontal" modules are of much less interest and may prove intractable to study, compared with the domain-specific or "vertical" modules such as language and face recognition.

We discuss these claims at length in Chapters 3 and 7. For present purposes, we point out that a serious problem with such claims regarding domain specificity is their failure to recognize that specificity may occur on (at least) five levels: *tasks, behaviors, representations, processing mechanisms,* and *genes.*

(a) *Specificity of tasks and problems to be solved.* We will define a "task" as a problem that the organism must solve in order to achieve some goal. Each task or problem can be defined in terms of a "problem space," a set of parameters that includes a well-defined goal (G), specification of the environment in which the organism must work (E), the resources and capabilities that the organism brings to bear on the problem (R), and a description of the sequence of operations that will lead to G given E and R. Of course such parameters are implicit in the situation; they are not necessarily represented explicitly anywhere in the organism or in the environment.

Our point for present purposes is this: Most problems are unique, and thus form the starting point for an analysis of domain specificity. For example, human languages emerged within a rich problem space that has little in common with the many other things we do. Put in the simplest possible form, languages represent solutions to the problem of mapping inherently nonlinear patterns of thought onto a linear sequence of signals, under a severe set of processing constraints from human perception, motor coordination and production, and memory. Of course all complex behaviors must be organized in time, but the temporal constraints on language may be unique due to the complexity of the information that must be conveyed and the multiplex of constraints on use of that information in real time—to say nothing of the problem of learning, addressed in more detail in Chapters 2 and 6.

(b) *Specificity of behavioral solutions.* Let us be clear on this point: language is "special," a unique problem unlike any other that we face, and unlike the problems faced by any other species. Similar claims can be made for other complex systems as well. This does not mean that every complex problem has forced special solutions all the way down to the genetic level. However, it is quite likely that specific problems will require specific solutions at the behavioral level. There is, for example, nothing outside of language that resembles past tense morphology in English, case marking in Turkish, nominal classifiers in Navajo, or relative clauses in German and Dutch. Grammars are complex behavioral solutions to the problem of mapping structured meanings onto a linear string of sounds. Any resemblance to other cognitive domains (real or fictional) is purely coincidental. But the same is true for domains that cannot conceivably be innate (e.g., tennis; chess; chicken sexing). The domain specificity of behavior does not in itself constitute evidence for the innateness of that behavior. The real arguments lie at the next few levels down.

(c) *Specificity of representations.* An individual who can produce domain-specific behaviors on a reliable basis must (within the framework we have outlined here) possess a set of mental/neural representations that make those outputs possible. We are persuaded that such representations must also have a high degree of domain specificity. But as we have suggested here, and shall elaborate in detail in Chapter 5, we are skeptical that detailed representations of any kind are innate, at least at the cortical level. So, in our view, the argument for "vertical modules" (Fodor's term for innate and domain-specific systems) must lie at another level.

(d) *Specificity of processing mechanisms.* Assuming that higher (and lower) cognitive processes require domain-specific representations, must we assume that a domain-specific representation is handled by a domain-specific processor, or that it is acquired by a domain-specific learning device? This is the crux of the debate about domain specificity, the point at which arguments about domain specificity and innateness cross in their most plausible and compelling form. This critical issue breaks down into two distinct questions: a "where" question and a "how" question.

The "where" question also breaks into two related issues (addressed in more detail in Chapter 5). Are the representations required for a specific domain coded in a particular (compact) region of the brain, or are they widely distributed across different cortical and/or subcortical areas? Do the representations required for a specific task occupy their own, unique, dedicated neural tissue, or must they share neural territory with other tasks?

It is obvious why these are related issues. On the one hand, if each cognitive function is localized to a specific, compact region, then it is not unreasonable to assume (or it could at least be true) that each region functions as a specialized processor, used always and only for a specific task (e.g., a face processor, a language processor, a music processor, a Grandmother cell, a yellow Volkswagen detector). As we have already noted, this claim is independent of the issue of innateness, since most of these specializations can be (and probably are) acquired through experience. In the limiting case, every concept (innate or otherwise) would be assigned to a specific neuron. With 10^{11} neurons to hand out to all comers, it would take a very long time to run out of neural capacity (see Churchland, 1995, for an enlightening discussion of this point).

On the other hand, if the representations required for a specific task are widely distributed across brain regions, then the case for specialized processors is necessarily weakened. Simply put, we would not have enough brain to go around if we handed out huge tracts of territory for the exclusive use of language, faces, music, etc. If the mechanisms that store (for example) linguistic representations are broadly distributed in the brain, then it is quite likely that those mechanisms are also used for other cognitive functions. That is why claims about localization and claims about domain specificity so often go hand in hand, and why the connectionist notion of distributed representations is not well received by those who believe that domain-specific (perhaps innate) representations must be handled by domain specific (perhaps innate) processing mechanisms.

The "how" question has to do with the nature of the operations that are carried out by a processing device. Assuming that we have a set of domain-specific representations (innate or learned), handled by a dedicated processor (probably local, since distributed but domain-specific processors are limited by their size), does it necessarily follow that the processor carries out unique and domain-specific operations? Can a general-purpose device (or, at least, a multipurpose device) be used to learn, store, activate or otherwise make use of a domain-specific representation? Or must the processing characteristics of the tissue be tailored to the requirements of a specific task, to the point where it really cannot be used for anything else? Even if the processor had rather plastic, general properties at the beginning of the learning process, does it retain that plasticity after learning is complete, or has it lost the ability to do anything else after years of dealing with (for example) language, faces, music, Grandmothers or yellow Volkswagens? The "big issues" of innateness, localization and domain specificity have clear implications for a fourth big issue addressed throughout this volume (but especially in Chapter 5): the issue of developmental plasticity.

We will insist throughout that connectionist models are not inherently "anti-nativist" and that they are certainly not the natural enemy of those who believe in domain specificity (in fact, a major criticism of current connectionist models is that most of them can do only one thing at a time; see Chapter 7, and Karmiloff-Smith, 1992a). However, these models do assume distributed representations, and those representations are usually distributed across processing units of considerable flexibility. To the extent that realistic developments can be modeled in a system of this kind, there is a prime facie case against the notion that domains like language, music, faces or mathematics *must be* carried out by dedicated, innate and domain-specific neural systems. Hence the issue of domain specificity at the level of processing mechanisms connects up with major controversies about innateness. This brings us to the sine qua non of "old-fashioned" nativism, the issue of genetic specificity.

(e) *Genetic specificity.* By definition, if we make a claim for innateness and domain specificity at any of the above levels, we mean that the outcome is ensured by and contained within the genome. But just how, and where, and when does this occur? There seems to be a widespread willingness to believe in single genes for complex outcomes (see Chapter 7 for a detailed accounting of the Grammar Gene Rumor). Things would be much simpler that way! But the evidence to date provides little support for this view. Alas, a complex cascade of interactions among genes is required to determine outcomes as simple as eye color in fruitflies or body types in earth worms. More often than not, the genes operate by controlling variables in timing. When the default schedule is followed (within some limits of tolerance), certain interactions between structures inevitably occur. There is no need for genes to encode and control those interactions directly. Instead, they follow from the laws of physics, geometry, topology—laws of great generality, but laws that have very specific consequences for the actors on that stage (D'Arcy Thompson, 1917/1968).

If this is so demonstrably true for the embryogenesis of simple organisms, why should things be different for the embryogenesis of the human mind? It is unsettling to think that our beloved brain and all its products result from such a capricious game. That is why the idea that single genes and detailed blueprints are responsible for what we are is attractive to many. Albert Einstein once insisted that "God does not play dice with the Universe." We sympathize with this view, but we remember Nils Bohr's reply: "Stop telling God what to do."

The shape of change

If developmental paths were always straight and always uphill, they would not be nearly as interesting as they are. One usually implicitly assumes a linear model of growth and change; we presuppose that, all things being equal, development will be gradual (no sudden accelerations or decelerations), monotonic (always moves in the same upward direction), and continuous (reflecting quantitative changes in some constant measure or dimension). This is the "garden variety" model of change.

So when we observe developmental phenomena which deviate from this model, we find it interesting. But more: We assume that these deviations reflect changes in the underlying mechanisms. So, for example, when we see a U-shaped curve in children's acquisition of the past tense (Brown, 1973; Ervin, 1964; Kuczaj, 1977), it seems reasonable to infer that the child's similar performance at the early and late stages arises from different sources (e.g., rote learning early on, use of rules at the later stage). If we see a child ignoring certain inputs which are attended to later (for example, the relative importance of mass vs. size when attempting to balance weights on a scale), we infer that something internal has changed or appeared which makes it possible for the child to process what was previously ignored. And when we see children able to learn languages which adults learners can only imperfectly master, we assume that some critical component of the learning mechanism which is present in childhood is lost to the adult.

As reasonable as these inferences are, we know now—and this has been one of the powerful lessons of connectionist models—that nonlinear change in behavior can be generated by mechanisms which themselves undergo no dramatic or nonlinear changes in operation whatsoever. Connectionist models illustrate how the same learning mechanism can give rise to behaviors which differ dramatically at various points in time. The changes need not be linear nor monotonic, nor do they need to be continuous. We describe networks with these properties in Chapters 2 and 3, and analyze the basis for their nonlinear behavior in Chapter 4.

Partial knowledge

The idea of partial knowledge is central to the mystery of development. If an adult knows something, and an infant does not, then—unless we assume instantaneous change—the state in between can only be characterized as a state of partial knowledge. And notions of innateness frequently invoke the concept of partial knowledge which is prespecified, to be supplemented by experience. But what exactly is meant by partial knowledge? What could it mean to have "half an idea?" As intriguing as this concept is, it serves no useful or explanatory function if left vague and unspecified.

In a few domains, it is easy to characterize partial knowledge. For example, a child who knows how to multiply and subtract may be said to have partial knowledge of how to do long division. One might even try to quantify just how complete that knowledge is and claim the child has 75% of the knowledge required. But such cases are few and far between. Knowledge of complex domains is more often *not* a matter of assembling pieces in a jigsaw puzzle together. Having 90% of an idea does not usually mean lacking only 10% of the pieces.

In the connectionist models we shall discuss, there are certain key concepts which give us a new way to think about partial knowledge. The idea of *distributed representations* is discussed in Chapters 2 and 3; we see what it means for knowledge to be distributed across a complex system, and the consequences of superposition of representation across the same processing elements. This makes it possible to talk about partial knowledge in situations where that knowledge is integrated across multiple domains. The properties of *gradient descent learning* and *nonlinear* processing are discussed in Chapters 2, 3, and 4. We see how it is possible for incremental learning to give rise to discontinuities in behavior; we can talk about "subthreshold knowledge."

The value of simulations

Although mathematical models are common in some areas of psychology, and computer simulations have a long tradition in fields such as artificial intelligence, many developmentalists may find the methodology of computer simulation of models to be strange. Yet such simulations play a central role in connectionism, and we deem them important enough that we have written a companion volume to this book in order to explain the methodology in detail. Why?

First, these simulations enforce a rigor on our hypotheses which would be difficult to achieve with mere verbal description. Implementing a theory as a computer model requires a level of precision and detail which often reveals logical flaws or inconsistencies in the theory.

Second, although connectionist models often appear simple—they are, after all, merely collections of simple neuron-like processing elements, wired together—their simplicity is deceptive. The models possess nonlinear characteristics, which makes their behavior difficult (if not impossible) to predict. The use of distributed representations also means that the models exhibit emergent behaviors which can usually not be anticipated. The simulation therefore plays the role of an empirical experiment in allowing us to study the behavior of our theory in detail. In the case of evolutionary simulations, there is the additional benefit that effects which normally unfold over hundreds of thousands, or millions or billions of years, can be explored in a speeded-up model which may run in hours or days.

Third, the model's innards are accessible to analysis in a way which is not always possible with human innards. In this sense, the model functions much as animal models do in medicine or the biological sciences. After a computer model has been trained to generate a behavior which is of interest to us, we can inspect its internal representations, vary subsequently the input to it, alter the way it processes the input, and so forth. With humans, we can usually only guess at the nature of the mechanism responsible for a behavior by inference, but with the simulation we can directly inspect the network in order to understand the solution. Of course, it remains to be demonstrated that the model and human that it simulates do things the same way; but the model can be a rich source of hypotheses and constraints which we might not have stumbled across in human experimentation. Indeed, the conceptual role played by these models, in giving us new ways to think about old problems, is for us one of the most exciting and profitable reasons to do connectionist simulations.

*   *   *

In the remaining chapters of this book we shall attempt to flesh out the perspective we have outlined above. We begin in Chapter 2 with a minitutorial on connectionism, selectively emphasizing those concepts which are especially relevant to developmental issues. We stress the importance of connectionism as a conceptual framework, rather than simply a modeling methodology. In Chapter 3 we present what we see as some of the major challenges and mysteries to be solved in development, and discuss a number of connectionist simulations which start to address these issues. Chapter 4 focuses on the different "shapes of change" and introduces notions of nonlinearity and dynamics. We then show how connectionist networks provide accounts for the various patterns of change and growth which are observed in development. Chapter 5 is about brains: what they look like, how they develop, and how they respond to injury. These data bear directly on the question about the plausible loci of constraints. In Chapter 6 we return to the theme of interactions and present simulations which illustrate how very specific outcomes can be produced through very indirect means. Finally, we sum things up in Chapter 7 and propose a new way of thinking about innateness.

We emphasize from the outset that the approach we take is not anti-nativist—far from it. Where we differ from some is that we view the major goal of evolution as ensuring adaptive *outcomes* rather than ensuring prior knowledge. The routes by which evolution guarantees these outcomes are varied and often very indirect. The mechanisms need not be optimal, nor tidy, nor easy to understand. It suffices that they barely work, most of the time. There is an evolutionary trade-off between efficiency and flexibility. We are prepared to call many universally recurring patterns of behavior—in languages, for example—innate, even though we find them nowhere specified directly in the genome. In this sense, our definition of innateness is undoubtedly broader than the traditional view. We also believe that it is richer and more likely to lead to a clearer understanding of how nature shapes its species. We hope, by the time you finish reading what we have to say, that you agree.

CHAPTER 2

Why connectionism?

A conversation

B: *A little bird told me you're writing a book about connectionism and development. I didn't believe it. You must be losing it! What on earth is a nice girl like you doing in such bad company? You must know that connectionism is nothing more than associationism in high tech clothing. I thought you believed in constructivism, interactionism, epigenesis and all that murky stuff.*

A: *Oh, I'm still a believer! But the connectionist framework allows me to come up with a much more precise notion of what all that stuff really means. It can inspire a truly interactive theory about developmental change. Associationist models rested on assumptions of linearity. The multi-layer nets connectionists use are nonlinear dynamical systems, and nonlinear systems can learn relationships of considerable complexity. They can produce surprising, nonlinear forms of change. They've made me completely rethink the notion of stages.*

B: *I don't believe my ears!* Connectionist nets *are simply reactive, they just respond to statistical regularities in the environment.*

A: *No, that may have been true of the earliest models, but the* more complex ones develop internal representations that go well beyond surface regularities to capture abstract structural relationships.

B: *Yeah, but how will you account for the rule-governed behavior of intelligent humans. Networks may be okay at the implementation level, but they can't represent rules of grammar and things like that, and that's what human intelligence is all about.*

A: *Another misconception!* <u>*The transformations that occur during pro-*</u> <u>*cessing in networks do the same work as rules in classical systems.*</u> *But what's interesting is that these* <u>*rules and representations take a radi-*</u> <u>*cally different form from the explicit symbols and algorithms of serial*</u> <u>*digital computers.*</u> *The representations and rules embodied in connec-* *tionist nets are implicit and highly distributed.* *They capture our intu-* <u>*itions about the status of rules in infants' and young children's*</u> <u>*knowledge.*</u> *Part of the challenge of modern research on neural net-* *works is to understand exactly what a net has learned after it has reached some criterion of performance.*

B: *From the sublime to the ridiculous! The next thing you'll say is that connectionism is compatible with nativist ideas too! Come on, it's tab- ula rasa personified! Anyway, though they claim to build nothing in, the connectionists sneak in the solutions by fixing the weights and con- nections or laying out the solution in the way they represent the input.*

A: *Wrong again! First there are lots of different kinds of connectionism and we're trying to make the case for a biologically- and developmen- tally-inspired connectionism. Second, connectionism's not incompati- ble with innately specified predispositions—just depends how you define the predispositions. You're gonna have to read the book! You're right that many early simulations assumed something like a tabula rasa in the first stages of learning (e.g., a random "seeding" of weights among fully-connected units before learning begins). This has proven to be a useful simplifying assumption, in order to learn something about the amount and type of structure that does have to be assumed for a given type of learning to go through. But there is no logical incompatibility between connectionism and nativism. The problem with current nativist theories is that they offer no serious account of what it might mean in biological terms for something to be innate. In neural networks, it is possible to actually explore various avenues for building in innate predispositions, including minor biases that have major structural consequences across a range of environmental condi- tions. As for sneaking in the solutions, we do a lot less of that than classical systems. Don't forget that connectionist networks are self- organizing systems that learn how to solve a problem. As the art is cur- rently practiced, the only one who fiddles with the weights is the sys-*

tem itself in the process of learning. *In fact in a simulation of any interesting level of complexity, it would be virtually impossible to reach a solution by "hand-tweaking" of the weights. As for the issue of "sneaking the solution into the input," don't forget that there are several examples of simulations in which the Experimenter did indeed try to make the input as explicit as possible—and yet the system stubbornly found a different way to solve the problem! Connectionist systems have a mind of their own and very often surprise their modelers.*

B: *Jesus, I need a drink! A mind of their own! What next?*

A: *Okay, I set that one up, but seriously, the way networks learn is surprisingly simple, yet the learning yields surprisingly complex results. We think that complexity is an emergent property of simple interacting systems—you see this throughout the physical and biological world.*

B: *So the next thing I'm gonna hear is that connectionism has some biological plausibility! Lip service to neurons! Spare us, please!*

A: *Well, connectionists work at many different levels between brain and behavior. In current simulations of higher cognitive processes, you're right, the architecture is "brain-like" only in a very indirect sense. The typical 100-neuron connectionist toy is "brain-like" only in comparison with the serial digital computer (and don't forget, old boy, that serial digital computers are wildly unlike nervous systems of any known kind). There are two real questions: First, is there anything of interest that can be learned from simulations in simplified systems, and second, can connectionists "add in" constraints from real neural systems in a series of systematic steps, approaching something like a realistic theory of mind and brain? You know, there are many researchers in the connectionist movement who are trying to bring these systems closer to neural reality. Some are exploring analogues to synaptogenesis and synaptic pruning in neural nets. Others are looking into the computational analogues of neural transmitters within a fixed network structure. The current hope is that work at all these different levels will prove to be compatible, and that a unified theory of the mind and brain will someday emerge. Don't tell me we are a long way off, that's obvious. But most connectionist researchers are really committed to ulti-*

mate neural plausibility, which is more than you can say for most other approaches. Anyway, what's really exciting is that it has launched a new spirit of interdisciplinary research in cognitive neuroscience, and why I got excited is that it has really crucial implications for getting a developmental perspective into connectionism and a connectionist perspective into developmental theorizing.

B: *Well, you've got a lot of arguing to do to convince me about that!*

A: *As I said just now, you'll have to read the book! So where's that gin and tonic?*

Nuts and bolts (or nodes and weights)

The first thing to be said about connectionist networks is that most are really quite simple, but their behaviors are not. That is part of what makes them so fascinating. We will describe some of these behaviors in detail in this chapter, particularly those which are relevant to the developmental issues we are concerned with in this book (e.g., innateness, modularity, domain specificity, etc.). Before doing that, however, we need to have some working knowledge of the nuts and bolts of the framework. We therefore begin with an overview of basic concepts which characterize most connectionist models.

Basic concepts

There is considerable diversity among connectionist models, but all models are built up of the same basic components: Simple processing elements and weighted connections between those elements. The processing elements are usually called *nodes* or *units*, and are likened to simple artificial neurons. As is true of neurons, they collect input from a variety of sources. Some nodes receive and send input only to other nodes. Other nodes act like sensory receptors and receive input from the world. And still other nodes may act as effectors, and send activation outward. (Some nodes may even do all three things.) Figure 2.1 illustrates several architectures. Here, we use the convention of a filled circle to represent a node, lines between nodes to indicate communication channels (conceptually, roughly similar to dendrites and axons), and arrows to indicate the direction of information flow.

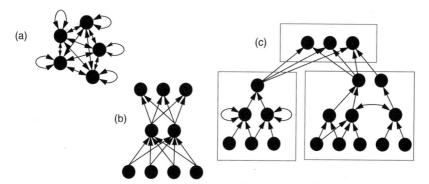

FIGURE 2.1 Various types of connectionist networks. (a) A fully recurrent network; (b) a three-layer feedforward network; (c) a complex network consisting of several modules. Arrows indicate direction of flow of excitation/inhibition.

We can expand this basic picture in a bit more detail by considering the dynamics of processing. A node receives input from other nodes to which it is connected. In some networks these connections are unidirectional (as in Figure 2.1b). In other cases, connections may be bidirectional (Figure 2.1a). The connections between nodes have *weights*. It is in the weights that knowledge is progressively built up in a network. These are usually real-valued numbers, e.g., 1.234, –4.284. These weights serve to multiply the *output* of the sending nodes. Thus if node *a* has an output of 0.5 and has a connection to node *b* with a weight of –2.0, then node *b* will receive a signal of –1.0 (0.5 × –2.0); in this case the signal would be considered *inhibitory* rather than *excitatory*.

A given node may receive input from a variety of sources. This input is often simply added up to determine the *net input* to the node (although other types of connections have been proposed in which products of inputs are taken, as in *sigma-pi* units). It is useful to develop a formalism for describing the manner in which the activity of a node is computed from other sources of activity in the network. Let us suppose that the single input coming from some node j is the product of that node j's activation (we'll call this number a_j) and the weight on the connection between node j and node i (we'll designate this weight w_{ij} where the first subscript denotes the receiving node and the second subscript denotes the sending node—a convention adopted from matrix notation in linear algebra). Then the single input from node j is the product $w_{ij}a_j$. The total input over all incoming lines is just the sum of all of those products (we'll use the symbol Σ_j to indicate summation from all node sources j). More compactly, we can write the net input to node i as

$$net_i = \sum_j w_{ij}a_j \qquad \text{(EQ 2.1)}$$

This is the total input received by a node. But like neurons, the *response* of the node is not necessarily the same as its input. As is true of neurons, some inputs may be insufficient to cause the node to "do" very much. What a node "does" is captured by the node's activation value. So we want to know what the *response function* of a node is: For any given input, what is the corresponding output (or activation)?

In the simplest case, if the node is a *linear unit*, the output activation is in fact just the same as its net input. Or the node may have a slightly more complex output function, emitting a 1.0 just in case the net input exceeds some threshold, and outputting a 0.0 otherwise. The Perceptron and the McCulloch-Pitts neuron both had this characteristic; they are examples of *linear threshold units*. (Note that, despite their name, they have an important nonlinearity in their response.)

A more useful response function is the logistic function

$$a_i = \frac{1}{1 + e^{-net_i}} \qquad \text{(EQ 2.2)}$$

(where a_i refers to the activation (output) of node i, net_i is the net input to node i, and e is the exponential). We have graphed the activation of a node with this activation function in Figure 2.2.

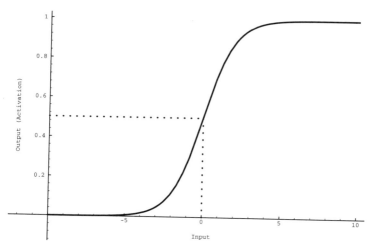

FIGURE 2.2 The sigmoid activation function often used for units in neural networks. Outputs (along the ordinate) are shown for a range of possible inputs (abscissa). Units with this sort of activation function exhibit an all or nothing response given very positive or very negative inputs; but they are very sensitive to small differences within a narrow range around 0. With an absence of input, the nodes output 0.5, which is in the middle of their response range.

This figure tells us what the output is (the vertical axis) for any given net input (the horizontal axis). For some ranges of inputs (large positive or negative ranges along the horizontal) these units exhibit an all or none response (i.e., they output a 0.0 or 1.0). This sort of response lets the units act in a categorical, rule-like manner. For other ranges of inputs, however, (close to 0.5) the nodes are very sensitive and have a more graded response. In such cases, the nodes are able to make subtle distinctions and even categorize along dimensions which may be continuous in nature. *The nonlinear response of such units lies at the heart of much of the behavior which makes such networks interesting.*

Note that in the absence of input (0.0 on the horizontal axis), the node's output is 0.5, which is right in the middle of its possible response range. Often this is a reasonable response, because it means that with no input, a node's response is equivocal. But sometimes times it is useful for nodes to have a default value other than 0.5, so that in the absence of input, they might be either "off" (output 0.0), "on" (output 1.0), or perhaps take some other intermediate value. This notion of different defaults is similar to the idea of varying thresholds, and can be accomplished by giving each node one additional input from a pseudo-node, called the *bias node*, which is always on. The weight on the connection from the bias node may be different for different units. Since each unit will always receive input from the bias unit, this provides a kind of threshold or default setting, in much the same way as humans exhibit default reactions to stimuli in the absence of additional data.

It is not difficult to see how simple networks of this kind might compute logical functions. For example, logical AND could be implemented by a network with two input units and a single output unit. The output unit would be "on" (have a value close to 1.0) when both inputs were 1.0; otherwise it would be off (close to 0.0). If we have a large negative weight from the bias unit to the output, it will by default (in the absence of external input) be off. The weights from the input nodes to output can then be made sufficiently large that if both inputs are present, the net input is great enough to turn the output on; but neither input by itself would be large enough to overcome the negative bias.

As this example makes apparent, part of what a network knows lies in its architecture. A network with the wrong or inappropriate input channels, with inputs which don't connect to the output, or with the wrong number of output units, etc., cannot do the desired job. Another part of a network's knowledge lies in the weights which are assigned to the connections. As the AND example shows, the weights are what allow the correct input/output relation to be achieved.

How do we know what architectures to choose, and what weights? In much of the connectionist work done in the early 1980's, and in an approach still pursued by many researchers today, networks are designed by hand and reflect theoretical claims on the part of the modeler. For example, in the word perception model of McClelland and Rumelhart (1981; Rumelhart & McClelland, 1982), there are separate layers of nodes which are dedicated to processing information at the *word, letter,* and *orthographic feature* levels (see Figure 2.3). The connections between nodes within and across layers reflect the non-arbitrary relationships between the concepts represented by the nodes. Thus, the node for the word "trap" receives positively weighted input from the letter nodes "t", "r", "a", and "p"; and it is inhibited by other word nodes (since only one word may be reasonably present at once).

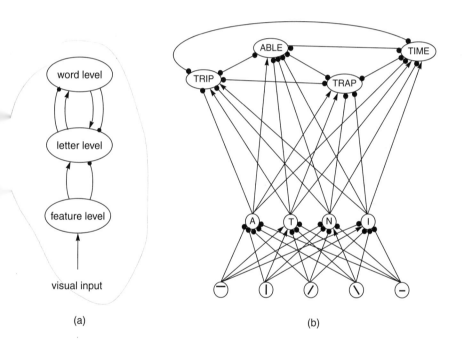

FIGURE 2.3 (a) Global view of word perception model of McClelland & Rumelhart. (b) Detailed fragment of the model. Connections ending in arrows are excitatory; connectionist ending in filled circles are inhibitory. (Adapted from McClelland & Rumelhart (1981).

This model instantiates a theory which McClelland and Rumelhart developed in an attempt to account for a range of experimental data. Even though the model was hand-crafted and relatively simple, it exhibits a number of behaviors which are not obvious simply from inspecting the model. As is often the case with these systems, the nonlinearities and high degree of interaction give rise to phenomena that were not themselves deliberately programmed in. (It would be wrong to say they were not programmed in, since they obviously do result from the specifics of the way the model has been designed. What is relevant here is just that they do not result from the modeler's intentional efforts to produce the behaviors.) Indeed, some of these behaviors are quite unexpected and make predictions about human behavior which can then be tested experimentally.

Learning

There are other significant aspects of models such as the McClelland and Rumelhart model about which we shall have more to say when we talk about representation. For now we note that the hand-wiring which is required may be problematic. As the degrees of freedom (these include, among other things, the number of nodes and connections) grow, so too does the number of ways of constructing networks. How do we know which is the best way? Or, equivalently, how do we know we have the right theory? There may be cases where we feel we have a good idea about the inputs and outputs which are relevant to some behavior—these can be observed directly. But what we want is to use the model to help *develop* a theory about the internal processing which gives rise to this behavior, rather than just *implementing* a theory we already hold. The question then becomes, is there some way by which networks can configure themselves so that they have the appropriate connections for a task? Can we use the networks for theory development?

Hebbian learning

One of the earliest ideas about how networks might learn came from Donald Hebb. Speaking of real nervous systems, Hebb suggested that

> When an axion of cell A is near enough to excite a cell B and repeatedly or persistently takes part in firing it, some growth process or metabolic change takes place in one or both cells such that A's efficiency, as one of the cells firing B, is increased. (Hebb 1949; p. 62)

There are various ways of capturing this notion mathematically, but Hebb is essentially proposing here that learning is driven by correlated activity. Thus, in an artificial system in which changes in connection strength plays the role of changes in synaptic potentiation, the rule for weight change might be

$$\Delta w_{ij} = \eta a_i a_j$$

(EQ 2.3)

(where Δw_{ij} represents the change in the weight on the connection from sending node j to receiving node i, η is a small constant of proportionality (the learning rate), and a_i and a_j are the activations of the two nodes). The Hebb rule has been studied extensively and is widely used in modeling today. It has at least two attractive virtues: there are known biological mechanisms which might plausibly implement it (e.g., Long Term Potentiation, Bliss & Lømo, 1973); and it provides a reasonable answer to the question, "Where does the teacher information for the learning process come from?" It is easy to believe that there is a great deal of useful information which is implicit in correlated activity, and the system need not know anything in advance. It simply is looking for patterns.

However, there is a serious limitation to Hebbian learning, which is that *all* it can learn is pair-wise correlations. There are cases where it is necessary to learn to associate patterns with desired behaviors even when the patterns are not pair-wise correlated, or may exhibit higher-order correlations. The Hebb rule cannot learn to form such associations.

The Perceptron Convergence Procedure

One solution to this dilemma is to base learning on the difference between a unit's *actual* output and its *desired* output. Two early techniques were proposed by Rosenblatt (1959, 1962) and Widrow and Hoff (1960). The approaches were very similar, and we will focus here on Rosenblatt's.

Rosenblatt proposed what he called the Perceptron Convergence Procedure (PCP; Rosenblatt, 1962). This is a technique which makes it possible to start with a network of units whose connections are initialized with random weights. The procedure specifies a way to take a target set of input/output patterns and adjust the weights automatically so that at the conclusion of training the weights will yield the correct outputs for any input. Ideally, the network will even generalize to produce the correct output for input patterns it has not seen during training.

The way that learning works is that the input pattern set is presented to the network, one pattern at a time. For each input pattern, the network's actual output is compared with the target output for that pattern. The discrepancy (or error) is then used as a basis for changing weights to inputs, and also changing the output node's threshold for firing. How much a given weight is changed depends both on the error produced and the activation coming along a weight from a given input.[1] The underlying goal is to minimize error on an output unit by apportioning credit and blame to the inputs (the intuition being that if you have made a mistake, you should probably play more attention to the people who were shouting loudest at you to go ahead and make that mistake).

This learning procedure (as well as the Widrow-Hoff rule) addresses some of the limitations of Hebbian learning, but it has its own limitations. It only works for simple two-layer networks of input units and output units. This turns out to be an unfortunate limitation. As Minsky and Papert (1969) showed, there are classes of problems which can not be solved by two-layer networks of percep-

---

1. There are actually several variations; in some the weight and thresholds change by a constant.

trons. The best known of these is the exclusive-OR (XOR) problem, but the basic problem is a very deep one which occurs in many situations. Two-layer networks are rather like S-R pairs in classical psychology. What is required is something between input and output that allows for internal (and abstract) representations.

Similarity in neural networks—a strength and a weakness

One of the basic principles which drives learning in neural networks is *similarity*. Similar inputs tend to yield similar outputs. Thus, if a network has learned to classify a pattern, say 11110000, in a certain way then it will tend to classify a novel pattern, e.g., 11110001, in a like fashion. Neural networks are thus a kind of analogy engine. The principle of similarity is what lets networks generalize their behaviors beyond the cases they have encountered during training.

In general, the similarity metric is a good rule-of-thumb. Lacking other information, it is reasonable that networks (and people) should behave in similar ways given similar situations. On the other hand, there are clearly cases where we encounter two patterns which may resemble each other superficially, but which should be treated differently. A child learning English must learn that although "make" and "bake" sound almost the same, they form the past tense in different ways. Even more extreme is the case where two words may be identical at the surface level. For instance, the word "did" can either be a modal (as in "I did go") or a main verb (as in "I did badly"). Despite their surface identity, we must learn to treat them differently. More prosaically, kittens and tigers may share many physical features but we do not want to treat them in the same way. The problem thus is that although similarity is often a good starting point for making generalizations, and lacking other information we do well to rely on similarity, there are many circumstances in which similarity of physical form leads us astray. We might still wish to invoke the notion of similarity, but it is an abstract or functional kind of similarity. Kitty cats, dogs, fish, and parakeets are dissimilar at the level of appearance but share the abstract feature "domestic animal"; tigers, rhinoceri, wild boar, and cobras do not look at all alike but are similar at the more abstract level of being "wild animals."

The neural network learning procedures we have described so far (Hebbian learning and the PCP with two-layer networks) have the important limitation that they can only learn to generalize on the basis of physical similarity. (To make this more precise, by physical similarity we mean here similarity of *input pattern*.) In fact, the PCP is actually guaranteed to discover a set of weights which give the right response to novel inputs, provided the correct response can be framed in terms of physical similarity (defined in a certain way). Conversely, when the correct response cannot be defined in terms of similar inputs yielding similar outputs, the PCP is guaranteed to fail!

One of the best known problems which illustrates the limitation of the PCP and simple two-layer networks is the Exclusive Or problem (XOR). XOR is a what is known as a Boolean function. (A Boolean function just means that we have to take some set of inputs, usually 1s and 0s, and decide whether a given input falls into a positive or a negative category. These categories are often denoted "true" or "false," or if we are dealing with networks, produce output node activations of 1 or 0.) In the case of XOR, the inputs consist of a pair of numbers (either one of which can be a 1 or a 0), and the task is to decide whether the pair falls into the "true" category or the "false" category. For a network this would mean taking inputs of the form shown in Table 2.1 and producing the appropriate output value.

TABLE 2.1

INPUT		OUTPUT
0	0	0
1	1	0
0	1	1
1	0	1

Why can't a two-layer perceptron network solve this problem? Let's imagine we have a network that looks like the one shown in Figure 2.4. It is not hard to see why this problem cannot be solved

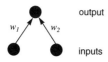

output

w_1 w_2

inputs

FIGURE 2.4 two-layer network that cannot learn to solve the XOR problem.

by such a network. Consider what we would be asking of the two weights (w_1 and w_2) in this network in order to produce the correct input/output mappings.

First, remember that these weights are multipliers on the inputs, so when we want an output to be "true" (i.e., 1) at least one of the inputs must be a 1, and at least one of the weights must be big enough so that when the two numbers are multiplied, the output node turns on. Now, the first two patterns (0,0 and 1,1) have to produce a 0 on the output. For that to happen, the weights connecting the inputs to the output node be sufficiently small that the output node will remain off even if both input nodes are on. That's easily enough obtained: just set both weights to 0. That way, even if both inputs are on (1,1), the output unit will remain off.

But this is at odds with what is required for the third and fourth patterns (01 and 10). In these cases, we need the weights from *either* input to be sufficiently big such that one input alone will activate the output. Thus we have contradictory requirements; there are no set of weights which will allow the output to come on just in case either of the inputs is on, but which will keep it off if both are on.

There is another way of thinking about this problem, and it's worth introducing now because it allows us to introduce a framework for thinking about representations in networks which we have found very useful. It is also a bit more intuitively understandable than simply thinking about weights and such.

We have referred to patterns such as 0,0 and 1,1 as vectors. A vector can be thought of either as a collection of numbers, or as a point in space. If our vectors consists of two numbers, as in the case of our inputs in this example, then the vector points in a two dimensional space, with each number indicating how far out along the dimension the point is located. We could thus visualize the four input patterns that make up the XOR problem in the two dimensional space shown in Figure 2.5.

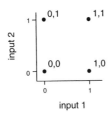

FIGURE 2.5 Spatial representation of the four input patterns used in XOR.

Thinking of these patterns as points in space now gives us a way of defining a bit more precisely what we mean by similarity. It is simple: we judge the similarity of two vectors by their (Euclidean) distance in space. As a first pass at understanding why XOR is difficult, notice that the pairs of patterns which are furthest apart— and therefore most dissimilar—such as 0,0 and 1,1 are those which need to be grouped together by the function. Nearby patterns such as 0,1 and 0,0, on the other hand, are to be placed in different groups. This goes against the grain.

The problem is actually a bit worse. Technically, the difficulty is that the input/output weights impose a linear decision bound on the input space; patterns which fall on one side of this decision line are classified differently than those patterns which fall on the other side. When groups of inputs cannot be separated by a line (or more generally, a hyperplane) then there is no way for a unit to discriminate between categories. This is shown in shown in Figure 2.6. Such problems are called *nonlinearly separable* (for the simple reason that the categories cannot be separated by a line).

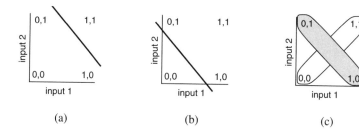

FIGURE 2.6 Geometric representation of the XOR problem. If the four input patterns are represented as vectors in a two-dimensional space, the problem is to find a set of weights which implements a linear decision boundary for the output unit. In (a), the boundary implements logical AND. In (b), it implements logical OR. There is no linear function which will simultaneously place 00 and 11 on one side of the boundary, and 01 and 10 on the other, as required for (c).

Although this example may seem a bit arcane, it is useful because it illustrates the idea of vectors as patterns in space, of similarity as distance in space, and of the problem that can arise when patterns which are distant in space (and therefore inherently dissimilar) must nonetheless be treated as though they were similar.

The problem, then, is how to take advantage of the desirable property of networks which allows similarity to have a role in generalization, while still making it possible to escape the tyranny of similarity when it plays us false. We want to be able to define similarity at an abstract and not only form-based level. (Or, to put it another way, we want to have our cake and eat it too.)

Solving the problem: Allowing internal representations

It turns out that this problem is relatively easy to solve. With an extra node or two interposed between the input and output (see Figure 2.7) the XOR problem can be solved.[2] These additional nodes

---

2. A two-layer network can actually solve XOR if the output is not a perceptron-like unit. More precisely, the output cannot have an activation function which is a monotonically increasing function of its input. A cosine activation function, for instance, will suffice.

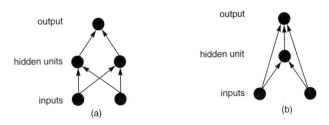

FIGURE 2.7 Two architectures for solving the XOR problem. In both cases, at least one internal ("hidden") unit is required.

are often called "hidden units" because they are hidden from the world. They are equivalent to the internal representations we invoke in psychological theorizing. The inputs and outputs to hidden units are confined to other parts of the network. Hidden units are extraordinarily powerful; they make it possible for networks to have internal representations of inputs which capture the more abstract, functional relationships between them. Indeed, it is tempting, and not entirely unwarranted, to think of input units as analogous to sensors, output units as motor effectors, and hidden units as interneurons. However, while this metaphor captures an important aspect of what units do, we stress that there are occasions in which inputs and outputs may be interpreted in very different ways and not correspond to sensory/motor maps at all. Input similarity still plays an important role, and all things being equal, the physical resemblance of inputs will exert a strong pressure to induce similar responses. But hidden units make it possible for the network to treat physically similar inputs as different, as the need arises. The way they do this is by transforming the input representations to a more abstract kind of representation.

We can use the example of XOR to illustrate this point. Once again the spatial interpretation of patterns proves useful. In Figure 2.8 we see in (a) what the inputs "look like" to the network. The spatial distribution of these patterns constitutes the intrinsic similarity structure of the inputs. The vectors representing these points in space are then passed through (multiplied) the weights between inputs and hidden units. This process corresponds to the

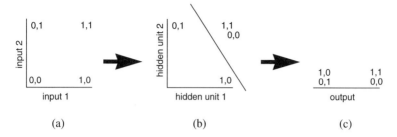

FIGURE 2.8 Transformation in representation of the four input patterns for XOR. In (a) the similarity structure (spatial position) of the inputs is determined by the form of the inputs. In (b) the hidden units "fold" this representational space such that two of the originally dissimilar inputs (1,1 and 0,0) are now close in the hidden unit space; this makes it possible to linearly separate the two categories. In (c) we see the output unit's final categorization (because there is a single output unit, the inputs are represented in a one-dimensional space). Arrows from (a) to (b) and from (b) to (c) represent the transformation effected by input-to-hidden weights, and hidden-to-output weights, respectively.

first arrow. The weights from input to hidden units have the effect of transforming this input space; the input space is "folded," such that the two most distant patterns (0,0 and 1,1) are now close in the hidden unit space. (Remember that hidden unit activation patterns are vectors too, so we can graph the location of the activations produced on the hidden layer by any set of inputs.) This reorganization sets things up so that the weights to the output unit (second arrow, between (b) and (c)) can impose a linear decision bound, the line cutting across the space in (b), and resulting in the classification shown in (c).

Having said that hidden units can be used to construct internal representations of the external world which solve difficult problems (in fact, Hornik, Stinchcombe, and White, 1989, have proven that a single layer of hidden units gives networks the power to solve essentially any problem), all might seem rosy. But there's still a problem. It is one thing for a network to be able to *solve* a problem, in principle; this merely says that there exists some set of weights which enable the network to produce the right output for any input. It is another thing for the network to be able to *learn* these weights.

So in fact, although it had long been known that more complex multilayered networks could solve problems such as XOR, the real challenge (which Minsky and Paper suggested probably could not be solved) was how to *train* such networks. The PCP works with two layer networks, but not when there are additional hidden layers.

Fortunately, there are several solutions to this problem. One of the best known is called backpropagation of error (Rumelhart, Hinton, & Williams, 1986). Because many of the simulations we describe in this book use this learning procedure, we shall spend a bit of time describing how it works, at least at a general level.

Backpropagation of error (informal account)

"Backprop," as it has come to be known, works very much like the PCP (or the Widrow-Hoff rule, which is also very similar; Widrow & Hoff, 1960). Recall that the approach of the PCP is to adjust the weights from input unit(s) to output unit(s) in such a way as to decrease the discrepancy between the network's actual output and its desired output. This works fine for the weights leading to outputs, because we have a target for the outputs and can therefore calculate the weight changes. When we have hidden units, however, the question arises: How shall we change the weights from inputs to hidden units? The strategy of credit/blame assignment requires that we know how much error is already apparent at the level of the hidden units—even before the output unit is activated. Unfortunately, we do not have a predefined target for the hidden units, only for the output units. So we cannot say what their activation levels should be, and therefore cannot specify an error at this level of the network.

Backprop solves this problem in a clever but entirely reasonable way. The first step is to figure out what the error is at the level of output units (just as in the PCP). This error is simply the difference between the activation we observe on a given output unit, and the activation it is supposed to have (commonly called the *target* or *teacher* value).

The second step is to adjust the weights leading into that unit so that in the future it is either more or less activated, whichever is needed in order to decrease the error. We can do this on all the weights leading into the output unit. And if there are more than one output units, we simply repeat the same two steps (error calculation; weight change calculation) for each one, using each unit's own target to determine the error.

Now we come to the question of how to change the weights leading into the hidden units. The procedure we used for the hidden to output weights worked because we knew the target values for output units. How do we determine target values for hidden units? We assume that each hidden unit, because it is the thing which excites (or inhibits) the output units, bears some responsibility for the output units' errors. If a certain output unit is very far from its target value, and has been highly activated by a certain hidden unit, then it is reasonable to apportion some of the blame for that on the hidden unit. More precisely, we can infer the shared blame on the basis of (a) what the errors are on the output units a hidden unit is activating; and (b) the strength of the connection between the hidden unit and each output unit it connects to. Hence the name of the algorithm: We propagate the error information (often called the error signal) backwards in the network from output units to hidden units. Notice that this same procedure will work iteratively. If we have multiple layers of hidden units, we simply calculate the lower hidden layers' error based on the backpropagated error we have collected for the higher levels of hidden layers.

Formal account

This informal account of backprop can be made somewhat more explicit and precise. In the remainder of this section we will attempt to provide a more rigorous treatment of the way the backpropagation learning works. Although not necessary for understanding how backprop works at an intuitive level, the formal account is worth at least perusing. The notation may seem daunting to those unfamiliar with mathematical formalism, but it is really just a shorthand to more compactly represent the concepts we have just discussed.

We said that the first step in the learning algorithm was to calculate the errors on the output units. We do this a unit at a time. Since the procedure is normally the same for all units, for the purposes of generality we will refer to an output unit with the index i. We will call the error for that unit e_i. The observed output for that same unit will be represented as o_i and the output unit's target value will be t_i. Computing the error is simple. It is just

$$e_i = t_i - o_i \qquad \text{(EQ 2.4)}$$

Now that we know the error that is being produced for a unit, we want to adjust the weights going into that unit so that in the future it will be more or less activated, but in a way that reduces the error we just calculated. Since weights connect two units, we will use the subscripts i to represent the unit on the receiving end of the connection, and the subscript j to represent the sending unit. The weight itself can be referred to as w_{ij} —so the first subscript always refers to the receiving unit and the second subscript to the sending unit. The change in the weight, which is what we want to calculate, is Δw_{ij}.

The weight adjustment should be such that the error will be decreased as the weights are changed. This notion is captured by a construct called a partial derivative, which tells us how changes in one quantity (in our case, network error) are related to changes in another quantity (here, change in weights). Therefore, letting network error be symbolized as E, and the symbols $\dfrac{\partial E}{\partial w_{ij}}$ to represent the partial derivative of the error with respect to the weight, then the equation for the weight change can be formalized as

$$\Delta w_{ij} = -\eta \frac{\partial E}{\partial w_{ij}} \qquad \text{(EQ 2.5)}$$

(where η is a small constant of proportionality called the learning rate).

We will not give the actual computations which can be performed on the partial derivative in order to convert it to a more usable form (see Rumelhart, Hinton, & Williams, 1986); we simply report the result that the right-hand side of Equation 2.5 can be calculated as

$$\Delta w_{ij} = \eta \delta_{ip} o_{jp}$$
(EQ 2.6)

The quantities on the right-hand side of Equation 2.6 are now in a form which we can identify and begin to use in training a network.

This equation says that we should change the weights in accord with three things (corresponding to the three symbols on the right-hand side). First, we are trying to find a set of weights which will work for many different patterns. So we want to be cautious and not change the weights too much on any given trial. We therefore scale the calculated weight change by a small number. This is the term η and is called the learning rate. Skipping to the third term, o_{jp}, we also make our weight change on the connection from the sender unit j to receiver unit i be proportional to j's activation. That makes sense; after all, if the sender unit hasn't contributed any activation to unit i, it won't have contributed to i's error. The second subscript p in the expression o_{jp} indicates that we are only considering the activation of unit j in response to input pattern p. Different input patterns will produce different activations on unit j.

Finally, the middle term, δ_{ip}, reflects the error on unit i for input pattern p. We have spoken previously of the error as being simply the difference between target and output; the term δ_{ip} includes this but is a bit more complicated. The reason for that just has to do with the calculation of the partial derivative in Equation 2.5. Although we will not attempt to explain the derivation here, there is one interesting consequence to the definition of these δ_{ip} terms which we comment on below.

For an output unit (i), this term is defined as

$$\delta_i = (t_i - o_i) f'(net_i)$$
(EQ 2.7)

which says that our error is, straightforwardly, the difference between the target value for unit i on this pattern and the actual output. This discrepancy is modulated by the derivative of the unit's current activation. As we said, we will not attempt here to justify the presence of this derivative; it simply has to do with the way Equation 2.5 is worked out. However, this term has an important effect on learning.

The derivative of a function is its slope at a given point. The derivative measures how fast (or slowly) the function is changing at that point. If we look back at Figure 2.2, which shows the activation function for sigmoidal units, we see that the slope is steepest in the mid-range and decreases as the unit's activation approaches either extreme. This has important consequences for learning which we shall return to several times in this book. When networks are first created, they are typically assigned random weights values clustered around 0.0. This means that during early learning, activations tend to be in the mid-range, 0.5 (because no matter what the inputs are, they are being multiplied by weights which are close to 0.0; hence the net input to a unit is close to 0.0, and the activation function maps these into the mid-range of the receiving unit's activation). There is a further consequence: When weight changes are computed, the error term is modulated by the derivative of the unit (see Equation 2.7). This derivative is greatest in the mid-range. Therefore, all things being equal, the weights will be most malleable at early points in learning. As learning proceeds, the impact of any particular error declines. That's because once learning begins, weights will deviate more and more from 0.0 and the net input to a unit will tend to produce its minimum or maximum activation levels.

This has both good and bad consequences. If a network has learned the appropriate function, occasional outlier examples will not perturb it much. But by the same token, it may be increasingly difficult for a network to correct a wrong conclusion. Ossification sets in. The interesting thing about this phenomenon, from a developmental viewpoint, is that it suggests that *the ability to learn may change over time—not as a function of any explicit change in the mechanism, but rather as an intrinsic consequence of learning itself.* The network learns to learn, just as children do.

We have just seen how the learning algorithm works for weights from hidden to output units. How do we compute the δ_{ip} for hidden units? We cannot use the difference between the target value and actual output, because we don't know what the targets for hidden units should be. But as we pointed out earlier, we can calculate a hidden unit's error in an indirect fashion from the error of the output units, because each hidden unit bears some responsibility for those errors. So for any hidden unit i, we collect the error of the k output units to which it projects, weighted by the connections between them, and use that sum:

$$\delta_i = f'(net_i)\sum_k \delta_k w_{ki} \qquad \text{(EQ 2.8)}$$

This procedure can be used for networks with arbitrarily many levels of hidden units.

In summary, backprop is an extremely powerful learning tool and has been applied over a very wide range of domains. One of the attractions is that it actually provides a very general framework for learning. The method implements a gradient descent search in the space of possible network weights in order to minimize network error; but what counts as error is up to the modeler. This is most often the squared difference between target and actual output (as described here), but in fact, *any* quantity which is affected by weights may be minimized.

Learning as gradient descent in weight space

Because the idea of gradient descent in weight space figures importantly in later chapters, we wish to be clear about what this means. A graphical representation is helpful here. Imagine a very simple network of the sort shown in Figure 2.9a.

There are two trainable weights in the network shown in Figure 2.9a. Let us suppose that we have a data set which we wish to train the network on. We might systematically vary the two weights through their possible range of values. For each combination of weights, we could pass the training data through the network and see how well it performs. We could then graph the resulting error as a function of all possible combinations of the two weights. A hypothetical version of such a graph is shown in Figure 2.9b. Regions of the surface which are low along the z axis

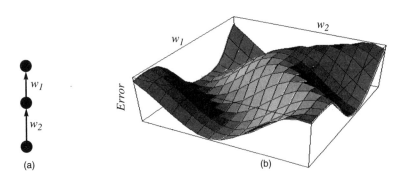

FIGURE 2.9 (a) Simple network with 2 trainable weights. (b) A hypothetical graph depicting the error produced for some imaginary set of input/output training patterns, for all possible values of weights w_1 and w_2. Low regions on the surface correspond to low error.

tell us that the combination of weights (x and y axes) produce low error and so these are weight settings which are desirable.

This same technique might be used, in principle as a form of learning. We could discover good regions of weight space empirically by sampling the entire space. But this would be cumbersome and in networks of any complexity, quite impractical. What back-propagation (and many other neural network learning algorithms) provides is a technique for starting at some random point on this surface and following the downward gradient. If we are fortunate, there will exist some combination of weights which solves the problem, and we will find it.

Is success guaranteed? Not at all. We know from theoretical arguments that any problem *can* be solved with a three-layer network (Hornik, Stinchcombe, & White, 1989) but we do not know *a priori* the precise architecture needed to solve that problem (how many hidden units, what pattern of connectivity, etc.). Furthermore, if in the process of following the gradient we take very big steps as we change our weights we might overshoot a good region of weight space. Or we might find ourselves trapped in a region which is locally good but not perfect (this is called a *local minimum*). Since the gradient at this spot points up—all weight changes lead to poorer performance—we may be stuck. In fact, one of the hypotheses of this book is that *evolution produces a developmental profile which interacts with learning in just such a way as to cleverly help us avoid such local minima.*

Other architectures and learning algorithms

Backpropagation and Hebbian learning are perhaps the best known and most widely used connectionist learning algorithms. There are a number of other alternative techniques for training networks, however. We chose not to delve into these largely because to do so would tax the patience of our readers, and there are many fine texts which give a more comprehensive presentation of network learning (see, for example, Hertz, Krogh, & Palmer, 1991 for an excellent mathematical presentation; and Bechtel & Abrahamsen, 1991, for a more cognitively/philosophically oriented review). Furthermore, the majority of the work which we will discuss employs either Hebbian or backpropagation learning, and so it is most important to us that readers be acquainted with these approaches. We simply wish to emphasize that the connectionist paradigm embraces a multitude of approaches to learning, and the interested reader should not imagine what we have presented here exhausts the range of possibilities.

Issues in connectionist models

As exciting as the early successes in connectionist modeling have been, there remain a number of important challenges. We would like now to turn to some of the issues which the field is currently focussing on.

Representing time

The networks we described in talking about backpropagation (e.g., the networks shown in Figure 2.4 and Figure 2.7) are called "feed-forward networks." The flow of activation proceeds from input through successive layers and culminating in output units. The flow of information is unidirectional, and at the conclusion of processing an input, all activations are wiped clean and the next input is received.

Processing in such networks is atemporal, in the sense that the activations of units reflects only the current input. The only sense in which time figures in is if the network is still undergoing training, in which case the weight changes implicitly reflect the time course of learning. We might think of these changes in connection strength as implementing semantic long-term memory, in much the same way that Hebb envisioned it.

Early models attempted to circumvent the lack of temporal history in processing through an ingenious trick. Input units would be conceptualized as being divided into groups, with each group processing one input in a series. In the simplest case, for example, 10 temporally ordered inputs might be presented having the first input unit respond to the first event, the second input unit respond to the second event, etc. All 10 events/inputs would then be processed simultaneously.

Although this approach has been used to good effect in a number of models (e.g., McClelland & Rumelhart 1981; McClelland & Elman, 1986; Sejnowski & Rosenberg, 1986), it has some very basic defects (see Elman, 1990, for review). This has led researchers to investigate other ways of representing time. An enormous part of the behavior of humans and other animals is clearly time-dependent, and so the problem is a serious one. What would seem to be lacking in the feedforward network is an analog to short-term or working memory.

Recurrent networks implement short-term memory by allowing connections from nodes back to themselves, either directly or indirectly, as in Figure 2.10. In this way, the network's activity at any point in time can reflect whatever external input is presented to it, plus its own prior internal state (where that state corresponds to activations at prior points in time). There exist a variety of algorithms and architectures which make such recurrence possible (e.g., Elman, 1990; Jordan, 1986; Pearlmutter, 1989; Pineda, 1989; Rumel-

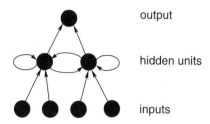

FIGURE 2.10 A recurrent network. The network has fully interconnected hidden units; each hidden unit activates itself and the other hidden unit. As a result, the current state (activation) of a hidden unit reflects not only new input from the input units, but the hidden units' prior activation states.

hart, Hinton, & Williams, 1986; Stornetta, Hogg, & Huberman, 1988; Tank & Hopfield, 1987; Williams & Zipser, 1989).

Scaling and modularity

The examples we have used so far have involved networks which are rather simple in design. The networks contain relatively few units; they tend to have uniform internal structure (e.g., they are organized into layers in which each layer is fully connected to the next layer); and they are trained on tasks which are usually very simplified versions of real-world situations.

No apologies are needed for such simplifications. In many cases the goal is to abstract away from irrelevant complexity in order to focus on the heart of a problem. In other cases, the simplification may be justified by the immaturity of the technology. We are still at a point where we are trying to understand basic principles. Much of the research is empirical, and it is useful to work with systems which are tractable.

There are also potential hazards associated with these simple architectures. Let us mention two.

First, there is the problem of scaling up to bigger and more realistic problems. For complex and large-scale problems, it is not clear that networks with uniform architecture are optimal. As a network grows in size (as it must, for larger-scale problems), the space of potential free parameters—weights—grows exponentially, whereas the size of the training data typically grows more slowly. Techniques such as gradient descent typically do not do well in searching out good combinations of weights when the weight space is very large relative to the training data available. In Figure 2.9 we showed a very simple network and its associated (hypothetical) weight space. Gradient descent relies on the information provided by training examples to find the set of weights which yield minimum error. Imagine how much more complicated that task may be in a search space containing hundreds of weights, particularly if we have a relatively small number of examples. We are likely to get "trapped" in regions of weight space which work for a few of the examples but not for others.

One way to address the problem of determining optimal network architectures involves the use of what are called "constructive algorithms." These are procedures which allow a network to be dynamically reconfigured during training through the addition or deletion of nodes and weights. One of the best known techniques was proposed by Scott Fahlman, and is called Cascade Correlation (Fahlman & Lebiere, 1990). We'll see a developmental example of this in Chapter 3. Omitting details, this procedure works with a network that initially contains no hidden units. It then adds new hidden units gradually in order to reduce error. A somewhat different technique has been suggested by Steve Hanson (Hanson, 1990). In Hanson's scheme, connections weights are not discrete, but instead have a mean and variance. As training progresses, if a weight's variance grows too large the connection may split in two (hence the name, "meiosis networks"). Dynamic configuration through pruning has also been studied. Rumelhart (1987) and Weigend (1991) found that in many cases, a network's ability to generalize its performance to novel stimuli was improved by gradually eliminating weights, as long as this did not lead to increased error.

In addition to scaling, there is a second risk which can arise with overly simple networks. This problem is subtler. Here the difficultly is not just that there are two few or two many units in the network. It is that they are not connected in the best way. For example, sometimes a problem which appears hard can be solved if it is first decomposed into smaller pieces. Then, rather than have a single network attempt to learn the entire problem as an undifferentiated whole, we might more efficiently use a network architecture which reflects the problem's functional decomposition. This is the insight which underlies a proposal by Jacobs and his colleagues (Jacobs, 1990; Jacobs, Jordan, & Barto, 1991; Jacobs, Jordan, Nowlan, & Hinton, 1991).

As an example of a difficult function which can be made easier if broken into pieces, Jacobs and colleagues point to the absolute value function. The absolute value of a number is simply its "positive value"; if the number is already positive, then that is also its absolute value. If the number is negative, then the absolute value negates that, making it positive. Formally, the function is defined as

$$f(x) = \begin{cases} -x & \text{if } (x < 0) \\ x & \text{if } (x \geq 0) \end{cases} \qquad \textbf{(EQ 2.9)}$$

This is a nonlinear function (because of the "bend" at 0), and can be learned by a single network which has at least one hidden unit. On the other hand, the function can also be learned by a network which has two modules (e.g., such as shown in Figure 2.11). Each module consists of a single linear unit, plus a simple gating network which decides which output to use, depending on whether the input is less than or greater than 0. Such a network should find it easier to learn the task because each subcomponent is linear and there are no hidden units.

Jacobs, Jordan and Barto (1991) have trained a network with such an architecture to do the "what/where" task. In this task, the network has to determine what an object is, and where it is in the visual field. They found that this modular architecture facilitated learning, compared with nonmodular networks. Furthermore, if one of the expert networks is composed of units with linear activation functions, that module always learns to carry out the "where" task.

We like this result, because it shows how it is possible for task assignment to be innately determined, not on the basis of the task

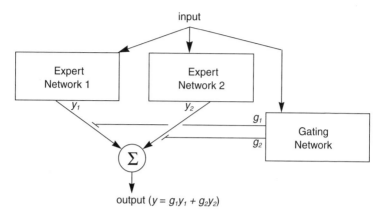

input

Expert Network 1   Expert Network 2

y_1 y_2

g_1
g_2

Gating Network

Σ

output ($y = g_1y_1 + g_2y_2$)

FIGURE 2.11 Modular network proposed by Jacobs, Jordan, and Barto (1991). Input is sent to both expert networks, each of which is specialized for one aspect of the task, and also to a gating network. Both expert networks generate output; the expert network decides the appropriate mixture, given the specific input.

per se, but rather the match between a task's requirements and the computational properties of network architectures. It is not necessary that a piece of network be committed to a task in an explicit and hard-wired fashion, with one set of "where" units and another set of "what" units. Instead, the intrinsic capabilities of the module simply select those tasks for which it happens to be suited. Thus, the way in which the task/architecture mapping is specified may be through biasing and not rigid assignment. The important lesson is that some problems have natural good solutions; they have computational requirements which impose their own constraints on how the problem can be solved. *Nature does not always need to provide the solution; it often suffices to make available the appropriate tools which can then be recruited to solve the problems as they arise.*

Where does the teacher come from? Supervised vs. unsupervised learning

Learning algorithms such as backprop depend crucially on a prior notion of what good performance is. The training environment consists of input/output patterns in which the output is a target or teacher for the network. The error is defined as the discrepancy between the network's output and the target output (supplied by the teacher pattern) which goes with the input. Is this assumption of a teacher pattern psychologically reasonable?

In some cases, it is. For example, let us say we carry out an experiment in which a subject is given some input, performs some action, and then gets feedback. This scenario clearly resembles the training regime which a network undergoes. The feedback is exactly equivalent to the teacher for the network.

But it is not necessary to be quite so literal-minded. There is no reason not to take a somewhat more abstract view of what a network model is capturing. For instance, consider the case of learning the various forms associated with different tenses, person, number, and mood of verbs. There is no obvious teacher which is provided to children (i.e., their learning experience typically consists simply of hearing correctly inflected forms). On the other hand, suppose we conceptualize the learner's task as one of *binding forms*. The child has to learn to associate a set of morphologically related forms with each other. Failure to learn the correct associations would lead the child to anticipate forms which are not confirmed by the actual input; these failures would then constitute a kind of indirect teacher. The teacher signal in this scenario is internally generated. In fact, Nolfi, Parisi, and colleagues (Nolfi & Parisi, 1993, 1994, 1995) have used an evolutionary approach in order to develop networks which provide their own internal teacher in just such a manner.

Nonetheless, forms of training which require a teacher of any sort—called "supervised learning"—clearly have a fairly restricted domain within which they can be plausibly applied. In many circumstances it is not reasonable to suppose that the kind of detailed feedback which is required is available, from any source. One important goal of connectionist modeling has been to find ways to overcome this limitation, while hopefully retaining some of the attractive aspects of gradient descent learning.

There are several ways in which the supervised learning paradigm may be altered to make it more realistic. One form of training involves what is called "auto-association." In this task, a network is given an input and is trained to reproduce the same input pattern on the output layer. What makes this a non-trivial problem is that such networks (such as the one shown in Figure 2.12) contain a narrow "waist" in the middle. This means that the network is forced to

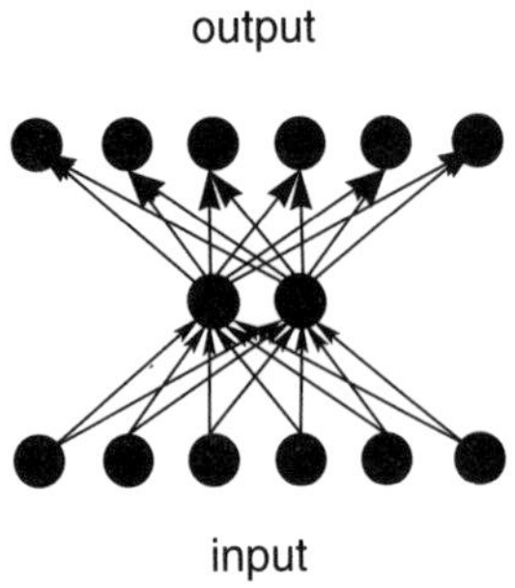

output

input

FIGURE 2.12 An autoassociator network. The network is trained to reproduce the input pattern on the output layer. The lower-dimensionality of the hidden units requires that the network find a more efficient encoding of the input patterns, and can be used for feature discovery.

find a lower-dimensional representation of the inputs. Often these internal representations capture interesting features of the inputs. For example, Elman and Zipser (1988) trained autoassociators to reproduce speech sounds, and found that hidden units learned to respond to different classes of sounds (e.g., vowels, consonants, and certain stops). This form of training also addresses the question of where the teacher comes from, since in autoassociation, the teacher is nothing more than the input itself. Minimally, all that is required is a short-term memory.

Another task which is similar in spirit to autoassociation is the task of predicting the future. The network shown in Figure 2.13 is what has been called a simple recurrent network (Elman, 1990), or SRN. (The network is simple in the sense that the error derivatives are propagated only one time step back into the past; this does not prevent the SRN from storing information in the distant past, but learning longer distance temporal dependencies may be difficult.). An SRN contains recurrent connections from the hidden units to a layer of context units. These context units store the hidden unit activations for one time step, and then feed them back to the hidden units on the next time step. The hidden units thus have some record of their prior activation, which means that they are able to carry out tasks which extend over time. (Note that the hidden units may continue to recycle information over multiple time steps, and also will find abstract representations of the time. So this sort of network is not merely a tape-recording of the past.)

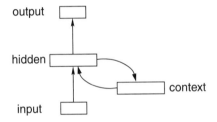

output

hidden

context

input

FIGURE 2.13 Simple recurrent network (SRN). Layers of nodes are shown as rectangles.

Insofar as there are interesting sequential dependencies in the training data, the SRN is often capable of discovering them through learning to predict. The task of prediction has a certain ecological validity, since there is evidence that anticipation plays a role in early learning in many domains. There are also biological mechanisms which are plausibly implicated in mediating learning through prediction (Cole & Robbins, 1992; Morrison & Magistretti, 1983). As is true for autoassociation, the prediction task requires no special teacher, since the target output is simply the next input. All that is required to be psychologically plausible is to assume that this processing can lag a few steps behind the actual input. These forms of learning might be called "self-supervised learning."

Finally, there exists a form of supervised training called "reinforcement learning" (Barto & Andanan, 1985; Barto, Sutton, &

Anderson, 1983; Sutton, 1984) in which the teacher takes a simpler form. Rather than being instructed on exactly which aspect of an output was right or wrong (remember that in backprop, each output unit gets an error signal), the network is simply given a scalar value reflecting its overall performance—a little like being told you're getting warmer (closer to the solution) or colder (further from the solution). The problem, of course, is figuring which part of the output is contributing to the error, and so it is not surprising that reinforcement learning is much slower than fully supervised learning. But it is certainly easier to believe that this sort of situation—in which we are only told how well we did, and not why—is a psychologically plausible one.

In addition to these weakened forms of supervised learning, there are training regimes which involve "unsupervised learning." Hebbian learning is a paradigm example of this. Here, the network weights are changed according to the correlated activity between nodes which are receiving input from the environment. None of that input need be instructive in any direct sense; the network may be thought of as more or less passively experiencing the environment and striving to discover correlations in the input. Many other forms of unsupervised learning have been proposed, including competitive learning (Grossberg, 1976; Rumelhart & Zipser, 1986), feature mapping and vector quantization (Kohonen, 1982; Nasrabadi & Feng, 1988), and adaptive resonance theory (Carpenter & Grossberg, 1987, 1988). These approaches are particularly relevant when the goal is to uncover latent features or categories in a set of stimuli.

Note, however, that although these training regimes may reasonably be called unsupervised, it is not exactly the case that these approaches are entirely theory-neutral or non-parametric. Each learning regime attempts to optimize some quantity, whether it be correlations, or mutual information, or harmony, etc. So there is still supervision; it's just folded into the learning rule itself rather than into the environment being learned. We point this out only as a reminder that no learning rule can be entirely devoid of theoretical content nor can the *tabula* ever be completely *rasa*.

Another promising approach which shares the goal of unsupervised learning has been proposed by Becker and Hinton (1992) In their multiple maps scheme, Becker and Hinton replace the external teacher with internally generated teaching signals. As they put it, "these signals are generated by using the assumption that different parts of the perceptual input have common causes in the external world" (Becker & Hinton, 1992; p. 372). Thus, *the principle of coherence functions as the teacher.* Practically, what this involves is having separate parts of the network receive different (but related) parts of the perceptual input. Each module then learns to produce outputs which provides maximum information about the way in which the other will respond to various inputs. Becker and Hinton have shown that such a scheme can be applied to the problem of extracting depth from random dot stereograms. In principle, one can imagine that the approach could be used to discover correspondences across modalities as well. This could be very useful in modeling infants' capacity for cross-modal imitation, for example. Other related approaches (related in terms of goals, while using different algorithms) have been proposed by Kohonen (1982) and Zemel and Hinton (1993).

A final note on Hebbian learning: Earlier, we talked about the computational limitations in what can be learned solely on the basis of correlated activity. Despite these limitations, we emphasize that Hebbian learning plays an extremely important role in many models, particularly those which are concerned with biologically plausibility. There are similarities between Hebbian learning and long-term potentiation (LTP; for example, in hippocampal neurons), according to some authors (McNaughton & Nadel, 1990; Rolls, 1989). Furthermore, despite the limitations, a great deal can be learned through correlations. For example, as we shall see, Hebbian learning provides a plausible model of how the visual cortex might self-organize to produce visual ocular dominance columns (Miller, Keller, & Stryker, 1989; see discussion below), orientation selective cells (Linsker, 1986, 1990), and to detect dilation and rotation (Sereno & Sereno, 1991). Hebbian learning also plays an important role in the models of object detection (O'Reilly & Johnson, 1994; O'Reilly & McClelland, 1992) and of synaptic pruning (Kerszberg, Dehaene,

& Changeux, 1992; Shrager & Johnson, in press) which we discuss in Chapter 7. (See Fentress & Kline, in press, for a collection of recent work involving Hebbian learning.)

Finding first principles

Connectionism comes in many flavors. In the past decade, there has been a stunning explosion of architectures, algorithms, and applications. Models have been developed of everything from inter-cellular interactions to lobster stomach muscles, expert systems, acquisition of grammatical gender in German, and plasma flow in nuclear reactions. As with any field, there are camps and factions which have very different goals. Given this diversity, is there anything shared in common? Are there any basic principles which underlie the connectionist approach, and which these different models have in common?

We think so. The differences are of course important, sometimes cut deep, and in fact represent a positive state of affairs. Such ferment and diversity are critical to the health of the field (one of our own goals in this book is to urge new directions for connectionists). But when all is said and done, we nonetheless note that there are recurring characteristics which appear in connectionist models and which are, sometimes tacitly, sometimes explicitly, valued by modelers.

We would like to make a stab at suggesting what some of these "first principles" might be. We do so not with the goal of establishing a catechism for determining who is a card-carrying connectionist, but simply because it is important to understand why the models work. To what extent does their behavior reflect superficial differences with other computational frameworks, and to what extent does it flow from underlying properties? Unless we have some notion of what these underlying properties are, such questions cannot be satisfactorily answered. There is another reason for trying to understand what makes these models tick. We have found the most useful aspect of connectionism to be the concepts it makes available. Thinking like a connectionist need not require doing simulations, in our view. What is more important is being able to use the conceptual toolbox.

With this perspective, we would like now to focus here on four aspects of connectionist models which seem to us to be particularly relevant to developmental issues: *the problem of control, the nature of representation, nonlinear responses,* and *the time-dependent nature of learning.*

Who's in charge? Eliminating the homunculus

One of the banes of cognitive science is the *homunculus*—the idea that our mental life is controlled by an inner being who observes the world through the retinal "screen," listens to sounds through the cochlear "earphones," and guides our actions by throwing switches that innervate our muscles. (If one were a preformationist, one would expect to find this fellow in miniature in the fertilized egg.)

The logical problem of such a scenario, with its potential for infinite regress, is self-evident. But although the homunculus is today not a popular fellow among most self-respecting cognitive scientists, he may often be found lurking in disguise in many theories. The majority of theories of planning and control, for example, presuppose some controlling entity which basically does what a homunculus does. It is easy to see why such a fellow would be useful. Without him, there's a real paradox: If something is not in control of behavior, then why is behavior not uncontrolled?

Connectionist models implement an appealing solution to this problem: Global effects arise from local interactions. Coordinated activity over the entire system thus occurs as an emergent property, rather than through the efforts of a central agency. This is not a new idea, certainly. The Belousov-Zhabotinsky reaction, in which entirely local oxidation reactions give rise to formation of waves over macroscopic distances, is a classic example of emergent behavior (see Chapter 3). Alan Turing (1952) was one of the first to show how such reaction/diffusion (RD) processes might be involved in cell patterning. More recently, RD models have been proposed for a number of developmental phenomena, such as the formation of the tiger's stripes (or a cheetah's spots, depending on initial conditions), or the development of visual ocular dominance columns (see below). *Connectionist models are attractive because they provide a computational framework for exploring the conditions under which such emergent properties occur.*

Computation is local in (most) connectionist models in two senses. First, nodes often have restricted patterns of connectivity; they connect only to some of the other nodes in the network. In this regard, they resemble neurons, which usually have a local area within which is there is dense interconnectivity and much sparser connections to distant areas. The activity which emerges over the entire assembly of units therefore reflects a complex process in which many local interactions, most involving relatively simple computations, yield a global state which is not the result of any single unit.

Second, many learning algorithms use only local information when changing system parameters (such as weights on connections). In the case of Hebbian learning, connections between units are modified in a way which reflects the units' joint activity. In the case of backpropagation learning, each output unit has a local sense of error and the weights into that unit are changed so as to reduce that single unit's error. (There are, to be sure, learning rules which require knowledge of global error, but such rules are often criticized for precisely this reason.) Let us give two examples of how local computation and local learning may produce globally organized behavior. The first example deals with the relatively low-level process by which visual cortex might become organized. The second example deals with the higher-level process by which the rules governing legal sequences of sounds in a language might be learned.

One of the notable characteristics of primary visual cortex in many species (such as cat, monkey, and human) is the presence of alternating patches of tissue which serve primarily one or the other eye. These ocular dominance columns are not present from birth, but in normal circumstances develop inevitably during the early part of life. Figure 2.14 shows a surface view of such tissue in the normal cat.

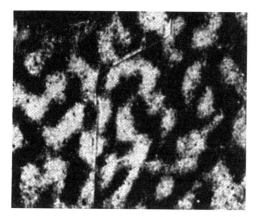

FIGURE 2.14 Ocular dominance patches in layer IV of the cat visual cortex. The image is produced from serial autoradiograph following injection of [^{3}H]-amino acid into one eye. From LeVay, Stryker, & Shatz, 1978.

Since the majority of this tissue initially receives input from both eyes, but the eventual preferential response is clearly affected by experience (deprivation of input from one eye may disrupt the process), an important question is how might this organization arise. Miller, Keller, and Stryker (1989) and Miller, Stryker, and Keller (1988) have shown by simulation one plausible explanation. In their model, layer 4 "cortical" cells receive excitatory input from both "retinal" sources (via the lateral geniculate nucleus) in such a way as to preserve retinotopic organization; synapses from the two eyes are nearly equal in strength. Intracortical connections also exist and are modifiable by a Hebb-like learning rule.

Initially, the randomly assigned connection strengths result in a nearly uniformly innervation from both eyes, as has been observed in new-born kittens. This is shown in the top left panel of Figure 2.15, marked T=0. As time progresses, there is a progressive differentiation of the response, leading to patches which resemble those found in the older kitten. These columns emerge as the result of naturally occurring differences in the degree of correlation between neighboring cells in each eye versus across eyes, along with competitive interactions among layer 4 cells. From their analy-

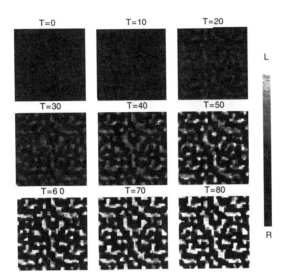

FIGURE 2.15 Development of ocular dominance columns in model by Miller, Keller, & Stryker (1989). Degree of innervation from each eye is shown by grayscale (right eye = white; left eye = black) at various time steps during learning, from T = 0 to T = 80.

sis of the model, Miller and his colleagues are further able to account for the precise form of the various pathologies which occur as a result of different types of abnormal experience.

A second example illustrates emergent behavior at a much higher level of cognitive phenomenon. The TRACE model (McClelland & Elman, 1986) was developed to account for a set of experimental findings in the area of speech perception. TRACE used an interactive activation architecture similar to that of the word perception model discussed earlier (Figure 2.3). Different layers of nodes carried out processing at the level of acoustic/phonetic feature extraction, phoneme recognition, and word recognition. Many of the effects the model attempted to account for involved the important role of context in speech processing, particularly as a solution to the high degree of variability observed in the signal.

An interesting by-product of the architecture was observed, however. The word-to-phoneme connections resembled those shown for the word-to-letter connections in Figure 2.3. This meant that the lexicon provided a very strong top-down influence on perception; that influence usefully compensated for degraded or missing input. But it also had another effect. The model could be given input corresponding to no known word. In that case, the correct set of phonemes would be activated, but of course no single word node would become active, although many partially similar words might achieve some activation. The pattern of phoneme activation was not unaffected by this activity at the word level, however, and noticeably different responses were obtained in the case of different sorts of non-word input. The sequence "bliffle," for example, was processed much better than the sequence "dliffle," in the sense that in the first case all phonemes were clearly activated; but in the second case the initial "d" and "l" phonemes were barely active.

In fact, this result is in close accord with the strong intuition English speakers have that the first sequence is better (i.e., more English-like) than the second. Linguists describe this tacit knowledge which speakers have about acceptable sound sequences with what are called phonotactic rules. English, for instance, is assumed to have a rule which marks word-initial sequences such as "dl-", "tl-", "bw-", "pw-" (among others) as ungrammatical. (Note that these rules must be language-specific and have nothing to do with articulatory difficulty, since many other languages happily tolerate sequences which are illegal in English.)

It is easy to see where the preference for grammatical sequences comes from. The non-word input activates many words which resemble it; the more word-like the input, the more word nodes first become active and then contribute top-down excitation to the phoneme level. In the case of very deviant input, few word nodes are activated and the phoneme activations depend solely on bottom-up input. (Subjectively, this is not unlike the experience of trying to identify the sounds in an unfamiliar language.) Thus what looks like rule-guided phonotactic knowledge arises simply as a result of the statistics which are present in the lexicon.[3] This is another example of emergent behavior in a connectionist network.

Connectionist representations

Early connectionist models, and also many current ones, adopt what is called a "localist" form of representation (more recently, these have been called "structured representations"). The word recognition model and the TRACE model are examples of models which use this type of representation.

Localist representations are similar in some ways to traditional symbolic representations. Each concept is represented by a single node, which is atomic (that is, it cannot be decomposed into smaller representations). The node's semantics are assigned by the modeler and are reflected in the way the node is connected to other nodes. We can think of such nodes as hypothesis detectors. Each node's activation strength can be taken as an indicator of the strength of the concept being represented.

The advantage of localist representations are that they provide a straightforward mechanism for capturing the possibility that a system may be able to simultaneously entertain multiple propositions, each with different strength, and that the process of resolving uncertainty may be thought of as a constraint satisfaction problem in which many different pieces of information interact. Localist representations are also useful when the modeler has *a priori* knowledge about a system and wishes to design the model to reflect that knowledge in a straightforward way. Finally, the one-node/one-concept principle makes it relatively easy to analyze the behavior of models which employ localist representations.

Localist representations also have drawbacks, and these have led many modelers to explore an alternative called "distributed representations." In a distributed representation, a common pattern of units is involved in representing many different concepts. Which concept is currently active depends on the global pattern of activity across the ensemble of units.

---

3. If this account is correct, it predicts that technically ungrammatical non-word sequences might still be perceived better than other grammatical non-word sequences, just in case the ungrammatical non-words happened to almost resemble many real words. This prediction was subsequently verified experimentally with human listeners (McClelland & Elman, 1986).

Figure 2.16 gives an example of distributed representations. The same group of units is shown in Figure 2.16a and Figure 2.16b, but the units have different activations. Concepts are associated, not with individual units, but instead with the global pattern of activations across the entire ensemble. Thus, in order to know which concept is being considered at a point in time, one needs to look across the entire pattern of activation (since any single unit might have the same activation value when it participates in different patterns). The information needed to decide which concept is being represented is distributed across multiple units.

FIGURE 2.16 Examples of distributed representations. Both (a) and (b) illustrate different patterns of activation of the same set of four units. Activation values for individual units are shown as numbers above each unit. Note that the second unit has the same activation value in both representations; in order to know which concept is represented, one has to look at the entire set of nodes.

This style of representation seems to be more consistent with the brain stores information than localist representations (e.g., Lashley, 1950). A note of caution, however: Most modelers who study higher-level cognitive processes tend to view the nodes in their models as equivalent not to single neurons but to larger populations of cells. The nodes in these models are *functional* units rather than *anatomical* units. So it is not clear how heavily to weigh what seems to be the somewhat greater biological plausibility of distributed representations.

The real advantages have to do with representational capacity. Localist representations impose a rigid framework on a model's conceptual contents. Although graded activations in a localist model allow one to capture probabilistic or partial knowledge, there is still a fixed and discrete inventory of concepts. Distributed representations tend to be richer and more flexible. This can be seen by spatial comparison of the two types of representation.

We pointed out earlier that patterns of activation can be thought of as vectors, in which the activation of each unit is the value of an element in the vector, and the entire vector represents the values of all the units in the pattern. As vectors, these activation patterns can also be represented geometrically. Each unit corresponds to a dimension (in "activation space"), and the unit's activation indicates where along that dimension the vector is located. Thus if we had activation patterns involving three units, we could represent the patterns in a three-dimensional space, using the x dimension to represent the values of the first hidden unit, the y dimension to represent the second, and the z dimension to represent the third.

If we use these three units in a localist fashion, then we are permitted exactly three distinct vectors, corresponding to the activation patterns $100, 010, 001$. The spatial representation of these vectors is shown in Figure 2.17a.

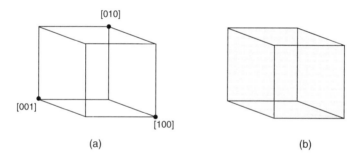

(a)   (b)

FIGURE 2.17 Localist representations in (a) pick out just three distinct vectors. Distributed representations in (b) fill the entire space occupied by the cube.

On the other hand, if we use a distributed representation, then our activation patterns may involve any combination of values on all the units. The activation pattern $0.5, 0.5, 0.5$ picks out a point in the center of the cube. This means we have at our disposal the *entire volume of space* represented in the cube shown in Figure 2.17b. Practically speaking, of course, there are limitations on just how precisely one might be able to distinguish vectors which are very close, but it is still clear that given the same number of units, the conceptual space is much larger for distributed representations than for localist representations.

There are several other outcomes from having available the entire activation space for representation, and these interact with another important observation. When we spoke earlier of the XOR problem, we said that networks operate on the principle that "similar inputs yield similar outputs." We defined similarity in terms of the spatial proximity of activation patterns and said that activation patterns—and the concepts they represent—are similar to the degree they are close in activation space. Vectors which are close, by Euclidean distance, are similar. Patterns which are distant in space are dissimilar.

Consider the implications of this for localist representations. The vectors in Figure 2.17a are all equidistant (they are also orthogonal to one another). There may or may not be relevant similarity relationships between the concepts they represent, but there is no way to capture this in concepts' representations. On the other hand, because the entire activation space is available with the distributed representations in Figure 2.17b, one can envision a range of possibilities. Things which are close in conceptual space can be represented by activation patterns which are close in activation space; the degree of dissimilarity can be measured by their distance. The space may even be organized hierarchically. Let us give a concrete example of this, from Elman (1990).

In this task, a simple recurrent network was trained to predict successive words in sentences. Words were input one at a time, and the network's output was the prediction of the next word. After the network made its prediction, backprop learning was used to adjust the weights and then the next word was input. At the end of each sentence the first word from the next sentence was presented. This process continued over many thousands of sentences.

The words themselves were represented in a localist fashion. That is, a word such as "cat" appeared as a 31-bit vector with one bit on (1) and the remaining bits set to 0. Because of the localist representations, words were equidistant in the activation space and there was therefore no similarity between them. This scheme was adopted quite deliberately, because it meant that the network was deprived of any clues regarding the grammatical category or meaning of the words which it might use in making predictions. Instead, the network had to rely entirely on the distributional statistics of the stimuli to carry out the task. (In much the same way, the acoustic form of a morpheme bears no intrinsic relationship to its meaning and one might imagine this as similar to the situation of the very young language learner, who has no prior knowledge of which meanings are associated with which words).

The network eventually learned to carry out the task, although not precisely as trained. For example, given the sequence of vectors corresponding to the words "the girl ate the...," the network activated *all* the output units which corresponded to the various words representing edible things, rather than predicting exactly the specific word which followed.

One might infer from this result that the network somehow developed the notion "edible." This category, and others such as noun, verb, etc., were not represented in either the input or output representations, so the only place remaining to account for the behavior would be in the hidden unit activations. The hidden units define a very high dimensional space (there were 150 hidden units, which map a 150-dimensional hypercube), and one might expect that the network would learn to represent words which "behave" in similar ways (i.e., have similar distributional properties) with vectors which are close in this internal representation space. Ideally, one would like to be able to visualize this space directly. But since the space is so highly dimensional, indirect techniques must be used. One of these involves forming a hierarchical clustering tree of the words' hidden unit activation patterns.

This is a fairly simple process. It involves first capturing the hidden unit activation pattern corresponding to each word, and then measuring the distance between each pattern and every other pattern. These inter-pattern distances are nothing more than the Euclidean distances between vectors in activation space we spoke of earlier, and we can use them to form a hierarchical clustering tree, placing similar patterns close and low on the tree, and more distant groups on different branches. Figure 2.18 shows the tree corresponding to hierarchical clustering of the hidden unit activations obtained from the sentence prediction task.

We see that the network has learned that some inputs have very different distributional characteristics than others, and forms hidden unit representations which places these two groups in different areas of activation space. These groups correspond to what we would call nouns and verbs. In addition to grammatical differences, the network uses the spatial organization to capture semantic differences (e.g., humans vs. animals). The organization is hierarchical as well. "Dragon" occurs as a pattern in activation space which is in the region corresponding to the category animals, and also in the larger region shared by animates, and finally in the area reserved for nouns.

This spatial framework allows some categories to be distinct and disjoint, but also makes it possible to have representations which lie between category boundaries. An example of this occurs in a phenomenon called sound symbolism. This refers to the common occurrence in which sequences of sounds have loose associations with meanings. In English, for instance, words which contain a final "-rl" ("curl," "unfurl," "burl," "whirl," "twirl," etc.) often evoke the image of circularity. The association is psychologically real in the sense that given a nonce word such as "flurl," speakers will generate definitions involving some circular aspect. These associations seem to lie somewhere between the level of sound (phonemes which by definition carry no meaning) and systematic meaning (morphemes). This is the sort of phenomenon which might be exploited by distributed representations.

Another aspect of distributed representations which makes them useful is their tendency to be context-sensitive. In this way they differ fundamentally from traditional symbolic representations, which are abstract and context-insensitive.

Context-sensitivity may be encoded by the precise location of a representation in activation space. Consider again the sentence prediction task described above. If one compares the position of the vector for a word across the many instances it occurs, one notices

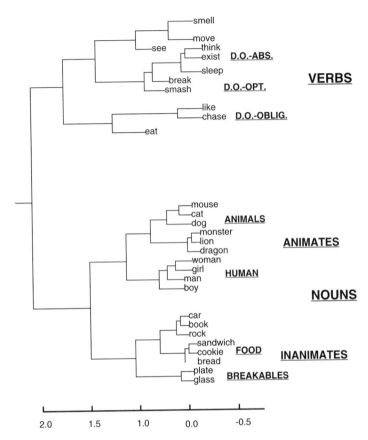

FIGURE 2.18 Hierarchical clustering of hidden unit activation patterns from the sentence-prediction task (Elman, 1990). The network learns distributed representations for each word which reflects its similarity to other words. Words and groups of words which are similar are close in activation space, and close in position in the tree.

that the location varies. Thus there will be many different vectors corresponding (for example) to the word "dragon"; the tree shown in Figure 2.18 was actually formed using vectors which averaged across context. (The reason why the hidden unit representations varies is because the internal representations are created by activation both from the word input itself, which is invariant, and the prior internal state encoded in the context units, which is variable.)

This proliferation of multiple *tokens* for different *types* might seem to be problematic. Given a specific activation pattern, how do we know which type it belongs to? As it turns out, the answer is easy, and it's the same one used to distinguish categories such as nouns from verbs: Tokens of the same type are all spatially proximal, and closer to each other than to tokens of any other type. The fact that they inhabit a bounded region of space is what tells us they are all the same abstract word.

There's a bonus as well. The spatial distribution of tokens turns out to be non-random. Instances of "dragon" when it appears as subject in a sentence are located in a different region of the "dragon" space than "dragon"-as-object. The relative positions of these two subclasses turns out to be identical across nouns! This means that the tokenization process is actually encoding grammatically important information, in a systematic way. *The internal representation of a word thus reflects not only its core identity, but provides a grammatically interpretable context-sensitive shading.*

Connectionist processing: The importance of nonlinearity

Earlier we remarked on the importance of the nonlinear activation function in neural networks. We return to this issue because we view it as one of the properties which gives connectionist models great computational power.

Essentially, the fact that nodes have nonlinear responses to their inputs means that there are some conditions under which they respond in a graded and continuous manner, and other conditions where their response is abrupt, discrete, and all-or-nothing. It simply depends which region of the node's activation function is being used.

The importance of this characteristic to development should be apparent. It has long been observed that there are periods during development where behavior changes slowly and progress is incremental. Such periods may be succeeded by spurts of activity in which change is dramatic and rapid. This phenomenon is also sometimes linked to the notion of "readiness," because during the time of rapid change it appears that the developing organism has a heightened sensitivity to inputs which previously elicited no apparent response. The organism is said to be ready to change in a way it was not before.

A reasonable interpretation of this developmental phenomenon might be that the organism has undergone a drastic internal reorganization, or that maturational factors have changed the organism in some fundamental way. Thus, dramatic—or nonlinear—changes in behavior are seen as diagnostic of dramatic, nonlinear changes in internal structure.

This interpretation is not the only one possible, and in Chapter 3 we discuss several connectionist simulations which exploit the nonlinearities in nodes' activation functions; in Chapter 4 we go into this matter of the shape of change in greater detail. The lesson will be that very small changes in internal structure may produce very big changes in observed behavior. Put another way, the same mechanism may produce discontinuous behavior over time without requiring drastic internal reorganization.

We have also briefly discussed another consequence of the nonlinearity, which we repeat here and which will figure in our account of the phenomenon we call "the importance of starting small," discussed in Chapter 6. Recall from Equation 2.6 that one of the terms which is used in computing weight changes during learning is the derivative of the activation function. This term, which measures the slope of the activation curve, modulates how much of the error signal is actually used to change the weights. Since the activation function is steepest in its midrange (see Figure 2.2), the effect of an error is greatest at this point. Because network weights are usually initialized with small random values with mean of 0.0, this means that when networks start off life they typically have net inputs close to 0.0, which is exactly the value that generates midrange outputs. The result, as we explained earlier, is that networks are more sensitive to error during early training. As training progresses there is a tendency for node activations to be pushed to their extremes, and this has the effect of slowing down learning. Whether this is good or bad, of course, depends on whether learning is successful! In any event, it provides a natural mechanism by which learning can be self-terminated. This too is a phenomenon which has been observed in development.

Final words: What connectionism is, and is not

Before we conclude this chapter, there are several issues outstanding which we wish to address. Several of these issues represent what we feel are misunderstandings about the nature of connectionist models. We will return to some of these points in later chapters, but feel it is useful to identify them clearly at the outset.

Just how *rasa* is the *tabula,* anyway?

There is a widespread belief that connectionist models (and modelers) are committed to an extreme form of empiricism; and that any form of innate knowledge is to be avoided like the plague. Since a basic thesis of this book is that connectionist models provide a rich framework and a new way to think about ways that things can be innate, we obviously do not subscribe to this point of view.

To be sure, it is not difficult to understand where the belief comes from. One of the exciting lessons from the connectionist work is that relatively simple learning algorithms are capable of learning complex things. Some of the results have demonstrated very dramatically, as pointed out before, that considerably more structure is latent in the environment than one might have guessed. In the past, a common argument in favor of innate knowledge has been the claim that the input to a learner is too impoverished—or the learning algorithms too weak—for the appropriate generalizations to be induced (e.g., Gold, 1967). So one very useful function which has been served by the demonstration that learning may be possible in cases where it was thought not to be is to encourage a bit more caution in resorting to claims of poverty of the stimulus.

It is also true that our species' ability (and need) to learn seems to be unparalleled in the animal kingdom. As Gould (1977) and others have pointed out, one of the most striking characteristics of human evolution is the dramatic increase in time spent during development. This prolonged period of dependence might be considered to be maladaptive, except that it increases the opportunity for learning and socialization. The connectionist emphasis on learning is thus highly relevant to understanding a major characteristic of our species.

Of course, the fact that something *can* be learned is not a sufficient demonstration that it *is*. The learning that takes place in a network simulation might in biological organisms occur through evolutionary mechanisms. And to the proof of the universal approximator capabilities of networks (Hornik, Stinchcombe, & White, 1989) must be counterposed another result, that when architecture is unconstrained, many classes of problems are NP-complete (i.e., for practical purposes, too hard to learn; Judd, 1990). So there are good reasons to believe that some kinds of prior constraints are necessary.

In fact, all connectionist models necessarily make some assumptions which must be regarded as constituting innate constraints. The way in which stimuli are represented, the parameters associated with learning, the architecture of the network, and the very task to be learned all provide the network with an entry point to learning. But there are other, even more interesting ways in which connectionist models allow us to take advantage of prior constraints and to understand how maturational factors may interact with learning. The net effect is to make possible highly complex interactions with the environment. To be innate need not mean to be inflexible or nonadaptive. A major goal of this book will be to explore and exploit this perspective.

Modularity

Modularity tends to travel with innateness on the Big Issues circuit. In fact, these issues are logically separable and it is unfortunate they are so often confounded with one another. In much the same way that connectionist models have been thought to deny any role for innateness, they are often thought to be anti-modular.

It is true that many models have worked from an assumption of minimal structure. This is not an unreasonable initial working hypothesis. Furthermore, many of the early models (such as the word-recognition model) focussed on phenomena in which interaction between various knowledge sources played an important role. The sort of strict encapsulation envisioned by Fodor (1983) seemed undesirable.

But nothing intrinsic to the connectionist frame precludes modularity, and we have already made the point that some degree of organization and modular structure appears necessary if models are to be scaled up. Work by Jacobs, Nowlan, Jordan, and others shows that connectionists take the challenge of modularity very seriously.

The real questions seem to us to be, first, to what extent is the modular structure pre-existing as opposed to emergent; and, second, what are the functional contents of the modules?

Answers to these questions will vary, depending on the modules involved. The retina is a module whose structure is highly pre-determined, and whose functional role is tightly coupled to a specific domain. Visual and auditory cortex, on the other hand, are modules which are partially pre-determined but in a highly indirect way. We know from results involving natural and induced pathologies (e.g., Hubel & Wiesel, 1963, 1965, 1970; Neville, 1991; Sur, Pallas, & Roe, 1990) that both the structure and content of these areas is highly dependent on appropriate input during development. To us the interesting question is not whether or not the brain is modular (it clearly is), but how and why it gets to be that way. There is a huge difference between *starting* modular and *becoming* modular. One of the important contributions of connectionist models has been in suggesting answers to these questions (e.g., Linsker, 1986, 1990; Miller, Keller, & Stryker, 1989; O'Reilly & Johnson, 1994). This will be an issue which will occupy much of our attention in the remainder of this book.

Do connectionist models have rules?

It is sometimes claimed that connectionist models show that systems are capable of productive and systematic behavior in the absence of rules. We actually do not believe this is true—but we do have great sympathy for what often underlies the claim.

To say that a network does not have rules is factually incorrect, since networks are function approximators and functions are nothing if not rules. So arguments about whether or not networks have rules really do not make much sense.

Others have tried to distinguish between behavior which is *characterized* by rules, and behavior which is *governed* by rules. Presumably, in the first case, the behavior only accidentally conforms to a rule, whereas in the latter case the rule has causal effect. Clearly, the behavior of a network is causally connected with its topology and connection weights, so ultimately this also is not an interesting distinction.

What we take as a more interesting question is, *What do the network's rules look like?* Are they merely notational variants of the rules one sees in more traditional approaches such as production systems or linguistic analyses? Or do they make use of primitives (representations and operations) which have significantly different properties than traditional symbolic systems, and which might capture more accurately—and with more explanatory power—the behavior of learning in humans?

It is important here to distinguish between theoretical behavior, in the limit, and behavior in practice, operating in the real world with real-time constraints. In principle, connectionist networks and traditional symbolic systems may be interconvertible. This is one reading of the proof of networks as universal approximators. But systems which are instantiated in the real world with space and time constraints have different properties than their idealized counterparts. The idealized digital computer may be a Universal Turing Machine, but no such machine exists in the real world, and no real neural network is a universal function approximator. So in reality we are dealing with systems which may have very different properties when they are placed in a real world context. In practice, certain sorts of generalizations and behaviors may be more readily captured in one system than in the other.

For instance, we know that it is possible to build a connectionist network which implements a LISP-style rule system (Touretzky & Hinton, 1985). But when it comes down to it, it is probably easier to write LISP programs in LISP. David Marr (1982) gives the example of computation with Arabic vs. Roman numerals. It is relatively easy to do numeric computation with Arabic numerals, but harder (especially multiplication) with the Roman system. In a similar manner, connectionist networks are not particularly well-suited to implementing truly recursive functions, nor to doing predicate calculus, nor to doing many basic mathematical operations. We are perplexed when people try to teach networks such things. They can be done, but at some cost and no gain.

On the other hand, connectionist networks do provide a natural formalism for carrying out many of the operations which are characteristic of biological systems. The criticism that connectionist models cannot implement strict recursion nor support syntactically composed representations (e.g., Fodor & Pylyshyn, 1988) are well-grounded, but the conclusion—that therefore connectionist networks are insufficient to capture the essence of human cognition—seems to us to reflect a profound misunderstanding about what human cognition is like. We believe that human cognition is characterized by interactive compositionality (or in van Gelder's terms, "functional compositionality," van Gelder, 1990) and that it requires exactly the kind of interactive and graded representations, and nonlinear operations which are the natural currency of connectionist models. So we believe that connectionist models do indeed implement rules. We just think those rules look very different than traditional symbolic ones.

Is connectionist neo-Behaviorism?

One concern that has been expressed is that connectionism is simply behaviorism dressed up in modern clothing. There is some irony in this worry, since Donald Hebb, whose insights have been a beacon for modern connectionism, was in profound disagreement with the behaviorist approach and saw connectionism (the term he used) as being distinctly anti-behaviorist.

Connectionism is apparently behaviorist insofar as connectionist models often involve inputs and outputs which play the role of Stimulus and Response. Behaviorists, however, eschewed attempts to speculate about the agency which mediated the S-R pairing. Behaviorism was both anti-physiological and anti-mentalist. Behaviorists were hostile to explanations which invoked unseen mechanisms and despaired of ever being able to relate mental function to brain function.

Connectionism, on the other hand, focuses precisely on the mechanisms which mediate behavior. Hidden units, for instance, play the role which behaviorists were unwilling to grant the brain. They allow models to form internal representations whose form and function may not be directly inferable from either input nor output. The resulting representations are abstract. Recurrent connections further enrich the system; they make it possible for the system itself to be both an input and an output of processing, and to generate activity which is not environmentally induced. Such endogenous activity is an essential component of thought.

The importance of biology

We end with this issue because it is one which lies at the heart of the this book. The question is how seriously one should take biological constraints.

First, we wish to make clear that we think that the connectionist paradigm is interesting in its own right, and that there are valid reasons to study connectionist models regardless of whatever biological plausibility they might or might not have. There are many routes to intelligent behavior. We see no reason to focus exclusively on organically-based intelligence and neglect (for example) silicon-based intelligence. Artificial intelligence may help us better understand natural intelligence. But even if it doesn't, artificial systems are fascinating on their own terms.

Having said this, we want to make clear our own bias. We *do* believe that connectionist models resemble biological systems in important ways. We believe that connectionist models will be improved by taking seriously what is known about how computation is carried out in neural systems. We also believe that connectionist models can help clarify and bring insight into why neural systems work as they do.

Certainly there is a large gap between models and reality. For instance, there is no known evidence of any biological system which implements backpropagation learning. (Hebbian learning, on the other hand, seems much more plausible.) Sometimes this gap is unavoidable; sometimes it is even desirable. Ten years ago, for instance, there was no biological evidence for the existence of multiplicative synapses of the sort described in Feldman and Ballard (1982) and Rumelhart, Hinton, and McClelland (1986). Nonetheless, such higher-order units (often called "sigma-pi units" because they sum products of inputs) have been shown to be very useful in simplifying circuitry (e.g., Durbin & Rumelhart, 1989; Giles, Griffin, & Maxwell, 1988; Mel, 1990; Poggio & Girosi, 1990) and in principle there seemed to be no reason why synapses with these properties might not exist. And indeed, such synapses have recently been discovered in the brain. In retrospect, it would have been a mistake to have rejected out of hand models which used sigma-pi units; there is obviously a great deal which remains unknown about nervous systems and one would not want modeling to always remain several paces behind the current state of the science.

There is another reason for being willing to tolerate a gap between the model and the meat. It is often difficult to know what functional role is served by specific neural mechanisms. As David Marr has pointed out, "trying to understand perception by studying only neurons is like trying to understand bird flight by studying only feathers: It cannot be done" (Marr, 1982). Models permit a level of analysis in which the emergent properties of a complex system may be revealed. Exactly which specific details of implementation are significant and which are not cannot always be predicted in advance. One might believe that some neural systems do gradient descent learning in a way which is *functionally* similar to backpropagation, even if one does not believe that backpropagation is the exact mechanism.

At the same time, we take seriously the goal of trying to build models which are informed by biological research. Nature's solution may not be the only one (and as Stephen Jay Gould points out, if it were possible to rewind the evolutionary tape and start over, even Nature would be likely to find a different solution on every rerun), but it is certainly an interesting one. And it works! So on pragmatic grounds alone, there are compelling reasons to try to reverse engineer nature.

More than this, however, we are interested in understanding nature's solution. So it makes no sense to ignore nature's lessons. We are willing to tolerate a reasonable gap between our models and the reality, particularly since our own interests veer toward high-level phenomena whose neural substrates are less well understood. But we take as our goal the development of models which are informed by the biology and at least roughly consistent with it.

Our view of what this entails is perhaps somewhat broader than might be first apparent. For us, a biological perspective involves not only the narrower view which focuses on developmental neuroscience in the individual, but also the broader perspective which views the individual as embedded in an evolutionary matrix. Marr's warning about studying bird flight is appropriate here again. Trying to understand individual traits without regard for the way they interact in the whole individual is a doomed enterprise; and so is trying to understand whole individuals without regard for the way they interact in societies and evolve over time. We are very interested in ways that things can be innate, and we do not see how this can be understood unless one takes an evolutionary perspective.

Of course, this broadens our brief considerably and we run the risk of over-reaching what may be reasonably grasped. But we think the risk is worth taking, and recent work which attempts to bring together connectionist models, the study of artificial life, and the use of evolutionary mechanisms represents exactly the kind of broad biological perspective we have in mind. Let us now elaborate that view in more detail across the following chapters.

CHAPTER 3

Ontogenetic development: A connectionist synthesis

Introduction

You couldn't think of a less experienced individual than a newborn. Yet, at birth babies already demonstrate many capacities. They show preference for face-like stimuli over other attractive visual stimuli. They can discriminate and imitate the facial gestures (tongue protrusion, mouth opening) of others. They exhibit categorical perception of different speech sounds. They discriminate between curved versus straight geometrical shapes. Also at birth, they discriminate linguistic input from other auditory inputs and at 4 days they have already learned enough about the sounds of their native tongue to be able to differentiate it from other languages. At 3 months they show surprise if two solid objects seem to occupy the same physical location. At 4 months they show surprise if a solid object seems to have passed through a solid surface. Also at 4 months their behaviors seem to demonstrate cross-modal perception, i.e., they can match speech sounds to the lip movements on the faces that produce them. At 6 months they exhibit talker normalization, i.e., they can recognize as equivalent different speech sounds from different talkers (male/female/adult/child). At 6 months too they can match the number of visual items that they see to the number of auditory beeps that they hear. At 7 months they expect objects moving on an ascending slope to decelerate, and objects on a descending slope to accelerate. Also at 7 months they expect an object whose center of gravity is not supported by another object to fall. At 8 months they are surprised if an object stops in mid-air before encountering a supporting surface. And so on. Numerous infancy researchers have provided us with these new and theoretically tantalizing data over the past decade or so (e.g., Antell & Keating, 1983; Baillargeon, 1987a, 1987b; Golinkoff & Hirsch-Pasek, 1990; Haith, 1980, 1990; Hirsch-Pasek et al., 1987; Johnson & Morton, 1992; Jusczyk & Bertoncini, 1988; Kuhl, 1983, 1991; Leslie, 1984; Mandler, 1988, 1992; Mehler & Fox, 1985; Meltzoff, 1988; Slater & Morison, 1991; Slater, Morison & Rose,1983; Spelke, 1991; Spelke, Breinlinger, Macomber & Jacobson, 1992; Starkey, Spelke & Gelman, 1990; Strauss & Curtis, 1981; Vinter, 1986; see full discussion in Karmiloff-Smith, 1992a).

Nativists have interpreted such data as examples of a face module, a speech module, a physics module, a number module, and so forth, and attribute innate representational status to these domain-specific competencies (Spelke, 1994; Leslie, 1994). Learning theorists, by contrast, have sought an escape from these nativist conclusions by advocating domain-general computations and stressing the role of 9 months in uterus during which the baby has constant (albeit filtered) experience of its native tongue and, during infancy, the massive early experience with faces, voices, objects, etc. We claim that in their strong forms, both these positions are misguided. Rather, we argue that some innate predispositions—architectural, chronotopic and, rarely, representational—channel the infant's attention to certain aspects of the environment over others. Our view is that these predispositions play different roles at different levels, and that as far as representation-specific predispositions are concerned, they may only be specified at the subcortical level as little more than attention grabbers so that the organism ensures itself of massive experience of certain inputs prior to subsequent learning. As we will argue throughout the book, at the cortical level, representations are not pre-specified; at the psychological level representations *emerge* from the complex interactions of brain and environment and brain systems among themselves.

For many non-developmentalists, development is seen as uninteresting and merely a series of landmarks on the road to becoming an adult mind. Only the end state is of interest. Some go as far as considering everything of interest to be built in to the infant mind from the outset. Thus typical nativist statements include:

Deep down, I'm inclined to doubt that there is such a thing as cognitive development in the sense that developmental cognitive psychologists have in mind. (Fodor 1985; p.35)

I, for one, see no advantage in the preservation of the term 'learning'. I agree with those who maintain that we would gain clarity if the scientific use of the term were simply discontinued. (Piatelli-Palmerini, 1989; p 2)

By contrast, throughout this book we advocate that a developmental perspective is essential to understanding the end state, and that the connectionist framework, with its focus on learning rather than on-line steady-state computations, is especially relevant to that endeavor. But what is development? And why is the human mind/brain of special interest from a developmental point of view?

Development is both fascinating and mysterious. Why should systems develop, rather than coming into existence full-blown? Some species are up and running immediately at birth (ungulates, e.g., cows, donkeys, horses and the like). Early mature behavior, then, is indeed one of Nature's options. What did Nature have in mind such that the higher species opt for a much longer period of maturation? Note, even some lower species undergo developmental changes (sometimes very dramatic, as in species which change morphology or sex). Did something go wrong? Or is a more progressive postnatal period of development somehow an advantage? What precisely does it buy an organism to start out with a far less than fully-fledged operating system? We claim that there is an underlying logic to development, involving a trade-off between plasticity and minimal predispositions which permits optimal adaptation to an ecological niche. For example, the plasticity of the brain is such that congenitally deaf children of normal intelligence have absolutely no problem in acquiring a fully fledged language in the visuo-manual modality, e.g., American Sign Language, British Sign Language, French Sign Language, etc. (Poizner, Klima & Bellugi, 1987; Pettito, 1987; Newport 1981; Volterra & Erting, 1992). The intricate interplay between predispositions and plasticity results in certain auditory circuits of the brain that process spoken language in hearing subjects becoming devoted to processing visuo-manual linguistic input in deaf children.

Compared to other species, the human child has a particularly protracted period of development during which numerous fundamental changes take place. In fact, the slow process of human ontogeny has several important functions. First, it makes it possible to simultaneously regularize traits while preserving maximal plasticity in the face of internal damage and/or environmental uncertainty. Second it allows phenotypic exploration of new solutions.

As with all systems that change, one can ask:

Why do they change?

What directs or determines the change?

With the exception of Piagetians, for many years developmentalists tended to explain developmental change solely in terms of the structure of the physical and socio-cultural environments. In the past couple of decades, however, the pendulum has swung the other way. Now many developmentalists explain development in terms of genetic specification and maturation. These extreme positions resulted in a polarized debate over nature or nurture—a definition of the logical alternatives. However, we argue throughout that an exclusive focus on either endogenous *or* exogenous constraints fails to capture the rich and complex interactions which occur when one considers organisms rooted in their environment, with certain predispositions, and whose interactions give rise to emergent form.

Development in terms of emergent properties

Below are some examples of structured events that occur in biology, physics and human invention. At first blush, each would seem to be a perfect illustration of prespecified structure decided on in advance by nature or by the mind of a creative architect. Yet in each case the structures turn out to be emergent products of the complex dynamics of many different interacting factors, none of which contain the design for the final form.

Consider, for example, the perfection of the soap bubble. As you blow air gently into the soap film, the bubble gradually expands as the air pressure inside increases. However, the surface tension created by the contact between the soap film and the air on the inside and the outside of the bubble forces the surface area of soap in contact with the air to reduce to the minimum possible amount. The more soap there is contact with the air, the greater the amount of total surface tension and hence the greater energy required. The shape that maximizes the volume of an object and minimizes its surface area is the sphere. The soap bubble packs a maximum amount of soap into the film that forms the bubble wall and reduces its surface contact with the air by assuming a perfect spherical shape. If more air is blown into the bubble, the air pressure inside increases providing extra energy to sustain the increased total surface tension when the bubble expands. The emergent shape of the soap bubble can thus been seen as a simple resolution of different forces acting at the molecular level between soap and air. The laws of physics take care of the end product.

Have you ever marvelled at the intricate pattern of the beehive? A perfect design of nature, you may have thought. In fact the form of the beehive is merely the result of the interaction of multiple physical constraints (D'Arcy Thompson, 1917/1968). Just as with soap bubbles, honey cells start their lives as spherical objects. In the hive these cells are packed together, each bee attempting to create as much space as possible for itself within an individual cell. Viewed along a plane, any single sphere can contact just six other spheres (see Figure 3.1). When honey cells are packed together, the wax walls will undergo deformation. Furthermore, the surface tension between two honey cells acts to create a flat plane surface. Since each cell contacts just six other cells, the surface tension of the wax and the packing pressure of the bees will force each sphere into a hexagonal shape. The hexagonal shape maximizes the packing of the hive space and the volume of each cell and offers the most economical use of the wax resource. Hence the labor of the bees and the forces of physics act in consortium to produce the hexagonal shape of the beehive. The bee doesn't need to "know" anything about hexagons. Its labors are vital to the structure of the hive but given

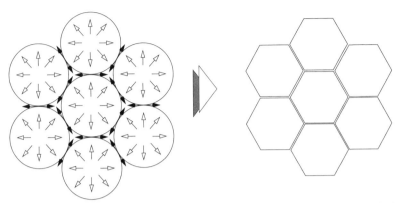

FIGURE 3.1 Building a beehive. Each spherical honey cell is surrounded by six other cells. The packing pressure resulting from the bees' activity is indicated by open arrows. The surface tension created by the contact between the wax surfaces is indicted by filled arrows. The packing and surface tension forces interact to deform the spheres into hexagons. The hexagon offers an efficient packing of the hive. Note that the exterior surfaces of the honey cells in the hive will maintain their spherical shape since there is no surface tension between two wax surfaces to stretch them into a plane. For convenience, we assume that the honey cells displayed in the figure are completely surrounded.

the materials used and the accommodation requirements of the hive's inhabitants, the shape of the resulting living quarters simply could not be otherwise. In other words, the resultant patterns are the emergent property of systems' dynamics, nothing more, nothing less.

Another oft-cited example comes from human artifacts. Many have stood in wonderment at the spandrels of St. Mark's Cathedral in Venice, Italy (Gould & Lewontin, 1978, and discussed by Clark, 1989). A spandrel is a particularly attractive triangular space formed by the intersection of two rounded arches. The illustrations worked into the spandrels seem so appropriate to the triangular shape that it is very tempting to view the overall structure of pillars and dome as the result of the need to create a triangular shape to receive the special design effects. In fact, this explanation is the exact reverse of the correct one. The spandrels are simply the emergent product of mounting a dome on rounded arches: a triangular shape necessarily emerges. What must have happened is that the artist recognized post factum the design potential of the resulting spandrels. But they were not part of the original design features.

The key point of these three examples is that something new seems to emerge from out of nowhere. In the above cases, the mathematical/physical principles that generate the emergent novelty are well understood. In the case of human development, however, we still do not understand the biological/psychological principles involved. Yet the order we witness in embryogenesis and ontogenesis can also be thought of in terms of emergent properties from systems dynamics. None the less, despite enormous advances in research, development is still shrouded in mystery. Inasmuch as development is largely a case of emergent properties, the problem of "reverse engineering" an emergent property by analysis is usually impossible. This is where the connectionist framework can suggest new ways to conceptualize the problem, as we will suggest throughout the book.

We will argue that despite such seeming discontinuity and stage-like changes in human development, the notion of emergent structure will apply in many domains and at different ages. Infants fail to solve some seemingly simple problems while demonstrating successful behaviors in other more complex ones. Children sometimes get worse at a task after a period of correct behavior. Young children spontaneously build theories and in doing so they frequently either ignore glaring counterexamples or invent data to meet their theoretical commitment. At times development appears to be continuous, and at other times an abrupt change occurs. Children move suddenly into a new type of reasoning that seems to bear no relation to their earlier thinking. Why do these changes arise? Where do they come from?

Nativism and empiricism give the same answer to these questions: the structures are not new, they pre-exist, either in the organism or in the environment. Constructivism offered a very different view, considering development in terms of self-organizing emergent structures arising from the complex interactions between both organism and environment. We believe that the biological-connectionist perspective opens the door to a new framework for thinking about development which embodies some aspects of Piaget's, Werner's and Vygotsky's constructivist intuitions, but which goes beyond them and provides a formalized framework within which to generate empirically testable questions. Previously vague suggestions about the fact that both organism and environment must be involved in ontogenesis can now be formulated in terms of exploring the nature of possible innate representational, architectural and chronotopic predispositions which evolution has made available to the human infant, tuned by the specificities of the human environment, to enable it to produce certain classes of behaviors.

For a long time, developmentalists accepted Piaget's domain-general view of the infant as assailed by chaotic input from competing sources which were gradually differentiated during development. Recently, however, exciting new paradigms for working with young infants have led us to discover many early competencies, as mentioned at the outset of this chapter. How to interpret these competencies without recourse to a strong nativist stance will be our connectionist-inspired aim throughout the book. Indeed, we believe that the connectionist framework can help us to rethink, rather than simply accept or reject, the notion of innately specified predispositions and to explore very different hypothetical starting points, be they chronotopic (i.e., temporal), architectural and/or representational. In the remainder of this chapter we discuss selected phenomena from human development and consider how they can be reinterpreted within a biological-connectionist perspective. We start with a discussion of children's sensitivity to certain social stimuli. We go on to consider language acquisition and simulations thereof, ending that section with a case study of the acquisition of the past tense as seen from different theoretical viewpoints. In the final section we explore children's sensitivity to events in the physical world and consider their simulation in a connectionist framework. Throughout the chapter, we draw out a series of basic principles governing the patterns of development which we believe can be particularly well explored in connectionist simulations. In cases where the connectionist simulations do not exist, we point to date or issues that can be usefully explored in the connectionist framework. In other cases, we describe existing connectionist models that simulate developmental phenomena in new and theoretical exciting ways. *In particular, we show how domain-specific representations can emerge from domain-general architectures and learning algorithms and how these can ultimately result in a process of modularization as the end product of development rather than its starting point.*

The child's sensitivity to faces

A basic survival mechanism in any species is to recognize its con-specifics. For human babies this involves recognizing human faces, voices, movement and so forth. We take as our first example infant face recognition. Developmentalists have either argued that face recognition is merely part of domain-general pattern recognition mechanisms (Slater, 1992) or that the details of the face are part of a genetically-specified social module (Brothers & Ring 1992). In this vein, nativists have used the existence of brain damaged adults suffering from prosopagnosia (an inability to recognize familiar faces) as a strong index of an innately specified face recognition module. However, within the connectionist framework and the way some developmentalists view ontogenesis, it can be argued that face recognition becomes modularized with time as the child gradually learns about faces and as circuits in the brain progressively become specialized for face recognition (Johnson & Morton 1991). Subsequent damage to these circuits in adulthood and the ensuing prosopagnosia cannot necessarily be used to argue for any face-specific genetically encoded information. If modules are progressively formed as a product of development (Bates, Bretherton, & Snyder 1988; Johnson & Vecera 1993; Karmiloff-Smith, 1986, 1992a; Kuhl, 1991), then they could be the result of relatively non-specific starting points, despite the fact that damage to any one of them *in adulthood* results in very specific impairments. However, studies of human neonates and other species suggest that species-specific recognition does not have a totally non-specific starting point. Some minimal face-specific predispositions give development a kick start in this domain.

Johnson and Morton (1991) carried out a series of experiments on face recognition in human neonates and young infants (see, also, the seminal work of Fantz, Fagan, & Miranda, 1975). They showed that neonates preferentially track a stimulus with three high contrast blobs (similar to a de-focused image of a face) over other stimuli (e.g., a symmetric face with the features scrambled, a blank face-like shape, etc.). Newborns show no consistent preference for a normal representation of a face with full features over the face-like blobs. Only later does the preference for the full face features emerge. It seems, then, that on the basis of a minimal predisposition, the human infant will initially pay particular attention to stimuli in its environment that resemble face-like objects. Only later does it learn about the specific characteristics of the face. Johnson suggests that the initial system is mediated subcortically (through structures such as the colliculus and pulvinar) and that, at around 2 months of age, the second learning system takes over via cortical structures which inhibit the earlier system (Johnson, 1990, in press). The subcortical system relies on an albeit minimal representational predisposition. The cortical system, however, requires no built-in representations, but it is primed for its subsequent learning by the preferential inputs set up by the first system. Evidence that the cortex acquires representations of faces as a result of experience comes from studies of neurons responsive to faces in the temporal cortex of mammals (Perrett, Rolls & Caan, 1982). In both monkeys and sheep the responses of these cells are dependent upon their exposure to faces. For example, sheep raised with others that have horns possess "horn-sensitive" cells (Kendrick & Baldwin, 1987), while those that are raised without seeing horns do not develop cells with such responses.

Once human infants have learned through experience about faces at around 2 months, they then behave differently than neonates and prefer stimuli that resemble real faces to the high contrast blob stimulus. Later still, at around 5 months, they preferentially attend to the movement of internal features of the face, suggesting that the infant is learning increasingly more about details concerning the facial characteristics of conspecifics from the subcortical predisposition that makes it attend to face-like objects in the first place.

A strong nativist position might argue for a relatively detailed template, as is posited to be the case for universal principles and parameters for language (Chomsky, 1986). Certainly Nature has had longer to build in details of human faces than of language, if her aim were to specify modules in detail. However, a further clue that the early predisposition is not a detailed template of the human face comes from work testing human infants and adults with the faces of other primates (Sargent & Nelson, 1992). Infants were habituated to a monkey face in a particular orientation. They were then shown either the same monkey face but from different orientations, or other monkey faces from the same orientation. While 9 month old infants successfully discriminated between the different monkey faces, adults could not. In other words, adults (and plausibly older infants) who have become very proficient at human face recognition, i.e., have filled in the full human details of the initial sub-cortical three-blob specification, can no longer make fine discriminations among the faces of other species without specific training.

The infancy work with primate faces suggest that recognition becomes species-specific *progressively with experience*—an argument against any strong nativist view. But the infancy face processing research does argue in favor of a sub-cortical head-start to the process of face recognition, because at birth infants consistently orient toward species-like faces in preference to other attractive stimuli. Thus, although the environment presents many potentially competing sources of information, specific biases channel the infant's attention preferentially to particular aspects of that environment, thereby ensuring proprietary inputs for subsequent learning.

Now, one might ask why it is necessary to build in any sub-cortical information about the human face. Nurture provides massive input to the infant and nature an appropriate learning algorithm. Surely a domain-general pattern recognition device would suffice? Yes, indeed, if the only thing the child (or a network) learns about is faces. But real environments are exceedingly rich; infants are not presented with faces neatly cordoned off from other visual or auditory input, with language neatly isolated from perceptual input, with syntax independent of meaning, or with physics input independent of number input, and so forth. Given such a potentially rich environment, we need to invoke some minimal predisposition that biases the newborn to pay special attention to faces. As yet we have no developmental simulations of either normal or abnormal face processing, though adult models of on-line face processing do exist (Burton, Bruce, & Johnston, 1990; Cottrell & Fleming, 1990). Of course, given the fact that networks are only presented with a subset of the data that the normal infant is faced with, a network simulation of the development of face processing would not need to build in any representational dispositions. However, what is clear from the human infant data is a complete account of the gradual development of human face processing should build into the starting state of simulations a minimal specification of three high contrast blobs in one of the networks and some architectural constraints enabling the setting up of two different face processing systems. Indeed, of relevance to the topic of the development of human face processing is a recent connectionist simulation of the developmental process of conspecific face recognition in the domestic chick. However, since our focus here is on human development, (and the chick simulation is discussed in Chapter 6) we will now turn to language acquisition and simulations thereof.

The child's sensitivity to speech and language

Just as with face processing, there's a fascinating infancy story to tell about speech and language, too. Infants have now been shown to display sensitivity to numerous speech and language contrasts. But as in other domains, here too in the case of language, we will invoke a view of development that embodies some predispositions together with a theory of the structure of the input and its effect on the developing brain.

As we mentioned at the outset of this chapter, newborns can distinguish human language from other auditory input at birth, and by 4 days they discriminate the prosody of their native tongue from other languages (Mehler et al., 1986; Jusczyk et al., 1993). It has also been shown that there are clear predispositions, though not restricted to human infants, with respect to the discrimination of different speech sounds. Patricia Kuhl and her colleagues have shown that human infants, chinchillas and monkeys exhibit categorical perception of different speech sounds, pointing to the non-linguistic evolutionary basis for this capacity (Kuhl, 1991). In the human case, the early capacity to discriminate all speech-relevant sounds is progressively lost and only those sounds relevant to the infant's native tongue are retained. Thus, Japanese children gradually lose the sensitivity to the /l/-/r/ distinction which they can discriminate early in infancy but which is not a contrast in the Japanese language. This also holds for infants' output. Babbling, whether vocal or manual (as in the case of deaf infants exposed to a sign language), is initially composed of multiple contrasts relevant to all human languages. But it progressively becomes reduced to those operative in the child's linguistic (speech or signing) environment (Pettito, 1993). Do these impressive capacities in the infant have to be built-in and specifically linguistic? Connectionist simulations suggest not.

Predicting the next sound

Elman (1990) discusses a network that discovers word boundaries in a continuous stream of phonemes without any built-in representational constraints (see Chapter 2 for a discussion of simple recurrent networks). Here we will again see an example of emergent form: the linguistically-relevant representations simply emerge from the processing of the input. Elman's network attempts to simulate the young toddler's task of identifying words from a continuous sequence of input phonemes. The network is fed one phoneme at a time and has to predict the next input state, i.e., the next phoneme in the sequence. The difference between the predicted state (the computed output) and the correct subsequent state (the target output) is used by the learning algorithm to adjust the weights in the network at every time step. In this fashion, the network improves its accuracy in predicting the subsequent state—the next phoneme. A context layer is a special subset of inputs that receive no external input, but feed the result of previous processing back into the internal representations (see Figure 3.2). In this way, at time

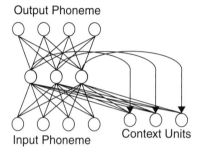

FIGURE 3.2 Simple recurrent network used by Elman (1990) to predict the next phoneme in a string of phonemes. The input to the network represents the current phoneme. The output represents the network's best guess as to the next phoneme in the sequence. The context units provide the network with a dynamic memory to encode sequence information.

2, the hidden layer processes both the input of time 2 and, from the context layer, the results of its processing at time 1. And so on recursively. It is in this way that the network captures the sequential nature of the input. These architectural constraints—particularly appropriate for sequential input—are built into the network. By contrast, network hypotheses concerning phonotactic constraints are not. They emerge from the processing, as we see below.

The input corpus consists of sentence-like strings made up of a sequence of phonemes. The phonemes themselves go together to make up English words and the words make up sentences. In attempting to predict the next phoneme in the sequence, the network must exploit the statistical regularities implicit in the phonotactics of the language. It is not given any explicit information about the structure of the language.

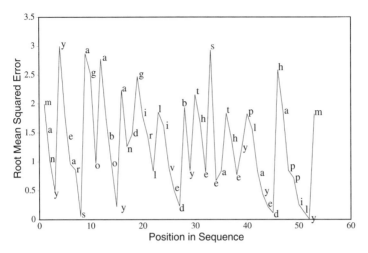

FIGURE 3.3 The error curve for a network trained on the phoneme prediction task (Elman, 1990). Error is high at the beginning of a word and decreases as the word is processed.

Figure 3.3 depicts the root mean squared error for predicting individual letters in the string:

Manyyearsagoaboyandgirllivedbytheseatheyplayedhappily.

Notice how the error tends to be high at the beginning of a word and decreases until the word boundary is reached. The error level for any phoneme can be interpreted as a measure of the level of confidence with which the network is making its prediction. Before it is exposed to the first phoneme in the word, it is unsure what is to follow. However, the identity of the first two phonemes is usually enough to enable the network to predict with a high level of confidence subsequent phonemes in the word. The time course of this processing is akin to Marslen-Wilson's proposal for a cohort model of word recognition (Marslen-Wilson, 1993; Marslen-Wilson & Welsh, 1979). Of course, when the input string reaches the end of the word the network cannot be certain which word is to follow so it cannot confidently predict the next phoneme. Consequently, the error curve for the phoneme prediction task has a saw-tooth shape with words falling into the teeth. The increased error at the beginning of a word (the start of the next tooth) shows that the network has discovered the word boundary.

Sometimes the network makes segmentation errors. For example, in Figure 3.3 the string of phonemes "aboy" is treated as a single unit by the network—the error continues to decline across the word boundary. This prediction is simply a consequence of the distributional characteristics of the article "a" and the noun "boy" in the language—they often occur together just like phonemes within a word occur together. In fact, there is considerable evidence in the child language literature that children make this type of mistake too. Peters (1983) and Hickey (1993) document the use of formulaic expressions in early first language acquisition. The following is a nice illustration of segmentation errors: young children often temporarily say: "the nelephant" having missegmented the indefinite article and the noun ("an elephant") or, as in one of the authors' children, the "ife" from missegmenting a "knife" (Bates, personal communication). Similarly, Plunkett (1993) shows how two Danish children produce overshooting solutions to the problem of lexical segmentation. Plunkett also identifies undershooting solutions in which children seem to postulate lexical units which are in fact shorter than the target words in the adult language. A typical equivalent in English would be "dult" instead of "adult."

Elman's network also produces these undershooting segmentation errors (this time based on orthographic missegmentation). For example, notice in Figure 3.3 that error curve for the sequence "they" rises after the "e" at the 39th position. The network recognizes the sequence "the" as a legal sequence and leaves the "y" stranded, unattached to any word. The "aboy" unit produced at the 13th position constitutes an example of an overshooting segmentation error.

The network can learn to rectify some of these segmentation errors. On exposure to further training examples where the indefinite article "a" combines with a wider range of nouns, it will eventually learn to split "aboy" into two separate words. In contrast, the network will continue to have difficulty deciding whether "the" should be continued into "they" since the former is a legal unit itself. In order to solve this problem, the network would require higher order information as to whether to expect a pronoun ("they") or an article ("the"). Since sentences can begin with either, there will be some contexts where it can never decide. Within a sentence, however, the network might eventually be able to make a better guess.

Connectionist models have been criticized for the way in which the inputs that they represent follow from the modelers own rule-based knowledge (Lachter & Bever, 1988). While this may be true for some models, Elman uses arbitrary localist input vectors (see Chapter 2 for a discussion of localist versus distributed representations) for the representation of all his inputs. Thus /p/ and /b/ are arbitrarily different, as are all phonemes. No part of the representation gives any indication of overlap of phonological features across phonemes or the phonotactics of phoneme combination. Each input vector is simply a long string of 0s with a single 1 at different arbitrary points. Phonotactic and lexical structure must be progressively inferred and represented in the hidden units as learning proceeds. Thus Elman's network receives inputs that are actually weaker in representational potential than those that the child receives. For example, we would expect a child to benefit from prosodic information in attempting to bootstrap its way into lexical segmentation.

Elman shows that, as with most connectionist networks using nonlinear functions, a long initial period is essential to learning. In the next section, we see this also to be true of vocabulary learning, both in real children and in network simulations. At first, Elman's network's predictions are random. However, with time, the network learns to predict, not necessarily the actual next phoneme, but the correct category of phoneme (vowel or consonant, etc.). Thus, the network progressively moves from processing mere surface regularities to representing something more abstract, but without this being built in as a pre-specified phonemic or other linguistic constraint. The phonotactic representations emerge from the discovery during processing of the structure inherent in the input. Thus, relatively general architectural constraints give rise to language-specific representational constraints as a *product* of processing the input strings. In other words, linguistically-relevant representations are an emergent property of the network's functioning. Elman's network will be further discussed in Chapter 6 with respect to the principles it embodies about how to get development off the ground, i.e., by starting with a small window on the input. But let's now turn to the child's output in language learning.

Vocabulary development

A phenomenon of language acquisition that has puzzled developmentalists for some time is why there is a lengthy period of slow vocabulary growth followed by a dramatic change during which there is a sudden increase in the number of words the child produces (Bates et al., 1988; McShane, 1979; Plunkett, 1986). Figure 3.4

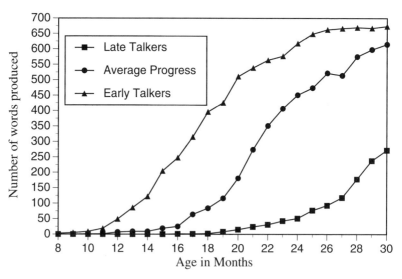

FIGURE 3.4 Profile of vocabulary scores for early talkers, late talkers and the 'average' child, taken from Thal, Bates, Goodman & Jahn-Samilo (1996).

illustrates this nonlinear growth (see Chapter 4 for an in-depth discussion of different growth shapes that characterize change). A connectionist simulation of the intra-domain and inter-domain processes involved in concept formation and vocabulary growth suggests once again that phenomena like the vocabulary spurt are in fact emergent properties of the dynamics of learning.

The network's task is to associate images with labels. The images are random dot patterns and are clustered in categories that have an internal prototype structure (cf. Posner & Keele, 1968). Pre-specified displacement lengths provide three different distortion levels corresponding to the distance between the image and its prototype. No prototype pattern has more than two dots in common with any other prototype. However, a small degree of overlap is permitted between high level distortions of one prototype and that of another. Hence, the network cannot categorize the random dot patterns purely on the basis of their perceptual similarity. The network must take account of the label assigned to each pattern in order to assign it the correct conceptual classification.

Thirty-two clusters of patterns are generated, resulting in a total of 192 distortions plus 32 prototype patterns. Before presentation to the network, the image patterns are pre-processed in a retina which serves to compress the representation of the pattern on the image plane, thereby forming a distributed representation of the random dot patterns with a corresponding retinal representation. Each cluster of images is associated with a discrete label. These consist of 32 bit vectors in which only a single bit is active for each label, similar to the input vectors in the Elman simulation discussed above. In other words, all label vectors are orthogonal to one another and are thus represented in a localist fashion. This means that there is no internal categorical structure built into the set of labels and that there is an arbitrary relationship between a label and its associated cluster of images. Figure 3.5 gives the network architecture. The task of the network is to reproduce at the output level the distinct retinal and label representations that are presented at the input level. This auto-associative architecture is quite different from the Elman network. So here we have a different architectural predisposition built into the network. Although the task may seem very simple, auto-association is non-trivial in that the input vectors are compressed through the hidden layers and thus have to be converted back to the original retinal/label representations at the out-

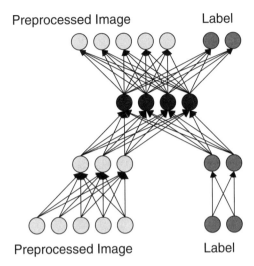

FIGURE 3.5 Network architecture used to autoassociate image-label pairs (see Plunkett, Sinha, Møller & Strandsby, 1992 for details). Note that the network is not drawn to scale—there are many more nodes in the model than is represented in the diagram.

put layer. This causes the network to progressively form representations of the relations between retinal and label inputs prior to reconverting them to outputs. The composite representation at the second hidden layer enables the network to produce a label on presentation of an image alone (roughly equivalent to the child's production of a vocabulary item when asked of an object, "What's that?") and to produce an image when presented with a label alone (roughly equivalent to the child evoking an internal image when told "That's an X").

Note in Figure 3.5 that each modality projects first to a separate bank of hidden units, which compresses the representation of that input, prior to converging on the second common set of hidden units where it is further compressed. Thus the network creates representations of each modality as well as representations of the relationships between them.

The network is trained in a three-phase cycle. First, a retinal representation is presented at the input units and activity is propagated through the network to the output units. The activity on the retinal output units is then compared to the initial retinal representations. Any discrepancy between the two is recorded as an error signal. As in the Elman model, a backpropagation learning algorithm is used. This then adjusts the weights on the image side of the network only. A similar trial is carried out for the label associated with the previously presented image, except that weight adjustment occurs this time only on the label side of the network. Finally, both image and label are presented simultaneously and error signals are used to adjust weights on both sides of the network. We can think of this alternation in terms of an attention-switching process. The three-phase learning sequence is successively and repeatedly applied to all image/label pairs in the training set.

There are many other details of the network that we will not go into here (see Plunkett, Sinha, Møller & Strandsby, 1992, for full description). The performance of the network in producing output labels and retinal representations is evaluated at various points during training, in terms of comprehension and production measures. We will focus here only on the network's production performance and its relation to the vocabulary spurt seen in child development. Given a retinal image, production is measured in terms of the network outputting the correct corresponding label. The network's performance is also evaluated by comparing its output on trained patterns with its output on prototype patterns which it has never seen. Correct production of a label for a prototype retinal representation assesses the network's capacity to generalize. A further production evaluation concerns over- and under-extension, by calculating the total number of distortions that elicit a given label, irrespective of the appropriateness of the label. Next, the number of correct output labels with respect to a given image are calculated. The comparison of these two calculations makes it possible to determine when a label is over- or under-extended in production.

The results of training the network exhibit several features that are characteristic of young children's early conceptual and vocabulary development. First, the performance on the prototypes which the network has never seen exceeds the performance on the distortions on which it has been trained. Thus, the network can be said to

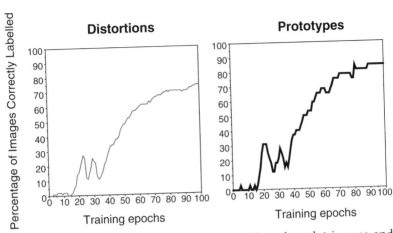

FIGURE 3.6 Overall performance on distorted random dot images and their prototypes as training proceeds in the network. Performance on prototypes exceeds that on distortions throughout most of training despite the fact that the network has never been trained on prototypes. Note also the vocabulary spurt that occurs around the 35th training epoch. Taken from Plunkett et al. (1992). Also see Chauvin (1988) for a similar model.

be extracting central tendencies from the constellations of the training distortions, grouping them together in clusters and responding most accurately to the central tendency of the cluster, i.e., the prototype pattern. Further, it is noteworthy that the network takes longer to discover the appropriate clusterings when labels are not included as input in the training. This suggests that the network not only uses the natural clusterings of the distortions in the image plane, but importantly that it also exploits the predictive power of the label in identifying category membership. This points to a rich inter-domain interaction between label and image.

Second, without biasing the input in that way, a vocabulary burst emerges simply from the nonlinear character of the processing. Production scores remain low for the first 20 to 30 epochs of training, but subsequently increase dramatically. There is no need to build in additional architectural constraints or to invoke changes in the input to explain the vocabulary spurt. It is simply an emergent function of the processing. The network also gives rise to over-

and under-extensions, the timing of which is typical of children's language (Barrett, 1986), in that under-extensions are witnessed during the early stages of development whereas over-extensions occur after the vocabulary spurt.

An important characteristic of the Plunkett et al. simulation is that it accommodates a variety of phenomena within a single explanatory framework. Elsewhere, these phenomena have been accounted for in very diverse ways. For example, the vocabulary burst has often been explained in terms of a naming insight (McShane 1979); over- and under-extension errors have been viewed as the inappropriate association of semantic features with labels (Clark 1973); prototype effects are regarded as emerging from children's pre-occupation with highly frequent tokens of objects/events in their experience (Barrett, 1986). By contrast, in the simulation, these three phenomena result from the same source—the gradual adjustment of connections strengths within the network to reduce the error on the output.

Important in the above simulation is the fact that the relationship between image and label grows progressively, is spontaneously generalized to prototypes not in the training set, and results in a sudden vocabulary spurt together with the over- and under-extensions typical of child language acquisition. Note that the computational predispositions are identical to the Elman network. But the architectural predispositions are different; in one case a recurrent net, in the other an autoassociative network. Both bear witness to emergent properties: new representations emerge from processing the input sets.

Although many connectionist simulations of child language acquisition are not necessarily well known outside connectionist circles, one language area has had a huge impact across connectionism, linguistics and psychology generally, i.e., the acquisition of the English past tense. It therefore seems worth devoting an entire subsection to this area, given the controversies and misunderstandings that continue to rage.

Learning the past tense

A well-documented phenomenon in the language acquisition literature is the occurrence of overregularization errors (Berko, 1958; Ervin, 1964; Marcus et al., 1992). For example, children may produce incorrect past tense forms of the verb like "go-ed" or "hitted." Typically, these forms are produced subsequent to correct performance on the same verbs. That is to say, children seem to unlearn what they already knew. This observation has often been interpreted as indicating that young children initially acquire the past tense forms of irregular verbs like "go" and "hit" through a process of rote learning. However, as they experience more verb forms, they discover the regular pattern for the past tense, i.e., add an /-ed/. In other words, the interpretation is that past tense formation in English involves the discovery of a rule. Children then apply this rule even to forms that do not obey the rule. Further progress in language acquisition is made when children discover that there are exceptions to the rules of English. This requires that children exploit two representational formats—one for dealing with the rule-governed forms (the regulars) and one for managing the exceptions (the irregulars). The phenomenon of overregularization (among other phenomena) has thus led to the proposal that the representation of language entails the existence of two separate mechanisms, a view that has been challenged by psychologists applying neural network models as a framework for understanding children's linguistic development.

The central issues at stake in the past tense debate are concerned with the argument from the poverty of the stimulus (in this case that the input to the child does not provide enough evidence for the apparent qualitatitive distinction between regular and irregular verbs) and the need for a dual mechanism (in this case, the need for a rule-governed device supplemented with a rote look-up table). However, the implications of this debate extend well beyond the particular details of English inflectional morphology. For example, much of current linguistic theory in the tradition of transformational generative grammar attempts to postulate highly general rules, such as *Move alpha* (move anything anywhere), in order to reduce the number of rules needed to characterize so-called Universal Grammar. In order to prevent these highly general rules from over-generating and producing sentences outside the target grammar, linguists propose the existence of specific linguistic constraints that constitute exceptions to general rules. Linguistic constraints are also taken to be part of the apparatus that implements Universal Grammar. In our terms, these appear to be representational predispositions of a specific nature. Appeals to dual mechanisms thus proliferate through much of current linguistic theory (e.g., Pinker & Prince, 1988). These architectural constraints are typically argued to be innately specified. By contrast, connectionist approaches to the past tense problem, several of which we discuss below, suggest that qualitatitive distinctions between regular and exceptional patterns of behavior can *emerge* through the operation of a *single* mechanism.

We will start with a very early and simple model of the past tense for expository purposes. But we will go on to look at the flaws in this model and to provide more recent and far more complex simulations of the acquisition of the past tense. Rumelhart and McClelland (1986)—henceforth R&M—were the first to demonstrate how a single architectural bias can be used to model children's acquisition of the English past tense. This mechanism takes the form of a single-layered feedfoward neural network, i.e., it has a set of input units connected directly in a simple feed forward fashion to a set of output units. These networks are often called single-layered perceptrons (see Chapter 2). Like the traditional account, it assumes that one of the problems facing the child in learning the past tense in English is to relate the stem of the verb to its past tense form. Therefore, the input to the model is a phonological representation of the verb stem and the output a phonological representation of the corresponding past tense form. Each of the input neurons is considered to represent some phonetic feature of the verb. Different groups of units represent different phonemes in the verb.

Like children, the network must learn to deal with both regular verbs and irregular verbs. So it must learn that the past tense of "grow" is "grew" while the past tense of "show" is "showed." The model learns through exposure to the different verbs: The verb stem representation is presented to the input units. Activation passes

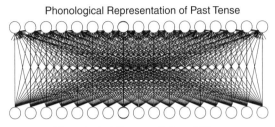

FIGURE 3.7 The learning network in the Rumelhart & McClelland (1986) model of the acquisition of the English past tense. The input is a distributed representation of the stem of the verb and output a distributed representation of the equivalent past tense.

along the connections to produce a pattern of activation on the output units. R&M used a probabilistic version of the continuous sigmoid function described in Chapter 2 as the activation function for the output units. The activation pattern on the output units is compared to the target activation pattern, i.e., the correct phonological representation of the past tense of the verb. Any discrepancy between the actual output pattern and the target output is used as an error signal to the learning algorithm to make small adjustments to the connections in the network. If the same stem were to be presented to the network again, then the output error would be smaller. Repeated presentations of the stem will eventually eliminate the output error, i.e., the network will have learned to produce the correct past tense form of the verb given its stem. The learning algorithm used by R&M was the Perceptron Convergence Rule (Rosenblatt, 1958—see Chapter 2).

It is also possible to evaluate the network's performance at various points in training by determining its current best guess as to what the past tense form of the stem ought to be. Initially, the output will not make any sense since the network knows nothing about the relation between the stem and past tense forms of verbs. However, as training proceeds, the network's attempt to construct the correct form can result in a variety of intermediary forms that can be compared to the forms that children produce en route to learning the English past tense.

Of course, the network has to learn more than just one stem/ past tense pair. In fact, R&M trained their model on 420 verbs. The set of verbs chosen was selected as a representative sample of regular and irregular verbs in English. Thus, most of the verbs chosen were regular verbs and the exceptions included an assortment of the irregular types found in English, e.g., "go/went," "hit/hit," "see/saw," "sleep/slept." On successive training trials, the network was presented with different verb stems. It was thereby required to learn many verbs at the same time since it was not permitted to completely learn a single stem/past tense pair before proceeding to the next. This type of training regime has important consequences for the behavioral characteristics of the network.

Consider a training cycle in which a regular verb is presented to the network. In order to produce the correct past tense form, the connections in the network must be adjusted so that the activity on the output units corresponds to the verb root plus an /ed/ form. Suppose on the next training trial another regular verb is presented to the network. Although a different stem should be activated on the output units, the network still has to activate the /ed/ form. Hence, the strengthening of the connections on the previous learning trial will facilitate the correct suffixing response on the current learning trial. And the learning that occurs as a result of the presentation of a new regular verb will strengthen the /ed/ response still further. In essence, verbs that belong to the same class will cause similar changes in the weight matrix of the network and so contribute to the learning of other verbs in the same class.

A critical feature of the English past tense is that not all verbs undergo the same type of transformation from stem to past tense forms. As noted above, the irregular verbs undergo a variety of changes, e.g., arbitrary change ("go"→"went"), no change ("put"→"put"), vowel change ("see"→"saw"), blending ("sleep"→"slept"). Let us suppose then that on our third hypothetical learning trial, the verb that is selected for learning is a no change, irregular verb, say "hit"→"hit." In this case, the network is once again supposed to activate the verb root on the output units but this time the /ed/ suffix should not be active. Since the network has only seen regular verbs so far, there will already be a slight tendency to turn on the /ed/ suffix. This tendency will be apparent on the output pattern for "hit" in the form of "hitted", i.e., there is likely to be quite a large discrepancy between the actual output on the suffix units and the target output. Importantly, this error signal will then be exploited by the learning algorithm to adjust the weights in the network so that the suffix units are turned off. If the regular verbs were presented again, the recent presentation of "hit" will have decreased the probability of the suffix units turning on. In other words, the adjustments to the connections in the network that arise as the result of presenting an irregular verb interfere directly with the configuration of connections constructed to deal with the regular verbs.

It is worth emphasizing here that these interference effects result from the fact that a *single set of connections* is being used to represent the stem/past tense relations of *many* verbs and that in English the regular and irregular verbs are in direct competition with each other in the manner in which they attempt to adjust connections strengths of the network.

Using this method of training, R&M succeeded in creating a network that could produce the correct past tense forms of the overwhelming majority of verb stems to which it was exposed. Hence, they were able to show that a single network mechanism was able to represent simultaneously both the irregular and regular verbs. There was no need for a dual mechanism. Furthermore, they questioned whether it was appropriate to characterize the network as having "learned a rule" for the past tense. If the network had learned an add /-ed/ rule for the past tense, then it should have applied the rule to all the stems in the training set—which it did not.

R&M also measured the performance of the network while it was being trained on the past tense task. Overall performance on regular and irregular verbs is shown in Figure 3.8. The regular verbs showed a gradual increase in correct performance throughout the training of the network. However, the irregular verbs showed a sharp decrement in performance around the tenth training epoch. At this point in training, the network started incorrectly to overregularize some of the irregular verbs like "go" and "hit" to produce

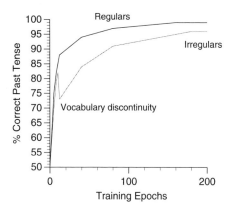

FIGURE 3.8 Performance on regular and irregular verbs in the Rumelhart & McClelland (1986) model of the acquisition of the English past tense. The vocabulary discontinuity at the tenth training epoch indicates the onset of overregularization errors in the network.

"go-ed" and "hitted." Furthermore, it was at this point in training that the network demonstrated a capacity to generalize to novel forms, i.e., add a suffix to verb stems that it had never seen before. As you might expect, it also had a strong tendency to regularize novel stems. As training of the network continued, the overall performance of the network on regular verbs continued to improve. By the end of training, performance on irregular verbs almost caught up with regular verbs.

It is instructive to consider why performance on the irregular verbs should recover once the network has begun to regularize. The explanation is to be found in the learning algorithm, i.e., a computational constraint built into the network. Weight changes are made in the network only if there is an error on the output. If there is no error for regular verbs, i.e., the network has learned them correctly, then no changes will be made to the weight matrix even during the training cycle. Weight changes will only occur for the irregular verbs for which there is an error. Hence, the irregular verbs will continue to improve in performance. Notice, however, that the changes to the weight matrix for irregular verbs may interfere with the performance on regular verbs.

The R&M model succeeded not only in performing the task of learning the past tense of English but en route to doing so, demonstrated some of the same phenomena that are observed in young children acquiring these forms. The R&M model signalled a major turning point in the way that psychologists think about language and development. However, although an interesting model in itself, there was a major problem with the R&M simulation as a model of language acquisition— the onset of overregularization errors coincided with a vocabulary discontinuity in the training set (Pinker & Prince, 1988). Consider again the learning curve for the regular and irregular verbs obtained in the R&M model as shown in Figure 3.4. The dip in performance on irregular verbs comes after 10 epochs of training. In fact, for the first 10 epochs of training, R&M used just 10 stem/past tense pairs. Of these 10 mappings, 8 were irregular. As you might imagine, a network of this size (460 input units and 460 output units) had no difficulty learning the training set. In fact, there were so many connections available that it could learn these 10 input patterns by rote memorization (recall the discussion of the constraints on learning in single layered perceptrons in Chapter 2). Indeed, no generalization to novel stems was observed after 10 epochs of training.

At this point in training, R&M increased the size of the vocabulary of verbs to 420. Of these 410 new stem/past tense pairs, the overwhelming majority were regular verbs. Since the network was now trained mostly on regular verb types, the weight changes resulted in the weight matrix moving to a point in the error landscape which was most suited to regular verbs—though not to irregular verbs. Consequently, the irregular verbs were overregularized. Continued training on the verbs gradually resulted in an improved performance on the irregulars—moving to another point on the error landscape.

R&M justified their tailoring of the training set on the basis of irregular verbs generally being more frequent in the language than most regular verbs. Furthermore, they noted that there is a vocabulary spurt in children learning language towards the end of their second year (see Figure 3.4 on page 124). However, children do not start typically to produce overregularization errors until the end of their third year. And the vocabulary spurt observed in the second year results from an increase in the number of nouns in their linguistic productions. Thus, the vocabulary discontinuity introduced by R&M does not reflect the conditions under which children learn the past tense of English. Consequently, their demonstration of rule-like behavior and U-shaped performance in a single mechanism appeared to be undermined.

There are other problems with the R&M model. Recall that the R&M simulation used a single layered network and the perceptron convergence procedure. In Chapter 2, we saw that such networks are only capable of learning problems that are linearly separable. Now it turns out that the past tense problem is not linearly separable. The easiest way to demonstrate this is to run the simulation in a network without hidden units for a sufficiently long time to see if the error reduces to zero. We know that the single-layered perceptron will guarantee to find a solution to the problem if one exists. Plunkett and Marchman (1991) ran a past tense simulation without hidden units and found that a residual error remained even after extensive training. This result pointed to the need for using hidden units and the backpropagation learning algorithm.

Plunkett and Marchman (1991)—henceforth P&M—showed how a network architecture that makes use of hidden units can reproduce the pattern of errors observed in children, without any need to introduce discontinuity in the training set, i.e., keeping the input closer to that experienced by real children. In essence, P&M argued that the conflict between regular and irregular verbs should alone be adequate to produce the required overregularization errors in the network. Just so long as there are competing verb types in the training set, interference between these different types will result in temporary erroneous performance since their individual weight adjustments will often be in conflict with each other. No manipulation of vocabulary size should be required. This prediction was borne out. Even though vocabulary size was held constant throughout training (the network was exposed to the complete set of verbs from the beginning of training), micro-U-shaped profiles of performance were observed (see Figure 3.9 for performance profile on different verb types). As we shall see shortly, this micro-U-shaped

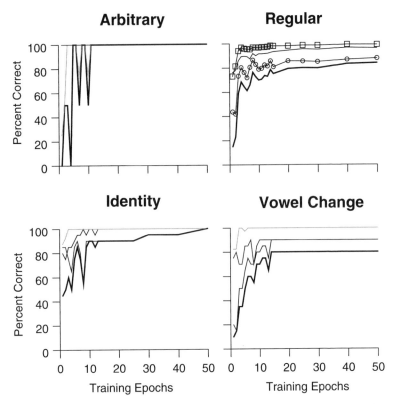

FIGURE 3.9 Performance on regular and irregular verbs in the Plunkett & Marchman (1991) model of the acquisition of the English past tense. The vocabulary size is held constant throughout training. Percentage word hit (solid line, no symbol) reflects overall performance for each verb type.

profile of performance is much closer to children's actual behavior than was originally supposed (Marchman, 1988).

Successful performance in the model was only achieved after suitable adjustments were undertaken relating to the frequency of the verb types in the training set. In the R&M model, the variable frequency of verbs in the language was modeled by introducing the high frequency verbs (mostly irregular) early in training and low frequency verbs later in training. Nevertheless, on any given epoch of training, each verb was presented only once to the network. P&M

found that presenting each verb token just once resulted in a disastrous performance outcome for the irregular verbs. The numerical superiority of the regular verbs swamped the network with a tendency to regularize from the very start of training. In order to overcome this regularizing effect, P&M increased the token frequency of the irregular verbs in the training set, again approximating what is known about the input to real children. In other words, on any given epoch of training, any individual irregular verb was presented more often to the network than any individual regular verb. As a consequence, the pattern of connections associated with an irregular verb was given an opportunity to establish itself before the regular verb patterns became too firmly established in the network. Thus, the high token frequency of the irregular verbs afforded them some degree of protection from regular verb interference.

Although the irregular and regular verbs are represented in a distributed fashion throughout the weight matrix in the network, the different verb types seem to have achieved some degree of functional modularity from each other. For example, Daugherty and Seidenberg (1992) show that a feedforward network trained on the past tense problem in exactly the same manner as the P&M simulations exhibits a frequency by regularity interaction. That is, reaction times (as measured by size of the residual error on the output units) are fast for high frequency irregular verbs and slow for low frequency irregular verbs whereas regular verbs show no such frequency effect. Marchman (1993) shows that regular and irregular verbs are differentially affected by lesions to such a network. Plaut and Shallice (1993) demonstrate a similar dissociation in the domain of reading (in Chapter 4 on the "Shape of Change," we explain why this occurs). These findings point to an important lesson—that *differentiation at the behavioral level need not necessarily imply differentiation at the level of mechanism. Regular and irregular verbs can behave quite differently even though represented and processed similarly in the same device.* The basic irony of the findings is that at some level a dual account of the U-shaped function is correct; we need competition between two types of verbs to generate micro-U-shaped during learning. But we neither need to represent nor to process the verbs by two different mechanisms. A single architectural and representational system suffices.

One noteworthy finding of the P&M simulations was that the particular configuration of frequencies associated with the different verb types required for successful network performance, closely followed the relative frequencies of the verb forms in English. For example, the so-called arbitrary verbs (like "go/went", "am/was") have a very high token frequency in the language, whereas the no-change verbs (like "hit/hit") are less frequent. Furthermore, no-change verbs are more numerous in the language than arbitrary verbs, i.e., no-change verbs ("hit/hit", "put/put") have a higher type frequency. Decreasing the token frequency of the arbitrary verbs in the simulations resulted in a serious deterioration in performance for those verbs, whereas the no-change verbs were more robust in the face of decreasing token frequency. Similarly, increasing the type frequency of the arbitrary verbs resulted in substantial performance decrements, while no-change verbs actually benefited from an increase in type frequency. Thus, arbitrary verbs could only be learned by the network when their token frequency was high and their type frequency low. No-change verbs were relatively unaffected by these frequency constraints.

P&M were able to show that the robustness of the no-change verbs derived from their shared characteristics with the regular verbs, i.e., like regular verbs the root of the no-change verb is preserved in the transition from the stem to the past tense form. Furthermore, all no-change verbs share the phonological characteristic of ending in a dental consonant, i.e., a /d/ or a /t/. The arbitrary verbs do not share any of these characteristics. Network performance on a particular verb was thus influenced by a complex interaction of its own token frequency, the number of other verbs in the same sub-class, the degree to which that sub-class shared features with other sub-classes and any phonological characteristics associated with particular verb classes.

These results suggested that the frequencies and phonological characteristics associated with different types of verbs in the language may play a crucial role not only for ontogenesis but also for the diachronic stability of the different verb classes in the actual language. High token frequencies may help children learn the irregular verbs in the language and thereby prevent the irregular verbs migrating to the regular class across successive generations. Similarly, phonological consistency within a class of verbs may help language learners identify a pattern for forming the past tense and so facilitate the learning and stability of that class.

It is also interesting to note that, in an attempt to model historical language change, Hare and Elman (1995) have trained successive generations of networks on the past tense problem. When the teacher signal for any given generation of the network was the actual output of the parent network (which was a distorted version of the original teacher signal), the partitioning of the verb classes in the language was observed to shift gradually. By introducing other characteristics associated with diachronic language change, e.g., loss of geminate consonants and vocalization of glides, Hare and Elman were able to model some of the known facts of verb migration in Old English. In particular, they showed how the weak class came to dominate the process of past tense formation. Hence, a connectionist approach not only seemed to offer a parsimonious account of the acquisition of the English past tense, it also provided an account of how the language got to be that way historically.

Another important finding from the P&M simulations concerned the profile of development associated with the learning of the past tense. In the classical view of the acquisition of the past tense (Berko, 1958; Ervin, 1964), the child is assumed to pass through a relatively circumscribed stage of development where all irregular forms are susceptible to overregularization errors. Indeed, the original R&M model captured this presumed stage-like performance rather well. In contrast, the P&M model did not exhibit a restricted period of development during which the overregularization errors were observed. Errors on individual verbs were observed to occur, recover and re-occur over protracted periods of training of the network. The pattern of errors in the P&M simulations resembled a micro U-shaped profile as opposed to a macro U-shaped profile, i.e. there is not a single circumscribed stage of development where irregular verbs are susceptible to overregularization errors, nor is there an indiscriminate application of the suffix to all irregular forms (also see Karmiloff-Smith (1983) for a similar distinction). As mentioned above, recent work with experimental elicitation tasks (Bybee & Slobin, 1982; Marchman, 1988) also indicate that overregularization errors occur for different verbs at different times in children up to the age of 9 years.

A recent reanalysis of naturalistic longitudinal data by Marcus et al. (1992) also indicates that overregularization errors occur over protracted periods of development. Marcus et al. demonstrated that the rate of overregularization is much lower than was typically assumed, i.e., children usually do not overregularize more often than 5–10% of the irregular verbs in their expressive vocabularies at any given time. Furthermore, the correct past tense form co-occurs with the incorrect version. The rates of overregularization of irregular verbs for four of the children in the Marcus et al. study is shown in Figure 3.10. In fact, the only robust characteristic retained from what turns out to be the erroneous classical picture of U-shaped development is the initial period of error-free performance. Even the oft-quoted examples of overregularization errors produced by children (/go-ed/, /hitted/) turn out not to reflect the actual data. For example, Marcus et al. demonstrate that high frequency irregular verbs (like /go/) are overregularized less often than low frequency irregulars (like /feel/). Also overregularization errors seem to be phonologically conditioned. Thus, low frequency no-change verbs (like /rid/) are less likely to be overregularized than other low frequency irregulars. Finally, Marcus et al. report only a small number of irregularization errors in their corpus. Irregularization errors occur when a regular verb is treated as though it were an irregular verb. For example, a child might use /pack/ as the past tense of /pick/ (presumably generalizing from the ring/rang, sing/sang subregularity). Other irregularization errors have been reported by Bybee and Slobin (1982), Marchman (1988), and Bates, Bretherton, and Bowerman (personal communication).

This careful reanalysis of actual child data is entirely consistent with a neural network model that attempts to represent regular and irregular past tense forms within a single mechanism. In particular, the micro U-shaped profile of development, the conditioning of errors by frequency and phonological shape, and the occurrence of

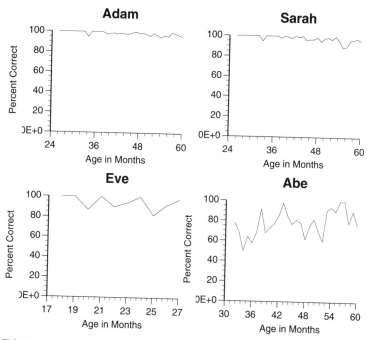

FIGURE 3.10 Percentage of irregular past tenses produced correctly by age for four of the children analyzed by Marcus et al. (1992). Abe had the highest level of overregularization of all the children reported in the study. Adam, Eve and Sarah exhibit more typical levels of overregularization of irregular verbs (data reconstructed from Marcus et al. 1992)

irregularization errors are directly predicted by the P&M simulations. However, the period of initial error-free performance is not predicted by these simulations. On the whole, performance on the past tense problem increases monotonically in the P&M simulations. In other words, early on in training the network performs relatively badly but improves as training is extended. Given the task with which the network is confronted this should come as no surprise. It has to learn the whole verb vocabulary from the very start of training. With a large verb vocabulary (500 verbs), the probability of achieving initial error-free performance is remote indeed—it would correspond to choosing an initial set of random weights that was very close to the solution of the past tense problem. Furthermore, if performance was completely error-free (in other words, if all 500 past tense forms were produced correctly), then there would be no error signal and no change in performance throughout training.

It is unlikely that children start off by attempting to learn large numbers of verbs in their language simultaneously. It would seem more likely that they focus on a gradually increasing set of verbs. In a later series of simulations, Plunkett and Marchman (1993) trained a multi-layered perceptron on the past tense problem where the initial number of training items started small and was then gradually expanded one verb at a time. The initial training vocabulary consisted of 20 verbs (10 regular and 10 irregular). New verbs were added to the vocabulary such that there was an 80% chance that each new addition was a regular verb. The goal of this work was to investigate whether a neural network would exhibit a similar pattern of overregularization errors to that found in children (see Figure 3.10) with an initial period of error-free performance (in contrast to the original P&M simulations). The error profile for this set of simulations is shown Figure 3.11. Comparison with Figure 3.10 reveals considerable convergence between the network performance and the profile of development for the children. In particular, overregularization rates remain very low and errors are not observed until the vocabulary has expanded to a size of around 100 verbs. Thus, the technique of gradually incrementing the size of the training set to which the network is exposed—a manipulation inspired by the assumption that children do not attempt to process all the verbs to which they are exposed in the input—results in a developmental profile in the network that closely resembles that observed in children.[1]

Further analysis of network performance revealed that the pattern of overregularization errors was closely related to the changing

---

1. It is worth noting that the comparison of the developmental profiles in Figure 3.11 is not entirely unproblematic: The input data to the network are not exactly identical to Adam's input. Furthermore, the child data are not fully longitudinal but based on regular samples. In the simulation we can observe network performance at *every* step in training.

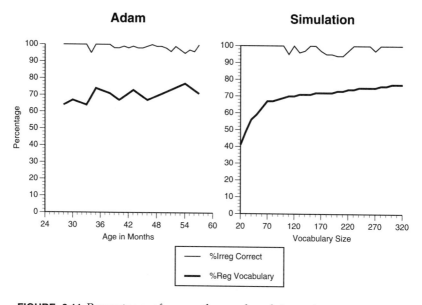

FIGURE 3.11 Percentage of correctly produced irregular past tenses from the Plunkett & Marchman (1993) simulation compared to those of Adam (Marcus et al. 1992). The level of overregularization (the micro-U-shaped profile) is similar in both cases. The thick lines indicate the percentage of regular verbs in the child's network's vocabulary at various points in learning.

generalization characteristics of the network. We can test for generalization by examining the response of the network to verbs that it has never been trained on—in effect, the network is exposed to a "wug" test (Berko, 1958). Figure 3.11 depicts the likelihood that a novel verb will be regularized by the network as a function of changing vocabulary size. It can be seen that during early training (vocabulary less than about 40 verbs) there is no systematic response by the network to novel verbs. This result indicates that the network has simply learned the training set by rote memorization. However, as verb vocabulary expands from 40 verbs to about 120 verbs, there is a dramatic increase in the rate of regularization of novel verbs—indicating that the network is reorganizing its internal representation of the past tense problem to reflect the dominant suffixation process. Beyond this range of vocabulary growth,

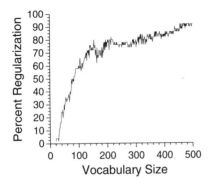

FIGURE 3.12 Network performance on novel verbs in the Plunkett & Marchman (1993) simulations. The graph plots the tendency to add a suffix to a novel stem.

the rate of regularization of novel verbs continues to increase but at a much slower rate. It should be emphasized that the nonlinear changes in the regularization rate of the network results from a gradual increase in the number of verbs in the training set and from a nonlinear increase in the proportion of types and tokens of regular verbs in the training set. Thus, the stage-like transition from a period of rote treatment of verbs to their more systematic treatment is an emergent property of the system, not one built in at the outset.

It is worth considering here why the onset of overregularization errors does not occur in the network earlier than is actually observed. Since the network already starts to regularize some novel verbs when vocabulary size has only reached 50 verbs, why do we not observe overregularization errors at this stage too? Recall that weight changes occur *gradually* in the network, determined by a parameter called the learning rate. So the effect of adding a new verb to the training set will not have its full impact until the network has had an opportunity to process the new verb a number of times, i.e., after several epochs of training. Most of the new verbs (approximately 80%) added to the network are regular verbs. The net result of vocabulary expansion will be a progressive strengthening of the regularization tendency of the network—though delayed in relation to the expansion process itself. Now when the network is tested on a novel verb, the response characteristics will reflect global characteristics of the training set, such as the relative proportion of regular to irregular verbs and the degree of similarity of the novel verb to other verbs in the training set. Consequently, the chances of a novel verb being regularized will increase as vocabulary expands. However, the chance of an irregular verb being irregularized is smaller. Note that an irregular verb is defined as such by the training set. Irregular verbs have had an opportunity to impact upon the connection strengths in the weight matrix (unless they are irregular verbs newly introduced to the training set). The regularization tendency of the network will be blocked if the connections associated with an irregular verb are sufficiently strong. In contrast, there is no such prior experience with the novel verbs to protect them from regularization. Hence, the onset of overregularization errors will be delayed with respect to the onset of the regularization of novel verbs. New in-depth analyses of child data point to very similar results (Marchman & Bates, 1994).

We have gone into much detail on this particular aspect of language acquisition, not of course because we believe the past tense to be the most crucial aspect of language that the child learns, but because it is a particularly rich example of the use of connectionist simulations for understanding the intricate interactions between mind and environment during development. Let us now move on from language to the case of the child's developing knowledge of the physical world.

The child's sensitivity to events in the physical world

According to Piaget, one of the most fundamental achievements of infancy is the attainment of object permanency, i.e., the infant's ability to maintain the identity of objects across changes in location and to represent the continued existence of objects despite their disappearance behind an occluder (Piaget, 1952). This, for Piaget, was the basis for all subsequent cognitive invariants, i.e., the conservation of matter, weight, volume, etc. in the face of perceptual changes.

Object permanence was presumed to develop progressively during the first 12 to 18 months of life. Piaget's measure of object permanence was the infant's successful retrieval of an object that she had seen occluded by a cover. Infants manage to do this at around 8–9 months of age, but if the object is hidden in one location (A) and is then moved to a second location (B), the young infant perseveres in looking for the object in the initially successful position. This has come to be known as the A$\overline{\text{B}}$ error which is only overcome in the reaching experiments many months later when, according to Piagetian theory, full and stable object permanence is acquired.

The last decade has witnessed a plethora of experiments attempting to show that object permanence is present in much younger infants. In order to demonstrate this purported capacity, researchers have used looking time measures, rather than reaching which is considered to be beyond the capacities of the young infant. If the infant looks significantly longer at a perceptually similar but conceptually dissimilar stimulus, this is taken to indicate that infants are sensitive to the conceptual principles being violated. Baillargeon (1987a) has demonstrated this with infants as young as 3.5 months. Baillargeon habituated 3–4 month old infants to the sight of a screen rotating 180°, until they showed boredom. Then, in full view of the infant, she placed a solid object behind the screen. Next, babies either saw the screen rotate to 45° (a normal event now that an object prevented the full rotation) or they saw the screen continue to rotate 180° (an impossible event, because the babies did not see that the object had been surreptitiously removed) (see Figure 3.13). As far as the visual input is concerned, the infants who saw the normal event were receiving new visual input (a 135° rotation), whereas the infants who saw the impossible event were receiving the same visual input as before (a 180° rotation). If infants' inferences are based on the representation of the continued existence of objects out of sight, and if they comply with the physical principle that two objects (the screen and the object behind it) cannot occupy the same space simultaneously, then they should show increased attention with respect to the impossible event even though the visual input has not changed. And this is exactly what happened. In other words, it would seem that 3–4 month old infants

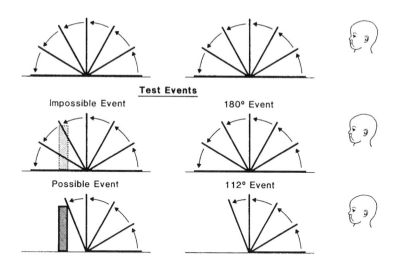

FIGURE 3.13 Schematic representation of the habituation and test events shown to the infants in the experimental and control conditions of Baillargeon (1987a).

are sensitive to the fact that even when an object is occluded by a screen, it continues to exist and should therefore block the screen's rotation. Baillargeon went on to use a similar rotating screen paradigm to demonstrate that 3–4 month old infants can compute the relation between the height of objects and the angle of rotation that they allow. Only later, but still during infancy, did infants show sensitivity to the rigidity or compressibility of objects.

In another experiment, Baillargeon (1987b) habituated young infants to a locomotive rolling down a slope on a rail, then disappearing behind an occluder, and then reappearing at the other side (see Figure 3.14). She then placed an object on the rail to block the locomotive's trajectory or behind the rail before lowering a screen in front of the block (see Figure 3.14b). In both cases, after the locomotive was released and had disappeared behind the occluder, it reappeared on the other side of the occluder. Again, very young infants showed surprise (significantly longer looking times) at the display if an impossible physical event occurred, i.e., when the locomotive reappeared having apparently passed through the block. It

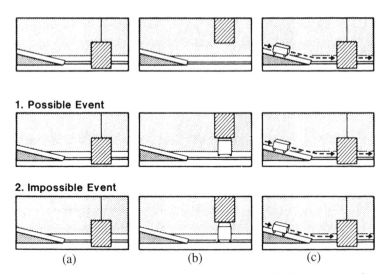

1. Possible Event

2. Impossible Event

(a)   (b)   (c)

FIGURE 3.14 Schematic representation of the habituation and test events used in Baillargeon (1987b). In (1b) a white object sits behind the track, and thus does not interfere with the movement of the locomotive in (1c). In (2b) the object has been shifted forward slightly and sits on the track, making the locomotive's reappearance in (2c) impossible.

is important to note that the child's increased looking is to an identical perceptual stimulus but which seems to be considered as new because of its different conceptual status.

Some theorists have interpreted these and similar results to suggest that the principles underlying object permanence are innately specified (Spelke, 1991). In our terms, they have invoked representational predispositions. However, that leaves open the question as to why young infants seem to "have" the object concept for looking tasks but do not use it for reaching tasks. One answer has been that young infants lack the means-end analysis enabling them to coordinate the necessary sequence of actions to successfully perform the reaching task although, the explanation goes, they do have the object concept (Diamond, 1991).

Connectionist-inspired alternatives to these interpretations have been proposed in terms of the principles of graded, embedded, experience-dependent learning (Munakata et al., in press;

Mareschal, Plunkett, & Harris, 1995; see, also, McClelland, 1992 and Russell, 1994, for discussion). Munakata et al., and Mareschal et al., attach particular importance to the notion of progressively strengthening representations (rather than representations present at birth as suggested by Spelke, 1991). Through increased processing experiences of the relevant stimuli, knowledge about the continued existence of objects accrues so that the internal representations of occluded objects become progressively strengthened. The argument is that a weak internal representation may have a lower threshold sufficient to guide looking behaviors but not reaching. Mareschal et al., argue that reaching also requires the infant to take account of more precise details of objects when reaching for them (distance, size, shape, etc.) and that this information needs to be coordinated with the infant's representation of the object's position. Mareschal et al. have shown that the need to coordinate information about the hidden object's identity with the hidden object's position delays the network's mastery of the task relative to pure tracking.

As a first step in the exploration of the object concept, Munakata et al. (op. cit.) trained a recurrent network with the backpropagation learning algorithm, similar to Elman's discussed above, to anticipate the future position of objects that disappear and reappear from behind an occluder. A recurrent network was used because it is particularly well suited to using past internal representations to influence subsequent internal representations which in turn affect behavior at the output level. The network sees sequences of inputs corresponding to several time steps of an occluder (a block) moving across a trajectory in a space which also contains a stationary object (a ball). At each time step, the input specifies the location of one or two objects (the ball and/or block). Each sequence simulates the movement of the occluder across the field at several discrete points along a horizontal axis, either from left to right or right to left. If the ball is present in a particular training set, it is stationary throughout a sequence. When the block passes the ball, the ball becomes temporarily invisible.

As with the Elman network for word segmentation, learning in the network is driven by discrepancies between the predictions that the network makes at each time step and the perceptual input it receives at the next time step. This enables the network to progressively learn to represent an occluded object, the ball behind the block, despite the fact that the perceptual input at that time step does not include the ball. These internal representations then serve as the basis for predicting the object's reappearance at a future time step. If the object does not reappear or reappears at the wrong time, there is a discrepancy between the predicted event and the actual event at the next time step. Munakata et al., argue that it is such a discrepancy to which the young infant is sensitive when showing surprise at unlikely or impossible events, rather than the understanding of a physical law.

The knowledge that subserves the network's predictions are stored in the form of connection weights, not in some explicit, symbolic representation of the occluded object. These internal representations are progressively built up over the course of learning. What is evident from Munakata et al.'s simulation is that the continued representation of an object that is no longer visible is a slow developmental achievement. This is consistent with the idea that looking behaviors may demonstrate earlier indications of some internal representation of the persistence of objects when tracking them behind occluders but that reaching behaviors may require much stronger internal representations to support their action. The Munakata et al. simulation suggests that in general, to speak in terms of not having, and then subsequently having, a concept is not the correct way to think about development which seems to be the result of *progressive* interactions at multiple levels of the network.

A similar task was performed by the network in the Mareschal et al. (1995) simulation. In addition to visual tracking, however, the Mareschal et al. network was required to initiate a reach for the object by generating an output representation that could be used to control a reaching response. The decision to reach was based on the identity of the object. Objects were predefined as very hot or cold. The network had to inhibit its reaching for hot objects to avoid "burning itself." Both the identity of the object as well as its position on the retina were important in the control of the reaching response. The network was required to learn spatially invariant representations to determine the object's identity wherever it occurred on the input retina and to learn to predict the next position of the object on the retina. These separate tasks were achieved by constructing two distinct processing channels in the network—a "what" channel and a "where" channel. Ungerleider and Mishkin (1982) provide evidence for the separation of these two pathways in the brain. The "where" channel (the Prediction module in Figure 3.15) performed the same function as the Munakata et al. simulation, i.e., keeping track of the object. However, information concerning details of the object were not maintained along this channel. This pathway was designed to keep track of an object but wouldn't know, say, if the physical features of the object had suddenly changed. The "what" channel (the Object recognition module in Figure 3.15) mapped the input retina to a set of complex cells that identified particular configurations of features at the input level. An object at the input level consists of a feature bundle, unique to that object. At any point on the input retina, up to 4 different features can be activated, allowing the definition of many different objects. The "what" channel learned to recognize object invariances at different points on the retina using an unsupervised learning algorithm (Foldiak, 1991). The extraction of these invariances permitted the network to maintain a constant representation of the object as it moves around the retina, reflecting the fact that individual objects were the same no matter where they appeared on the retina. In order to initiate a reach for an object, the network had to coordinate the information flowing along the "what" and "where" channels. The integration of this information was carried out by the Response Integration module in Figure 3.15. The complete network architecture depicted in Figure 3.15 summarizes the overall architectural constraints built into the model (though see Footnote 2). Representational biases emerge from these architectural predispositions.

Figure 3.16 summarizes the performance of the network when it attempts to reach for an object or track an object with or without an occluder. When there is no occluder, the network learns quickly both to track and reach for a desired object. In the presence of an occluder, however, the error rate for predicting the reappearance of the object from behind the occluder reaches an acceptable level much faster than for reaching for the object while behind the

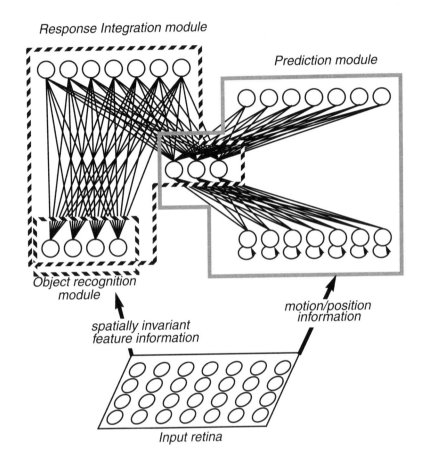

FIGURE 3.15 Network used by Mareschal et al. (1995) for simultaneous tracking and object recognition. The network learns to predict the next position of the object and to generate a motor command to drive a reaching response. The prediction function operates in ignorance of the object's featural definition. However, reaching is elicited only for specified objects and therefore requires both information about where the object is, and featural information about the object's identity.

occluder. This is because the network must learn to coordinate the developing internal representations of the invisible object's position with the separately developing internal representations of the invis-

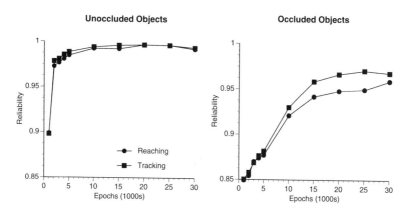

FIGURE 3.16 Performance on the Mareschal et al. (1995) network for reaching and tracking with and without an occluder. The network learns quickly to track and reach for desired objects in the absence of an occluder. Both reaching and tracking in the presence of an occluder exhibit a developmental lag in comparison to the unoccluded condition. However, this lag is more pronounced for reaching than tracking. The error discrepancy between reaching and tracking in the network resembles the developmental lag observed in infants between reaching and tracking skills.

ible object's identity. Although the distinct modules of the network have had equal amounts of experience with objects on the input retina, the integration of the internal representations constructed by the modules requires additional training to become sufficiently coordinated to support a reaching response when the object is out of sight. The model was thereby able to reproduce the behavior of young infants both when tracking objects and reaching for objects behind occluders.

The important point to note about both the Munakata et al. and Mareschal et al. simulations is that the basic mechanisms of change are constant throughout the development of these tracking and reaching skills. In the Mareschal et al., simulation the onset of reaching for hidden objects does not correspond to the sudden maturation of a new motor system. The basic potential has been present throughout the training of the network.[2] However, the realization of this potential must await the construction of successful representations for sub-tasks of the reaching response. Early in training, the necessary representations are not particularly well-developed when objects are out of view. The difficulty in coordinating these early representations will undermine a successful reaching response. Successful reaching must await more accurate *coordination* of the representations in the "what" and "where" channels.

Fast Learning

Many recent infancy experiments demonstrate precocious abilities in new-borns. One interpretation of these findings is that no real learning has occurred—the ability must be hard-wired (Baillargeon, et al. 1990; Spelke, 1994). However, more recent training studies with young infants (Baillargeon, 1994) and connectionist simulations have shown how precocious behaviors can emerge very rapidly in development as a result of very early learning, with no need to invoke detailed innate specifications. For example, Elman has extended his work with simple recurrent networks using abstract bit strings and language-like sequences to other domains in which change occurs over time. We will now briefly describe two examples from the domain of perceptual learning.

The first simulation involves a rotating cube with the following scenario. A cube was allowed to rotate for 30 seconds at a randomly chosen velocity and a randomly chosen angle of rotation. The cube was transparent, but there were lights at the 8 vertices. As the cube rotated, it cast an image on a 1000x1000 retina. Because the cube was transparent, all that was seen were the 8 points. These were represented by 8 X,Y coordinates. After 30 ticks of the clock the network began to rotate at a new speed and along a new axis. The task of the network was to track the rotation and predict where the verti-

---

2. Recall that the network in the Mareschal et al. simulation is *created* with separate "what" and "where" processing channels. However, the implication that these channels must be innately pre-wired for the infant is questionable. For example, we saw in Chapter 2 how a network, modularized to perform distinct tasks can start off life without commitments to specific tasks. Modularization according to function *emerges* from the network discovering that different learning algorithms are more suited to some tasks than others (Jacobs et al., 1991).

ces would be at the next moment in time. There were 16 inputs/output units and 20 hidden/context units respectively.

The network learned to predict the motion of the vertices quickly (within about 1000 30-second trials). There are several aspects of this simulation that are particularly noteworthy:

- What the network sees is a 2-dimensional projection of a 3-D object.

- It is presumably the 3-D object which provides the best "account" of the motion. That is, if one knew *in advance* that the point- light display was of a rigid 3-D object, and that the display was collapsed onto 2 dimensions, then one could easily solve the prediction task.

Most important is the fact that the network solves the task without being given any such information in advance. Its rapid learning suggests that representing 2-dimensional information in terms of the fact that coherent motion of all points must involve a 3-dimensional object turns out to be the best way of solving the problem. We considered 1000 30-second trials as rapid learning, but connectionist simulations of course remain challenged by what is presumed to be one-off learning in the human case.

A second example leads naturally from the previous simulation and has implications for our processing of biological as opposed to mechanical motion. Well-known in the psychological literature is a series of experiments on adults perception and labelling of human motion such as dancing, hopping, walking, etc. (Jansson & Johansson, 1973) when only point light displays on the main articulatory parts of the body (e.g., wrists, elbows, hips, knees, ankles etc.) are available. Elman's simulation displayed a person walking across the stage. There were "lights" attached at 11 joints (neck, shoulders, elbows, wrists, knees, ankles) and these were the only stimuli that the network "saw." Each point was input as an X,Y pair. The task of the network was to predict the motion of the person.

Again, the network performed extremely well. This result is interesting in that some might have been tempted to interpret Johansson's and other similar results as implying special (innately specified) knowledge about biological motion (e.g., Leslie, 1994;

Premack, 1990). In the network, the 11 points move together in a manner which is coherent only given the peculiar facts about how bodies are wired together and the biomechanical constraints on different gaits such as walking. This "function" is quite arbitrary in the grander scheme of things. The fact that the network was indeed able to do the prediction task without prior knowledge suggests that built-in representations about bodies are not required but can be inferred rapidly from the input as the best solution to accounting for the coherent motion.

We have mentioned these various simulations in the light of the tendency in the last decade for many researchers to seek younger and younger competencies in infants and to jump to the conclusion that they must therefore be innate. The simulations demonstrate that, although this is a possible conclusion, it is far from a necessary one. Indeed, again and again we rediscover with Gibson (1969) more structure in the environment than often suspected which, together with very simple learning algorithms and architectures, can give rise to very rapid learning resulting in surprisingly sophisticated internal representations and behavior. Apart from learning about whole objects, persons and movement, children also learn about various subtle properties governing the behavior of objects.

Readiness

Take a look at Figure 3.17. Which objects will topple? Without thinking about it, rapid computations allow you to see immediately which ones will fall. And, surprisingly, young infants also seem to know which of the symmetrical objects will fall. That is, they look longer at the displays which violate the physical principles of gravity and the law of torque (Baillargeon & Hanko-Summers, 1990). Yet they cannot immediately generalize this to the asymmetrical objects; this they have to learn from experience during the months which follow. It might be tempting to interpret these data as suggesting that infants have some innately specified knowledge about gravity which supports development through subsequent learning. It comes as no surprise to the reader that our connectionist perspective would suggest a more complex story (see also Baillargeon

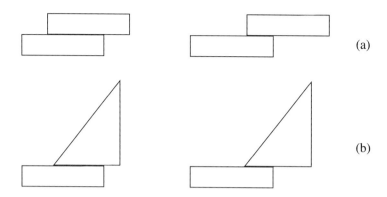

FIGURE 3.17 Stacks of blocks: Two are stable, two will fall. Our brains have no trouble computing which is which extremely quickly. Infants can solve this problem quickly too (after Baillargeon & Hanko-Summers, 1990).

1994). But, given the paucity of simulations of infant abilities in this area, our connectionist examples this time will come from more complex reasoning in older children.

A tidy description of development is that it proceeds in a series of stages, involving qualitative change at the representational and/or computational level (Brainerd, 1978; Piaget, 1955; Siegler, 1981). The simulations we have discussed so far call this into question. We have shown that abrupt behavioral change is not necessarily accompanied by abrupt representational change. We now discuss a model with a similar architecture and learning algorithm to the vocabulary simulation, which captures key features of children's progressive understanding of the balance beam. But first briefly the child data.

Children are shown a balance beam, as in Figure 3.18, with varying weights at varying distances from the fulcrum. They are asked to judge whether the beam will balance and, if not, which side will go down. Children are not given feedback with respect to their response, but simply questioned on a new item. According to Siegler (1981), children pass through a series of discreet stages determined by a sequence of rules which focus on weight or distance or a combination of both. At stage 1, children judge only on the basis of weight. Thus the side with a greater number of weights

FIGURE 3.18 Balance beam.

will go down, irrespective of the position of the weights on the pegs. At the second stage, children take distance into account. but only provided the weights on each side of the fulcrum are equal. The third stage is characterized by correct responses when either weight or distance differ, but children guess when both weight and distance differ. Finally, some but not all adolescents succeed in using the law of torque, taking weight and distance both into consideration and quantifying the differences. These four abrupt changes of focus of attention are thought to characterize the stage-like developmental pattern in this domain between 4 and 12 years of age.

In order to explore the mechanisms underlying this developmental pattern, McClelland and Jenkins (1991) trained a network on the balance beam problem. As in the vocabulary simulation discussed in an earlier section, architectural constraints are built into the network. The network's architecture is divided into two channels, one receiving weight-relevant input and the other receiving input relevant to the distance of weights from each side of the fulcrum (see Figure 3.19). A 5-place input vector encodes in a localist fashion 5 possible weight values and 5 distance values for each object presented to the network. The network is not given explicit information about which units correspond to weights or distances nor which represent the right or left objects on the beam. These are the representations that the network must create during training.

As can be seen from Figure 3.19, each bank of input units projects to a separate bank of two hidden units. These 4 hidden units then project to two output units which represent either balance or one side going down. If the activation of the left-hand unit is greater than the right-hand one by some criterial threshold, then

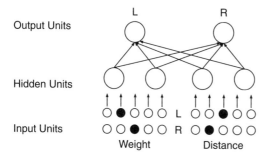

FIGURE 3.19 Network architecture for balance beam.

the beam is predicted to go down to the left, and vice versa for the right-hand side. If neither output unit is greater, then the beam is predicted to balance. The network's training involved a subset of the possible 625 combinations of weights and distances on the beam through the successive presentation of individual balance problems, similar to the questions posed to children in Siegler's task. However, contrary to Siegler's task in which children were not given feedback, the network's output is compared at each trial to the correct output and the discrepancy between the two is back-propagated to adjust the connection strengths in both the weight and distance channels. At regular intervals the network is tested on trained items as well as on generalization to novel weight/distance combinations. For purposes of this simulation, network training is taken to be equivalent to children's pre-test experience on balance problems.

Two characteristics are essential to the network's performance. First, the architecture is such that weight and distance are processed via separate channels. Second, the network initially learns more about weight than distance. The early focus on weight rather than distance is achieved by selecting training examples in which the range of weights experienced is wider than the range of distances. This is justified by the authors by the fact that children seem to have more experience with weight than distance in determining factors like heaviness. Had we justified it by the fact that weight is a one-place predicate whereas distance is more complex because it involves a two-place predicate, then we would have opted for different built-in representations. In this simulation, the input representations for weight and distance are equivalent. The weights in the network are initially random, and at first the network makes no systematic decisions, i.e., there are no systematic differences between the two output units. This is likely to be true of children who have not yet encountered balancing problems of any kind and may expect two objects that they place in any physical contact to balance (Karmiloff-Smith, 1984). Once training starts, the weight matrix moves rapidly away from its random state, and systematic output patterns start to emerge. Over time, these reflect the rules that Siegler argued underlie the stages through which children pass in coming to understand the balance beam. 85% of the network's early performance falls into Siegler's category 1. The network's early predictions involve only weight differences. Later on in the training, distance is taken into account if weights are equal on each side of the beam. Finally correct performance is achieved provided the weight and distance dimensions are not in conflict, i.,e. that they do not predict opposite results. The calculation of both unequal weights and unequal distances, i.e., the law of torque, is not reliably reached by the network, as it rather infrequently is in ontogenesis.

Important for our present concerns is the fact that the transition from one stage to another is abrupt, as in child development. Yet, unlike the explanations that have been invoked to explain the human case, the architecture, learning algorithm and input representations of the network remain unchanged throughout training. Consider again the network architecture for the balance beam problem depicted in Figure 3.19. The network's response to a given weight/distance input is evaluated by comparing the activities of the output units. If the left- hand output unit is criterially more active than the right-hand output unit, then the network is interpreted as predicting that the left-hand side of the beam will go down and vice versa. The activity of the output units is determined by a nonlinear integration of the net input from the hidden units (see Equation 2.2 in Chapter 2, page 52). Input can be either excitatory or inhibitory depending on the strength of the connections from the hidden units to the output units. For example, consider the connections in the weight channel from the two hidden units to the output units. Assume that the left-hand hidden unit grows strong and positive connections from the input units representing the value of the weight on the left-hand side of the beam. Likewise, assume that the right-hand hidden unit grows strong and positive connections from the input units representing the value of the weight on the right-hand side of the beam. This pattern of connections between the input weight units and the hidden weight units force a complementary pattern of connections from the hidden weight units to the two output units. In particular, the left-hand hidden weight unit will grow strong excitatory connections to the left-hand output unit and strong inhibitory connections to the right-hand output unit. The right-hand hidden weight unit will form connections to the output units which are asymmetrical with those of the left-hand hidden weight unit. Consequently, when the weights on either side of the beam are equal, the activity feeding into the output units from the hidden weight units will cancel each other out. An analogous distribution of connections in the distance channel of the network will permit the hidden distance units to influence the output units in a complementary fashion. Figure 3.19 gives a diagram describing the pattern of connectivity.

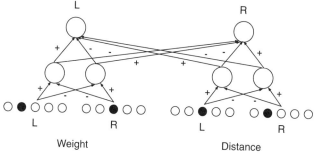

FIGURE 3.20 Network connectivity.

The task of the network is to discover this matrix of connections such that each channel operates in an internally consistent fashion and such that the two channels work together so that the dimensions of weight and distance are integrated in the network's response characteristics. The stage-like behavior of the model is directly attributable to the network discovering solutions to each of these problems at different points in its training, without introducing any changes in architecture or representation of weight and distance inputs.

First, recall that the training set contains more frequent variation of weight values than distance values. Obviously, the greater frequency of variation in weight values offers the network more opportunities to extract predictive correlations between weight values and balance beam solutions. In other words, the network experiences that it is more likely to reduce the error on the output units for a greater number of input patterns if it pays particular attention to the weight channel. The network achieves this focus by strengthening (in both an excitatory and inhibitory fashion) the connections between the input units, hidden units and output units on the weight channel. Changes are also made to the connections on the distance side of the network. However, these changes are less frequent since the distance dimension varies less frequently and appears to be less valuable to the network in predicting the correct output.

Connections in the weight channel increase rapidly after 20 epochs of training whereas the connections in the distance channel change relatively slowly throughout training. Changes during the first 20 epochs are slow because the initial random state of the network's connections inhibits the discovery of the relation between weight value and correct output. Some learning occurs, but is slowed down in much the same fashion as learning on the distance channel is slowed down by the low frequency variation in distance values. However, even small changes in the connections on the weight channel improve the network's efficiency in exploiting the high degree of correlation between weight value variation and correct output, resulting eventually in a sudden change in the magnitude of the strength of connections in the weight channel. Once the strength of the connections in the weight channel is adequate to produce the criterial activity differences on the two output units, the network can be considered to have represented the weight dimension. At this point in learning, the network has mastered the first category of response identified by Siegler (1981), i.e., the weight values exclusively determine the response of the network.

During the next phase of training (20th to 40th epochs), the strength of the connections on the distance channel are too weak to influence decisively the relative magnitude of activity on the output units, even when the weight values are identical. During this period, the performance of the network remains stable in relation to the type of solution it offers to the balance beam problem. That is to say, the network's performance remains characteristic of Siegler's first stage. Nevertheless, throughout this period of training, the distance connections continue to be gradually strengthened. The transition to the second stage occurs when the distance weights are strong enough to produce criterial activity differences on the output units when the weight values are equal. The network now pays attention to distance under those conditions when the dominant dimension of weight is equal on both sides of the balance fulcrum (Rule 2). Note here something crucial that we can derive from the model but not from studies of real children (until, say, we have fine enough measures of brain activity). In the network, although for a long period of time there is no sign of distance affecting the *output*, distance is progressively already affecting the *internal* representations. The stage-like changes at the output level are abrupt; the changes at the hidden level are continuous. It is therefore possible, even probable, that children's internal representations register knowledge that is not apparent for a long time in their output available to the observer.

Network performance again remains stable with respect to this category of response just so long as the strength of the distance connections do not approach the strength of the weight connections. However, with further training the distance connections become stronger such that the distance dimension influences network performance under conditions of unequal weight values (Rule 3). In some cases, the strength of the connections in the two channels may be sufficiently coordinated that the network can solve balance beam problems in which weight and distance are in conflict with each other (Rule 4).

The stage-like behavior of the McClelland and Jenkins model can thus be seen to arise from two crucial assumptions. First, the network's architecture is pre-structured to process the dimensions of weight and distance separately before integrating them at the output level. Second, input to the network is systematically manipulated to simulate attentional differences between the relevant dimensions. The interactions between the structural assumptions and input assumptions result in varying patterns of behavior. Initially, the network discovers the role of the weight dimension. The connections in the weight channel are systematically strengthened. Next, the network discovers the role of the distance dimension as it strengthens the connections in the distance channel. Finally, the two dimensions are integrated through the fine tuning of the connections between the hidden and the output units. Each of these representational changes correspond to individual stages in the development of the model. Representational changes in the network are continuous, i.e., weight changes are continuous. Furthermore, the structural and input assumptions remain constant throughout training. Thus, the network simulation suggests a different explanation in terms of quantitative change from that found in the literature in terms of qualitative rule change. (Chapter 4 explores these differences in explanations of change in detail.)

According to Plunkett and Sinha (1992), this network exemplifies the biologically-inspired notions of Piaget's epistemological theory: Assimilation of the input to the existing state of the network, accommodation of the network to the new input, and a process of gradual equilibration between the two (see also Oyama, 1985). Piaget interpreted the resultant behavior in terms of qualitative changes in representational stages, but the simulations suggest that this is not a necessary assumption. By contrast, the functional processes of assimilation, accommodation and equilibration invoked by Piaget seem to be very well captured in the dynamics of the network's architecture and learning algorithm.

Clearly this initial simulation of the balance beam task is not entirely satisfactory in that it has built-in architectural distinctions that may not be part of the real child's brain and environment. But the simulation does serve as a useful heuristic for rethinking the notion of stages and rules by providing a telling example of how domain-general architectures and computational devices (recall the same architecture and learning algorithm was used for the vocabulary simulation) can give rise to domain-specific representations. It is also important to stress that although the weights and connections are initially random in these simulations, i.e., there are no built-in rules and representations, this does not imply a tabula rasa starting point. Note that the architectural differences between the vocabulary and balance beam simulations on the one hand, and the Elman simulation of lexical segmentation on the other, point to different architectural starting points to learning. Similar differences can be explored at the computational and representational levels also. This is precisely why the connectionist framework is such a useful heuristic in exploring developmental questions.

The Siegler data and the simulation that ensued were based on the fact that children focus first on weight and only subsequently take distance into account when faced with the balance beam. Of course, many children come to a balance scale experiment with no prior experience whatsoever of balance scales. But that doesn't mean that they bring no relevant knowledge to the task. They may focus on weight in tasks using the traditional balance scale because weights are what the experimenter more obviously manipulates and/or because biases in the child direct his or her attention to properties of objects rather than to their locations. But in other tasks that also involve balance and the law of torque, e.g., balancing on a support a series of blocks with evenly and unevenly distributed weight, children ignore weight and focus solely on length (Karmiloff-Smith & Inhelder, 1974; Karmiloff-Smith, 1984). Obviously if the McClelland and Jenkins network were presented with problems where children first take distance into account, then it would behave differently to the simulation discussed above. Clearly, children come to balance tasks already having learned something about how rulers fall from tables, how children balance on see-saws, and so forth. But note that the see-saw is not a pure balance scale. It does not have a neat line of equidistant pegs on which children of absolutely equal weight can be placed one on top of the other! Development is not simply task-specific learning. It is deriving knowledge from many sources and using that knowledge in a goal-oriented way. The initial simulations have been crucial in clarifying the relationship between abrupt behavioral change at the output level and gradual and continuous representational change at the hidden level, an issue discussed in more detail in Chapters 2 and 4. But future connectionist simulations will need to set up richer input vectors and to model internal interactions in order to fully explore the ways in which real children learn in real and complex environments.

All of the simulations that we have discussed so far use a static network architecture to learn a specific task. This does not mean, however, that connectionist approaches to development need necessarily remain committed to such architectures. For example, Shultz et al. (1995) have used a cascade correlation network (Fahlman & Lebiere, 1990) to model the development of seriation in children (Inhelder & Piaget 1958). In cascade correlation, the network starts out life without any hidden units. The connections run directly from the input to the output units. Training proceeds in the normal error-driven fashion either until the network solves the problem or until it gets "stuck" at a particular level of error. In the latter case, the network recruits additional internal resources by gradually constructing an architecture involving hidden units. Cascade correlation recruits hidden units (initially introduced with random connections from the input units) whose activations correlate with the behavioral error that the network is experiencing. Newly recruited hidden units receive input from the network's input units and from any previously installed hidden units. Because high level hidden units receive both raw descriptions of inputs and interpreted descriptions from previous hidden units, they permit ever more sophisticated interpretations of the domain being learned. Cascaded hidden units permit the construction of increasingly powerful knowledge representations that were not available to developmentally earlier instantiations of the network.

Cascade-correlation goes through two recurrent phases: The network begins in what is called the output phase, reducing behavioral error based on environmental feedback by adjusting output-side weights (those connections leading into output units). When this error reduction stagnates, the network enters the so-called input phase, where the focus shifts to building new hidden units. The input phase adjusts input-side weights to candidate hidden units so as to maximize the correlation between network error and candidate unit activation. When these correlations level off, the candidate unit with the highest absolute correlation with network error is installed into the cascade, just beyond the last hidden unit. Then the algorithm reverts to the output phase, in which the network must adjust to this new representation of the problem domain by again training the output-side weights. Training continues until the residual error is eliminated or the process of hidden unit procurement has to be reinitiated.

In their simulation, Shultz et al. (1995) presented the network with an input pattern corresponding to an unordered sequence of different size sticks. The activation of each input unit represented the size of each stick. These input units projected to two separate banks of output units. One bank of output units was trained to indicate *which* of the sticks presented at the input should be moved and the second bank of output units was trained to indicate *where* the selected stick should be placed. Correct responses (and hence the teacher signal) were defined by Inhelder and Piaget's (1958) operational procedure, i.e., move the smallest stick that is out of order to its correct place. The information generated by the output units was used to drive a notional motor response that moved the chosen stick to the chosen location, ready for the network to select its next response. Using the Cascade-Correlation algorithm, Shultz et al. (1995) were able to demonstrate how this type of network architecture was able to capture many of the stages in the development of seriation observed in children.

Simulating Cognitive Development

All of the simulations discussed have involved error-driven learning. Some have questioned the thesis that progress is solely driven by error (Karmiloff-Smith, 1979, 1992a; Siegler & Munakata, 1993), but it is possible that the connectionist framework will suggest a more subtle understanding of the error-driven nature of dynamical systems where the notion of error goes far beyond a mismatch between inputs and outputs.

Before concluding this chapter, let us be very clear about the important distinction between implementation and theory building. It is of course relatively easy to implement developmental outcomes in connectionist models. Some of the models we have discussed are obviously still at a very early stage of development and go little beyond the implementation of behavioral outcomes. As yet they are obviously far from capturing all the subtleties of human development. There is, for instance, a crucial difference between modeling the fact that an organism can continue to hold a representation of an object in a certain location after it disappears from sight behind an occluder, on the one hand, and the conceptual knowledge that objects are permanent, on the other. The point of this chapter has not been to claim that connectionist simulations have solved developmental theory! On the contrary. What we do wish to claim is that connectionist simulations can make one think about development in new ways, drawing out distinctions that have hitherto been ignored.

Our discussion of developmental simulations is obviously not exhaustive. However, the simulations that we have discussed should give the reader a flavor of how connectionist thinking constrains developmental interpretations. Throughout the chapter we have identified a number of principles that guide connectionist modeling and that seem particularly relevant to explaining human development. In particular, domain-general architectures and learning algorithms can give rise to abstract domain-specific representations which ultimately become modularized as the product of learning, not its starting point. But the other side of the coin is to focus on how developmental theories and data pose new constraints on connectionist models. This will be part of our purpose in Chapter 7 when we look at the challenges that future connectionist researchers will have to meet in order to offer a more thorough account of child development. In this chapter we have focused on the current state of simulations that have attempted to capture the behavioral data available in the developmental literature. But the behavioral level clearly needs to be supplemented by specific issues raised by what we now know about biological constraints (see Chapter 5) and by in-depth analyses of why connectionist networks behave the way they do, i.e., of the dynamics of change, to which we now turn.

CHAPTER 4

The shape of change

Children change in thousands of ways, every day. There are changes in body size, attitudes, food preferences, style of dress, sense of humor, and a host of mental abilities. Sometimes change is slow and gradual. Sometimes it happens so fast that the child is a different person from one hour to the next. And sometimes consistent progress is followed by retreat. Out of all the myriad patterns of change that could be studied in a human child, the field of child development has traditionally focussed on a relatively narrow subset. Simply put, some forms of change are more interesting than others, because they present an important challenge to theories of development and (above all) to the peripatetic issue of nature, nurture and their interaction.

In this section, we want to unpack the term "interesting" into a set of primitives, presenting a taxonomy of growth patterns and an analysis of the theoretical assumptions that accompany each one. This taxonomy has a very specific function in a volume on connectionism and developmental psychology, because connectionism offers alternative explanations for some of the most interesting and exotic patterns of growth on the developmental landscape.

To some extent, the interest value of a particular growth pattern comes from its content rather than its shape. For example, with some exceptions (e.g., Gesell, 1929; Rochat, 1984; Thelen, 1986, 1994), developmentalists have not focussed on physical growth or motor development—except, perhaps, as metaphors for mental and/or behavioral change (e.g., Lenneberg, 1967, on proposed parallels between motor and language milestones; or Chomsky, 1980, who likens the language faculty to a mental organ). We will not dwell here on the inherent interest value of specific content domains, except to note that many of our own preferred examples come from the perceptual, linguistic, cognitive and social domains that motivate other developmental theories. Our focus here will be on the shape of change *within* these domains, along three nested dimensions: *linearity* (i.e., linear vs. nonlinear), *direction* (monotonic vs. nonmonotonic) and *continuity* (continuous vs. discontinuous). It is our contention that most developmentalists adopt an implicit theory of these three dimensions, a theory in which the least interesting phenomena lie at one end (change that is continuous, monotonic and linear) while the most interesting phenomena lie at the other extreme (change that is nonlinear, nonmonotonic and discontinuous).

In this context, the word "interesting" refers to the nature, number and transparency of the causes that must be invoked to explain each form of change. In particular, it is usually assumed that phenomena at the "uninteresting" linear end can be explained by a variety of simple and transparent causes—so many that their explanation poses no real theoretical challenge (i.e., these domains are unconstrained). By contrast, phenomena at the "interesting" nonlinear end of the spectrum have proven much more difficult to explain, requiring theorists to seek complex and highly-constrained sources of causation that often bear a very indirect and non-obvious relationship to the final product. In the most interesting cases, it seems as though a different set of causal factors may be required at different points along the developmental trajectory. Hence these phenomena provide a greater challenge to developmental theory, and their resolution can be viewed as a substantial victory. *One of our goals here is to illustrate how the number of causes or mechanisms required to explain complex patterns of change can be reduced within a connectionist framework.* In particular, there are cases in which several distinct causal mechanisms can be subsumed by a smaller set of explanatory principles. Of course not everyone finds this kind of unification aesthetically pleasing ("*De gustibus non est disputandum*"). But for those who value parsimony and theoretical austerity, this should be viewed as an advance.

Dynamical systems theory will figure importantly in what follows, and for the sake of completeness, we wish to hail the recent trend, both in developmental theory (e.g., Smith & Thelen, 1993;

Thelen & Smith, 1994; van Geert, 1994) and in cognitive science more generally (e.g., Port & van Gelder, 1995) to use dynamical systems approach in understanding behavior. Some of this work (Smith & Thelen, for instance) focuses on motor development, whereas other work (for example, the Port & Van Gelder collection) looks at cognitive and perceptual processes as well. This approach has demonstrated that it is often possible to capture complex patterns with models that have relatively few degrees of freedom.

We now consider six different forms of change, discussing their shapes and the formal models which give rise to them.

1. Linear change. Figure 4.1 illustrates simple linear patterns of change over time, including cases of linear increase (Figure 4.1a) and linear decrease (Figure 4.1b). Hypothetical examples of Figure 4.1a might include the gradual growth of vocabulary that is usually observed from 18 years of age to adulthood (Bates & Carnevale, 1993). Similar claims could be made for any domain of cultural knowledge that depends (or so it is assumed) on accumulative experience, and/or perceptual-motor skills that are characterized by incremental gains. Hypothetical examples of Figure 4.1b might include a gradual loss of speed in motor skills after age 30, or gradual loss of accuracy on memory tasks after age 50 in normal adults. Other examples (inverting the usual developmental convention of plotting upward growth) might include a drop-off in the errors observed on spelling tests for grade school children from 7–12 years of age.

Investigators are rarely surprised to find such patterns of linear change, because it is assumed that (1) they are quite common, and (2) they can be easily explained by a host of simple additive mechanisms (experiential or genetic). Although the second point is undeniably true, the first is questionable. In fact, there are good reasons to believe that truly linear patterns of gain or loss are relatively rare in the study of biology or behavior, and many putative cases of linear change be artifacts of sampling. For example, changes in shoe size may fit a linear pattern if we restrict our attention to children between 5–11 years of age. However, we would see some compelling nonlinear increases in shoe size if the age range were expanded

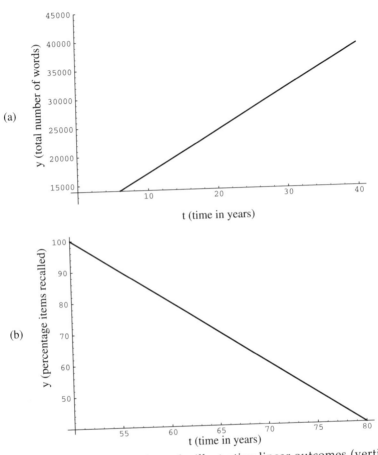

FIGURE 4.1 Hypothetical graphs illustrating linear outcomes (vertical axes) as a function of time (horizontal axes).

to include adolescence. Another source of artifactual linearity comes from summing across individual patterns of growth in group studies. For example, the burst in shoe size displayed by many individual children around adolescence might be lost if we were to graph cross-sectional data from 9–16 years of age in a group of 1,000 children from different cultural, socioeconomic and ethnic groups. To be sure, these data would fit a linear pattern (that is, we have not made a mistake); however, it would be a mistake to assume that this linear pattern reflects a linear (incremental, accumulative) mechanism.

This brings us to a critical distinction between cause and effect, i.e., between the mechanisms responsible for growth and the patterns of growth that are the output of such a mechanism. Invariably, we are forced to infer the former from the latter—and that is where we often go astray. To make this point, we need to distinguish between two different aspects of change: the **outcome function** (i.e., the state of the system as a function of time) and the **dynamic equation** (i.e., the rule governing the rate of change of the system). These two aspects of change have equally important but logically distinct implications for theories of development, particularly when we move on to cases of nonlinear outcomes, with or without nonlinear dynamics.

By definition, a relationship between two variables is linear if it can be fit by a formula of the type

$$y = at + b$$

(EQ 4.1)

where y and t are variables, and a and b are constants that are independent of y and t. Any relationship that cannot be fit by a formula of this kind is, by definition, nonlinear. The outcomes (the y's in Equation 4.1) illustrated in Figure 4.1 can each be described by a linear equation (which defines the ascending slope in outcomes over time in Figure 4.1a, and the descending slope in outcomes over time in Figure 4.1b).

Let us make the example in Figure 4.1a more concrete. We will assume that the t in Equation 4.1 refers to time. The y stands for quantities of some measurable behavior. In the specific relationship illustrated in Figure 4.1a, t (on the horizontal axis) stands for an age range from 6–40 years and y (on the vertical axis) represents the estimated number of words in the vocabulary of the average English speaker, from 14,000 words at age 6 (Templin, 1957; see also Carey, 1982) to an estimated average of 40,000 words at age 40 (McCarthy, 1954). What of the constants a and b in Equation 4.1? The constant a represents the rate of change (which is why we want to multiply it by the age, t). The constant b is the starting number of words (which is why we add that to the other term). It turns out that values of 730 words per year (approximately 2 per day) for *a* and 9620 words for *b* (necessary to produce the estimated value of 14,000 words at age 6) work well.

Although often it is the outcome function one is interested in, it is also often useful to be able to focus on the way the outcome changes over time. In the example above we say that the rate of change is represented by *a*; in the example, the rate remains constant. We can describe the rate of change using the following linear dynamic equation:

$$dy/dt = a \qquad \text{(EQ 4.2)}$$

(The term *dy/dt* is a special notation, called a derivative, which in this case does not refer to division. Instead, it is to be read as "the rate of change in *y* per unit of change in *t*." The *d*'s here have no other meaning). As we said, the symbol on the right-hand side of the equation is constant, so if we graphed the rate of change over time, we would expect to find the horizontal line shown in Figure 4.2. This simple example illustrates an important thing to

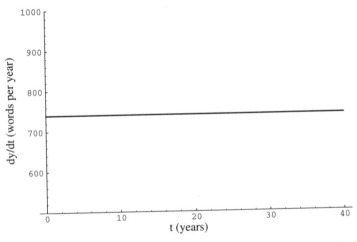

FIGURE 4.2 Dynamic equation underlying the outcome graphed in Figure 4.1a. Because the rate of change (dy/dt) is constant over time, the graph is flat.

remember, which is that although the actual outcome defined in Equation 4.1 changes over time, the rate of change remains the same over time.

What do equations like these have to do with cause and effect? Of course these equations are purely descriptive, but they describe (or so we will assume) the shape of change (outcome functions) and the principles that generate that change (dynamic equations). Hence, loosely speaking, the outcome equations that we will provide here and below are offered as a description of effects, while dynamic equations are offered as a description of the laws governing a causal agent. Sometimes investigators postulate layers of causality that might require an intermediate level of description. For example, the linear drop in performance described in Figure 4.1b may be attributed, in turn, to a linear drop in capacity. In this case, the relationship between performance and capacity is isomorphic, and the two are described by equations from the same family. Indeed, in this hypothetical case the same figure (Figure 4.1b) and the same equation can be used to describe both. But this pushes us another level down: what kind of equation would best describe (generate) the inferred loss of capacity? If the loss of capacity takes the linear form illustrated in Figure 4.1b, then we are assuming a rate of capacity reduction that is constant over time—which means, once again, that we are assuming an underlying cause that gives a constant rate of change as in Figure 4.2. This division into causal layers is not particularly interesting in the present case, but it will take on more importance later on.

More complicated dynamical systems could be represented by an equation of the form

$$\frac{dy}{dt} = f(y, t)$$

(EQ 4.3)

where $f(y,t)$ gives the rule that governs how fast the system, represented by y, is changing in time. This may depend on the state of the system at the time in question, so f depends explicitly on y. It may also depend on the age of the system in some way not directly related to the evolution of the system but rather is prescribed or forced to occur at fixed times. These prescribed rule changes could be due to some internal clock of the individual or be external and just related to the rate at which the individual is exposed to new information. We could represent such prescribed or externally controlled changes in the rule by including an explicit dependence of f on time. The case in which there is no explicit time dependence is called autonomous. In that case f depends only on y, in other words, the rate of learning at any given time depends only on the state of the system at that time. This may be appropriate if the influx of new words is a constant in time.

In the present context, we could consider the system to be the language learning system of the individual, and say we represent the state of the system by the number of learned words, y. We could postulate that the rate of learning words is given by some efficiency, ε, times g, the number of words presented to the child per unit time. For example, if the number of new words presented to the person in a year is 1000 but the person's learning system has an efficiency of 20% then the number of words learned would be $\varepsilon g = 200$. To simplify matters, let us also assume that the number of new words presented per unit time is constant. Thus we can represent the evolution of the system by

$$\frac{dy}{dt} = \varepsilon(y, t)g \qquad \text{(EQ 4.4)}$$

Such an equation is capable of representing as complicated an evolution as we like simply by specifying a complicated evolution of the efficiency with time. Thus changes in rates of learning could simply be prescribed to occur at different times. On the other hand, it may be that the efficiency of learning evolves mainly due to the internal dynamics of the learning system and is not governed or regulated, at least in any critical way, by outside processes. Such a model would be written as

$$\frac{dy}{dt} = \varepsilon(y)g \qquad \text{(EQ 4.5)}$$

That is, the efficiency of learning here depends explicitly only on the state of the system at any given time. As we shall see, this system too can give rise to complicated behavior, but now any changes in the efficiency of the system can only arise out of the natural evolution of the system subjected to a constant input of data, as opposed to the case where the changes are explicitly prescribed.

In what follows, we will consider some simple models for $\varepsilon(y)$ that can give interesting behavior for y. Let us turn now to some more interesting cases.

2. Nonlinear monotonic change with linear dynamics.

Now we can move into patterns that have played a more important role in developmental theory. We will start with patterns of growth that follow an "interesting" nonlinear trajectory, patterns that have inspired complex explanations involving two or more distinct causal mechanisms. As we shall see, however, the apparent complexity of these nonlinear outcome patterns is illusory, because they can be parsimoniously explained by a simple linear dynamics. (And this is why we made the distinction above between outcome equations and dynamical equations.)

Figure 4.3 illustrates the nonlinear relationship between vocab-

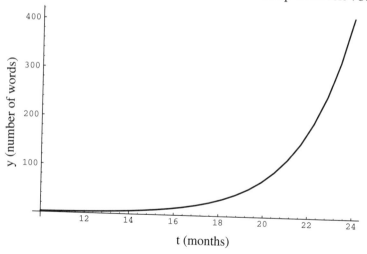

FIGURE 4.3 Hypothetical relationship between a child's expressive vocabulary and age.

ulary and age that has been observed in the earliest stages of language development (from Bates & Carnevale, 1993). In this case, t represents age from 10 to 24 months, and y represents the number of words in the child's expressive vocabulary. This graph is an idealization, but it is patterned after vocabulary growth functions that have been observed in diary studies of real live human children (e.g., Dromi, 1987; Nelson, 1973).

In contrast with the graph of adult vocabulary growth in Figure 4.1a, the infant data graphed in Figure 4.3 illustrate a nonlinear outcome. It appears from a cursory examination of this pattern that there is a marked acceleration somewhere around the 50-word level in the rate at which new words are added (occurring here around 20 months). For example, the child learns only 24 new words between 10 and 14 months, but she learns 328 new words between 20 and 24 months. This nonlinear pattern can be described by the nonlinear function

$$y = y_0 e^{b(t - t_0)} \qquad \text{(EQ 4.6)}$$

Here y_0 (=1 in this case) is the number of words the child knows at 10 months of age (Fenson et al., 1994). The constant b is called the exponential growth rate (which in this case is 43% per month), and e^x is the exponential function. Actually all this function e^x means is that the constant e (e is a constant frequently used in mathematics and its value is approximately 2.718) is raised to the power x. For example, e^2 is approximately 2.718^2 or approximately 7.388, etc.

"Burst" patterns like this one have been reported many times in the language acquisition literature. They are common in vocabulary development between 14 and 24 months of age, and similar functions have been reported for aspects of grammatical development between 20–36 months of age. How should such bursts be interpreted? A number of explanations have been proposed to account for the vocabulary burst. They include "insight" theories (e.g., the child suddenly realized that things have names, Dore, 1974; Baldwin, 1989), theories based on shifts in knowledge (Zelazo & Reznick, 1991; Reznick & Goldfield, 1992), categorization (Gopnik & Meltzoff, 1987), and phonological abilities (Menn, 1971; Plunkett,

1993). Although these theories vary greatly in the causal mechanisms brought to bear on the problem, they all have one thing in common: The proposed cause is located at or slightly before the perceived point of acceleration (i.e., somewhere around 50 words). In a sense, such theories assimilate or reduce the data in Figure 4.3 to a pair of linear relationships, illustrated in Figure 4.4, i.e., two linear functions whose cross-point indexes a sudden, discontinuous change in the rules that govern vocabulary growth.

However, if the function illustrated in Figure 4.3 is accurate,

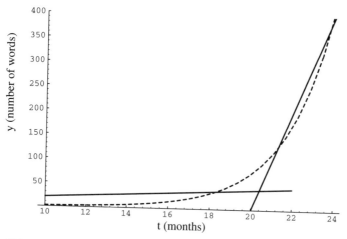

FIGURE 4.4 A piece-wise linear approximation (solid lines) of the data shown in Figure 4.3 (dotted curve).

this perceived point of discontinuity in slope is really an illusion. Figure 4.3 represents a function with a constant fractional rate of increase and can be generated by a linear dynamical equation of the form

$$\frac{dy}{dt} = by$$

(EQ 4.7)

which tells us that the increase at any given moment is always proportional (the constant b is the percentage increase per time) to total vocabulary size (given by the variable y). This linear dynamical relation between the rate of change and y is graphed in Figure 4.5.

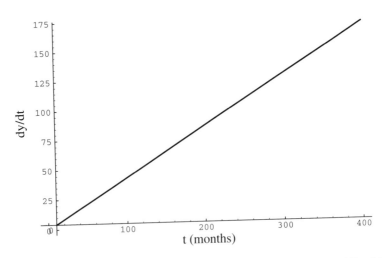

FIGURE 4.5 Graph of the linear dynamic equation Equation 4.7, which underlies the dotted curve in Figure 4.4.

As van Geert (1991) has argued in a paper on the equations that govern developmental functions, we do not need to invoke intervening causes to explain this kind of growth. The real cause of the acceleration that we commonly observe between 50–100 words may be the growth equation that started the whole process in the first place, i.e., conditions that have operated from the very beginning of language development. Of course this does not rule out the possibility that other factors intervene along the way. Environmental factors may act to increase or reduce the rate of gain, and endogenous events like the "naming insight" and/or changes in capacity could alter the shape of learning. Our point is, simply, that such factors are not necessary to account for nonlinear patterns of change.

The most general linear dynamical (autonomous) equation in one dynamical variable, y, is

$$\frac{dy}{dt} = by + c \qquad \text{(EQ 4.8)}$$

Here b and c are constants independent of time and y. Again, by linear dynamics, we mean that the relationship between the rate of change of y and y itself is linear. This relationship when plotted is simply a straight line like that in Figure 4.5. If $y = 0$, then the rate of word accumulation is the constant c. In other words, this dynamics assumes that even when you have no words you have the ability to learn some. The rate at which you learn those first words would be c, the product of a nonzero efficiency times the number of words per unit time to which you were exposed. If b is positive, the term by implies that the more words you know, the easier it is to accumulate more. This one term by can be thought of as the net of an increase in ability to learn words resulting from previous accumulation and the rate of loss of words due to forgetfulness which may also be proportional to the total number of words. In a sense, the two terms on the right hand side of Equation 4.8 can be thought of as two mechanisms that compete with each other at all times. The term that dominates depends on the size of y which changes in time. Thus the behavior of the system can have different characteristics at different times although the same mechanisms are always operating. In such autonomous models all of the complexity is assumed to come from the natural dynamics of the system itself under a constant environmental forcing.

The general solution of Equation 4.8 is

$$y(t) = \left(y_0 + \frac{c}{b}\right)e^{b(t - t_0)} - \frac{c}{b} \qquad \text{(EQ 4.9)}$$

The solution is simply an exponential plus a constant. Whether the solution increases or decreases with t depends on the sign of $by_0 + c$. Four examples of the shape of this function are shown in Figure 4.6.

This differs from the simpler solution Equation 4.6 in that one can have growth even if the initial number of words is zero. Also, if b is negative, this solution can represent the growth or decay of the number of words to some constant value (explicitly, y tends toward $\frac{-c}{b}$, as t increases indefinitely).

The simple vocabulary burst example in Figure 4.4 illustrates how a two-cause theory can be replaced by a theory based on a single mechanism (a linear evolution equation). More complicated data may suggest more mechanisms. But, as in the general linear dynamical case, it is interesting to consider the possibility that all

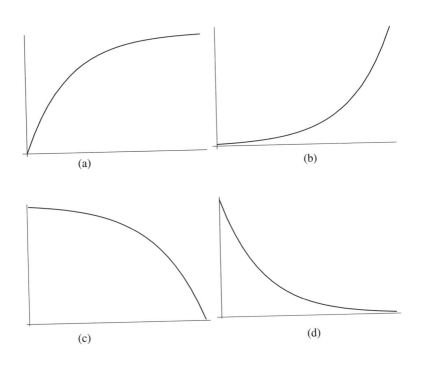

FIGURE 4.6 Hypothetical graphs illustrating nonlinear outcomes (vertical axes) as a function of time (horizontal axes).

the mechanisms are always operating and their relative importance depends only on the state of the system given by y, rather than invoking some ad hoc change prescribed at a certain time or age.

As an interesting example of this possibility, we consider the result found by Johnson and Newport (1989), who have focussed on second language learning by children and adults. Johnson and Newport set out to test the critical period hypothesis, i.e., the idea that there are maturational constraints on language learning that make it easy for children but difficult for adults to acquire a second language (we discuss a connectionist model which makes similar assumptions in Chapter 6).

To test this hypothesis, Johnson and Newport studied a sample of Chinese-English bilinguals who arrived in the United States (and began the process of second language learning) at various points from infancy to the early adult years. To overcome some of the methodological problems encountered in earlier studies comparing child and adult second language learners, they focussed entirely on the end state of language learning in adult bilinguals, approximately thirty years after arrival. In adult bilinguals who arrived between 0–16 years of age, they report a significant negative correlation between age of arrival and accuracy scores on a test of sensitivity to English grammar ($r = -.87$). In adults who arrived somewhere between 16–40 years, there was no significant relationship between age of arrival and grammar scores ($p < -.16$), although as a group the late-arrivers performed at a lower level than those who arrived before age 16.

Because these two regression lines are so different, Johnson and Newport conclude that different forces are operating in these two periods of development. Specifically, Johnson and Newport supposed that there is a change in maturational state, from a plasticity or readiness for language learning under age 16 (a state that undergoes a linear decrease from birth) to a steady state of limited success in second language learning after age 16.

But this is not the only possible explanation for these data. In Figure 4.7, we have replotted the data from Johnson and Newport on a single graph that spans 0–40 years.[1] The two straight dotted lines in Figure 4.7 represent the two regression lines, for 0-16 years and for 16–40 years, respectively. These two lines illustrate the "change of state" view presented by Johnson and Newport.

However, Figure 4.7 also includes a single curvilinear function (solid line) that we have fit to the same data, spanning the arrival range from 0–40. This curvilinear outcome function is a solution of the linear dynamical equation Equation 4.8, given by Equation 4.9. Here y is the percentage of correct performance. The simple relationship between the rate of change and y is illustrated in Figure 4.8. In other words, the "two-state" maturational effects described by Johnson and Newport can be fit by a theory in which

---

1. We thank Jacqueline Johnson for generously making these data available to us.

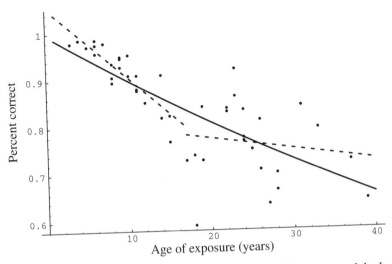

FIGURE 4.7 Data from Johnson & Newport (1989), fit by two models. In the two-stage model proposed by Johnson and Newport, language learning below the age of 16 is facilitated by a state of plasticity which is lost after the age of 16. The two phases are modeled by two different linear regression equations (graphed with dotted lines). The mean percentage of variance accounted for by these two lines is 39.25%. Alternatively, the same data may be fit by a single nonlinear function, shown by the solid line (the nonlinearity is slight, so that visually, within the range of values shown, the curve looks linear). The percentage of variance accounted for by the single stage nonlinear model is 63.1%

two simple mechanisms constantly compete to produce a smooth curve of changing slope. The changes are all determined by the direct response of the learning system to the constant environment and the periods of rapid versus slow change are determined by the state of the learning system itself, and not by any system external to it. Of course these are not the only facts in the Johnson and Newport data. One must also deal, for example, with a marked increase in the variance around the curve after 16 years. Our point here is not to dispute the conclusions offered by Johnson and Newport, but to point out that a simple linear dynamical alternative is available to account for the same nonlinear outcomes.

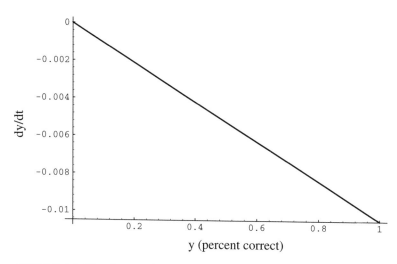

FIGURE 4.8 The linear dynamic equation underlying the outcome graphed in the nonlinear model of the Johnson and Newport data shown in Figure 4.7

It turns out that a similar analysis applies to another domain in which apparently abrupt changes in learning have been interpreted as evidence for two-state models: the acquisition of bird-song. Although birds and humans are clearly different, tantalizing similarity in pat- terns of acquisition of song by young birds have long been cited as evidence for the inherent plausibility critical period models.

Marler and Peters (1988) presented new data which they argue bolsters the claim that certain species of birds undergo a critical period of plasticity during which exposure to adult models will result in the acquisition of normal song. If young birds are deprived of the necessary input during the first 100 days after hatching, the percentage of correct song structure learned drops dramatically.

In Figure 4.9 we have graphed the data from Marler and Peters, along with two ways to model the data. The model Marler and Peters propose is shown by the dotted lines, which illustrate a very extreme form of the two state hypothesis. During the first state (roughly the initial 75 days of a bird's life) learning is at a high level. If learning is delayed until this initial critical period is passed,

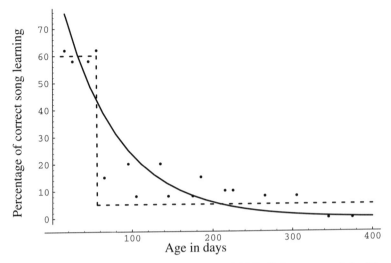

FIGURE 4.9 Data from Marler and Peters (1988), fit by two models. The two-stage model (dotted line) assumes an abrupt loss of plasticity at approximately 75 days; this model accounts for 59.9% of the variance in the data The solid curve corresponds to the single-stage model, which assumes a nonlinear drop in learning; this model accounts for 73% of the variance.

then performance drops to a very low level. This two state model accounts for 59.9% of the variance. However, the same data can be accounted for by an exponential function plus a constant graphed in the solid line. This single-stage nonlinear function accounts for 73% of the variance. Once again, we see that a single nonlinear function arising from a simple linear mechanism or dynamics may produce effects which look as if they arise from multiple state systems.

On the other hand, there are excellent reasons to believe that the linear dynamical equation shown in Figure 4.4 tells only part of the story (see also van Geert, 1991). Let us suppose for a moment that vocabulary growth continued to follow this dynamic function for a few more years. At this rate of growth, our hypothetical child would have a vocabulary of approximately 68,000 words at 3 years of age, 12 million words at four years, and 2 billion words by the time she enters kindergarten! Since there are no known cases of this sort, we are forced to one of two conclusions: (1) some exogenous force intervenes to slow vocabulary growth down, or (2) the initial acceleration and a subsequent deceleration were both prefigured in the original growth equation. We will turn to this point shortly, in a discussion of nonlinear dynamics.

(3) Nonlinear monotonic functions with nonlinear dynamics. Figure 4.10 illustrates a relatively simple nonlinear pattern known as the logistic function. Functions of this kind are quite common in

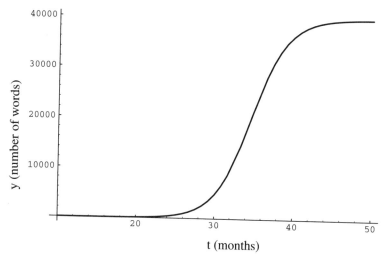

FIGURE 4.10 A nonlinear form of change in vocabulary (y axis) as a function of time (x axis).

behavioral research, and they are also common in neuroscience (where they can be used to describe the probability of firing for a single neuron or population of neurons, assuming some threshold value).

The particular example in Figure 4.10 reflects a hypothetical example of vocabulary growth from approximately 10–48 months of age. The first part of the graph from 10–24 months is almost identical to the growth function in Figure 4.6b (although that is difficult to see because of the change in scale). In contrast with the ever-increasing exponential burst in Figure 4.6b, Figure 4.10 does have a true inflection point (half-way up the curve), defined as the point at which the rate of change stops increasing, and starts to slow down. This pattern can be described by the equation

$$y = \frac{y_0 e^{b(t-t_0)}}{\left(1 + \left(\frac{y_0}{y_{max}}\right)(e^{b(t-t_0)} - 1)\right)}$$

(EQ 4.10)

Although this is a more complicated equation than the ones we have seen so far, the only new term here (in addition to the ones introduced in the previous example) is the constant parameter y_{max}, which stands for an estimated upper limit on adult vocabulary of 40,000 words. (We do not have to assume that the child knows this upper limit in advance; instead, the limit might be placed by the available data base or by some fixed memory capacity). The nonlinear dynamic equation that underlies this nonlinear pattern of change is

$$\frac{dy}{dt} = ay^2 + by$$

(EQ 4.11)

where a is defined as $\frac{-b}{y_{max}}$.

This dynamic relationship between rate of growth and the variable undergoing change is graphed in Figure 4.11. The main thing to notice here is the changing relationship between ay^2 and by which explains why the initial acceleration and subsequent decline in growth are both contained in the same equation. To obtain the shape in figure Figure 4.10, we make b positive and a negative. Early in the evolution, say from 10 to 24 months, by is much larger than $-ay^2$, and so the evolution during that period is almost identical to that in Figure 4.6b. As time proceeds, and y increases in size, ay^2 becomes closer in size to by. Because the growth rate is defined as the difference between these two terms of opposite sign, the rate of growth approaches zero at the specified vocabulary maximum, and that vocabulary maximum, $y_m = \frac{-b}{a}$, is predicted by balancing

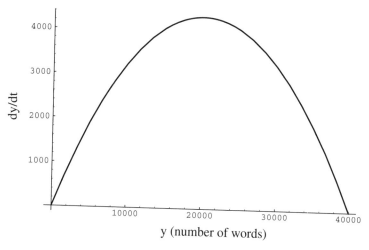

FIGURE 4.11 The graph of the nonlinear dynamic equation (Equation 4.11) underlying the outcome shown in Figure 4.10.

the two terms. Here we see the two different "stages" in development are given by a continuous change in the relative magnitude of these two terms.

This dynamic equation provides a better overall match to the child vocabulary growth data than the exponential in Figure 4.3 (i.e., the function that predicts a 2 billion word vocabulary when the child enrolls in kindergarten). However, it is still grossly inadequate. In Figure 4.10, our hypothetical child is already close to adult vocabulary levels at 4 years of age—unlikely under any scenario. In other words, as van Geert also concludes in his analysis of lexical growth patterns, the symmetrical properties of the logistic function cannot capture the apparent asymmetries in rate of growth evidenced across the human life time. Additional mechanisms are required to explain the asymmetries that are actually observed in infant vocabulary development, in particular the asymmetric damping or slowing down of vocabulary learning between 2–4 years of age. For example, Dromi (1987) has proposed that this damping occurs because the child has switched her attention from vocabulary to grammar. In this case, we have formal justification for adopting a more complex developmental model. This contrasts with the vocabulary burst example described above, where a simple linear dynamic is sufficient to handle all the data. The solution proposed by van Geert was to use a delay equation, that is, the rate of change of the system depends on its state at an earlier time. An alternate solution would be to introduce a slightly more complicated version of the dynamical Equation 4.11. In that equation the rate of change is zero if $y = 0$. It is more reasonable to assume that there is a mechanism for learning even if you have no words to start with. This can be achieved by adding a constant to the right hand side of Equation 4.11. Thus we have

$$\frac{dy}{dt} = ay^2 + by + c \qquad \text{(EQ 4.12)}$$

and the general solution to this evolution equation is

$$y(t) = y_m \frac{1 - e^{2a(y_m - y_i)t}}{1 - \frac{y_m}{2y_i - y_m} e^{2a(y_m - y_i)t}} \qquad \text{(EQ 4.13)}$$

where y_m is the maximum number of words that the curve tends toward, y_i is the inflection point where the rate of growth stops increasing and begins to slow down. Both y_m and y_i are defined by the values of the constants a, b, and c. Also a can be defined in terms of y_m, y_i, and c, which is the initial slope of the solution curve. The formula for this is

$$a = \frac{c}{2y_i y_m - y_m^2} \qquad \text{(EQ 4.14)}$$

Thus with this dynamics, one can choose the initial slope of the graph, the inflection point, and the maximum number of words all independently to match the data. Figure 4.12 shows this solution with these parameters all chosen to make a reasonable fit to typical vocabulary growth over a lifetime.

We can also write the rate of change as an efficiency of learning

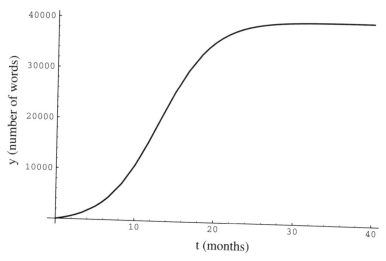

FIGURE 4.12 Modified model of relationship between change in vocabulary (y axis) as a function of time (x axis).

$\varepsilon(y)$ times the rate of input of new words g. That is,

$$\frac{dy}{dt} = \varepsilon(y)g$$

(EQ 4.15)

where

$$\varepsilon(y) = \frac{ay^2 + by + c}{g}.$$

(EQ 4.16)

The c term corresponds to the mechanism that permits learning at a constant rate for the first words. The term by, proportional to the number of words known, increases the efficiency of learning because of the familiarity with other words. The term ay^2 is negative, thus it decreases the efficiency of learning. A possible mechanism that would require such a quadratic term would be interference between words. Since the number of pairs of words that can interfere increases as $y(y-1)$ or approximately y^2, it is reasonable to include such a quadratic term in the dynamics.

(4) Nonmonotonic change. Now we come to the patterns that have always provided the biggest challenge to theories of development. These involve nonmonotonic outcomes, in which behavior moves in one direction for a while, and then (perhaps only temporarily) reverses.

The first example, shown in Figure 4.13 is actually quite com-

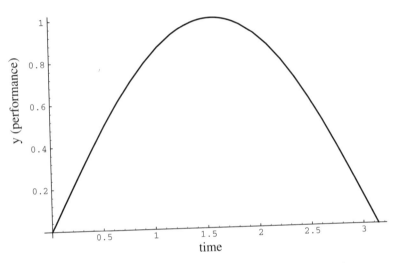

FIGURE 4.13 Hypothetical performance curve in which performance increases early during life, peaks, and then declines during the remainder of life.

mon if we take the whole human life-span into account: An increase in outcomes (performance, size, ability, etc.) in the first half of life, followed by a symmetrical decline in the second half. The outcome function shown in Figure 4.13 is actually

$$y(t) = \sin bt \qquad \text{(EQ 4.17)}$$

The dynamical equations of the type that we have been dealing with take the form

$$\frac{dy}{dt} = f(y)$$
(EQ 4.18)

that is, autonomous equations for the rate of change of a system described by only one dynamical variable y. These kind of equations can only represent monotonic behavior. This is because the rate of change is a function only of y. Imagine at two different times the function $y(t)$ has the same value, for example once on the increasing side of Figure 4.13 and once on the decreasing side. Since the value of y is the same at both of these points, the function $f(y)$ will be the same at both of those times, thus $\frac{dy}{dt}$ must be the same at both points, but then they must both be points where the curve is increasing or decreasing because the slope must be the same at two points that have the same value of y.

One way to deal with nonmonotonic functions would be to introduce explicit time dependence into the equation,

$$\frac{dy}{dt} = f(y, t)$$
(EQ 4.19)

that is, to make the equation nonautonomous and change the slope by prescription. For example, we could generate the $\sin bt$ curve from the equation

$$\frac{dy}{dt} = b\cos bt$$
(EQ 4.20)

(the cosine function has positive/negative values just where the slope of $\sin bt$ is positive/negative). But this would be a poor model of the possible interesting dynamics of the system that produce the evolution. This kind of equation would be better reserved for cases where the systems evolution really is controlled in detail by external driving.

Another way to approach the problem is to realize that using one dynamical variable to describe a complex system may be inadequate. In reality the learning system is highly complex and many components would be needed to accurately describe its behavior. We would then want to represent the evolution of a dynamical system with a vector equation

$$\frac{d\hat{y}}{dt} = f(\hat{y}) \qquad \text{(EQ 4.21)}$$

where

$$\hat{y} = (y_1, y_2, ..., y_n) \qquad \text{(EQ 4.22)}$$

is a vector of n variables that represent different components of the system. Even in linear dynamics, then, nonmonotonic change is possible.

For example, the dynamical equation which generated the curve in Figure 4.13 is

$$\frac{dy_1}{dt} = by_2$$
$$\frac{dy_2}{dt} = -by_1 \qquad \text{(EQ 4.23)}$$

where b is a constant. This is a system of coupled linear equations. It is linear because no higher than the first powers of the dynamical variables are involved, and the two equations are coupled because the growth of y_1 depends on the value of y_2 and vice versa. For example, in this case if b were positive, the first equation says that y_1 would be increasing if y_2 were positive and decreasing if it were negative. Similarly, the second equation says y_2 is increasing/decreasing in time when y_1 is negative/positive. The result is that $y_1 = \sin bt$ and $y_2 = \cos bt$.

Dynamical systems can be modeled at many different levels. For example, the y_i may be some macroscopic variables such as the number of words accumulated and the mean word length of utterances. At the other microscopic extreme the system may be described by its most fundamental units. For example, y_i could be the level of activation of neuron i in the brain. Actually, by increasing the number of components contributing to the behavior, arbitrarily complicated behavior could be represented even with just linear dynamics. By including the possibility of nonlinear interactions extremely complicated behavior may appear even for systems with as little as three interacting components such as in the case of the Lorenz attractor that we will describe in a later section of this chapter.

The important insight here is that we need no external cause to change the course of development halfway through life. The rise and fall of this developmental system are prefigured in the same dynamic equation, set into motion at the beginning of its life. Of course we would not want to insist that this is the necessary explanation for the unhappy events in Figure 4.13. Alternative accounts are possible, but they would necessary involve external influences to impinge on the learning system rather than an uninterrupted interaction between the system and the data presented.

Now we come to the parade case of nonlinear development: the so-called U-shaped functions in which things get worse for just a little while before they get better. An outcome function of this kind is illustrated in Figure 4.14.

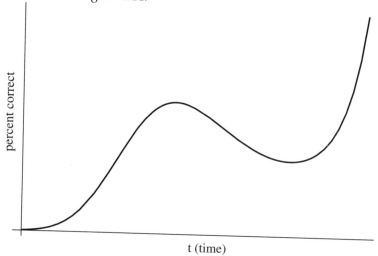

FIGURE 4.14 Hypothetical U-shaped curve, in which an initial improvement in performance is followed by a temporary decline, followed by mastery of the task.

This particular function was chosen to approximate the famous U-shaped pattern for production of the English past tense (Berko, 1958; Brown, 1973; Bybee & Slobin, 1982; Ervin, 1964; Marchman, 1988; Marcus et al., 1992; Pinker & Prince, 1988; Plunkett & Marchman, 1991; Rumelhart & McClelland, 1986), discussed at length in Chapter 3, and which will figure prominently in our arguments.

A graph such as this could be obtained either by making our two variable model nonlinear, or adding a third variable, y_3, in a linear dynamical system. A simple nonlinear model could be constructed as follows:

$$\frac{dy_1}{dt} = h(ay_2^2 - by_2 + c)y_1$$

$$\frac{dy_2}{dt} = h$$

(EQ 4.24)

The solution to this system of equations is

$$y_1(t) = y_{10}e^{\frac{at^3}{3} - \frac{bt^2}{2} + ct}$$

$$y_2(t) = ht$$

(EQ 4.25)

where we have assumed $y_2(0) = 0$. The idea behind the dynamical model is that y_1 represents the capability for getting the past tense correct and y_2 is some measure of the comprehension of the regular rule. In this crude model we just assume y_2 keeps growing with time. Now if we assume y_1 should grow at a rate proportional to the examples of the past tense that the child has already absorbed, we would write

$$\frac{dy_1}{dt} = ry_1$$

(EQ 4.26)

But we also want to build in some loss of ability when the knowledge of the regulars starts destroying some of the irregular patterns that had originally been correctly incorporated. This happens as y_2 grows. Finally we want y_1 to have a positive growth rate again when the knowledge or the regular rules has reached a sufficiently high level. This is accomplished by writing the growth rate as

$$r = h(ay_2^2 - by_2 + c)$$

(EQ 4.27)

Thus when y_2 is small y_1 grows at the rate $h = dc$, at some intermediate value of y_2 the $-by_2$ term dominates and makes r negative and y_1 decreases, then when y_2 is sufficiently large, the growth rate is again positive. So this is one way to model such behavior.

Now, let us consider for a moment the usual treatment of outcomes like the U-shaped development of the English past tense. The basic story (grossly oversimplified) is something like this:

(1) Children begin to talk about actions by using an unmarked form, usually in the present tense (e.g., "Daddy come").

(2) The child's first systematic efforts to explicitly mark the past tense usually involve high frequency irregulars, produced correctly (e.g., "Daddy came").

(3) After a period of correct production, the same child begins to produce errors in which the regular past is over-generalized of the irregular past (e.g., "Daddy comed"). Note that these errors are usually relatively rare (under 20% of all past tense forms), and they coexist with continued usage of the correct form for a period that can range from weeks to years. In the same period, some investigators report the existence of a very rare error in the opposite direction, i.e., "irregularizations" in which an irregular pattern is over-generalized to regular and/or to other irregulars (e.g., "pick"→ "pack", perhaps by analogy to "sit"→ "sat").

(4) Eventually these errors of over-generalization phase out, and the child enters into the mature state of correct regular and irregular verb marking.

When these phenomena were first described by Roger Brown and his colleagues (Brown, 1973), investigators proposed an explanation based on two developing mechanisms: rote memory (which is responsible for the early mastery of high frequency irregulars), and rule extraction (which is responsible for the subsequent extraction of the "-ed" rule and its over-generalization to other forms). Presumably, these two mechanisms compete across the period in which over-generalization errors occur, until the boundary conditions are established for each one.

This behavioral example and its two-mechanism explanation are familiar to most developmental psychologists. Indeed, this may be the best-known and most-cited case in the field of language acquisition. It is perhaps for this reason that the famous Rumelhart and McClelland (1986) simulation of the past tense had such a dramatic impact, inspiring a wave of criticisms (Marcus et al., 1992; Pinker & Prince, 1988 and counter-attacks (Daugherty & Seidenberg, 1992; Hare & Elman, 1995; Marchman, 1988, 1993; Plunkett & Marchman, 1991, 1993). Rumelhart and McClelland were able to replicate many of the details of past tense learning in a neural network with no explicit division between rote and rule. Although this initial work made a number of important points, the model contained several flaws, a fact that is acknowledged by most connectionists today. But subsequent studies by other investigators have overcome the initial weaknesses in the R&M model, and it now seems clear that the basic pattern in Figure 4.14 can be obtained in a homogeneous neural network.

So where does the U-shaped pattern come from, if a single mechanism is responsible for the entire developmental trajectory? As P&M have pointed out, the U-shaped function reflects a dynamic competition between regular and irregular mappings within a single system. As the number of verbs in the competition pool expands across the course of learning, there are shifts in the relative strength of regular and irregular forms. The U-shaped dip in the learning curve occurs around the point in development in which there is a change in the proportional strength of regular "-ed" compared to other mappings.

Here is the basic irony that we want to underscore. At some level, as pointed out, the two-mechanism account of the U-shaped function is correct. That is, to produce the pattern in Figure 4.14, we need two competing terms in the equation. The difference between this account and the classic two-mechanism view lies in the fact that our two competing terms are contained within the same equation, and hence (by implication) within a single system. In other words, there is fundamental difference between the classic account and the connectionist view at the level of cognitive architectures. However, the basic intuition that "two things are in competition" is necessary in both accounts.

(5) True discontinuities. In the examples presented so far we have tried to illustrate that changes in the behavior of a system can be due to competition between different mechanisms within the system and do not need to be prescribed ahead of time nor imposed by changes in the environment. Depending on the kind of nonlinearity employed the autonomous dynamical system described by

$$\frac{dy}{dt} = f(y)$$ (EQ 4.28)

can exhibit behavior with sharp changes in the rate of growth. A simple way to introduce such changes into the models is to use nonlinear functions that change rapidly for some small variation of the dynamical variables. The logistic function is such a function.

Even more extreme is the step function, $H(y)$. This function is 0 when its argument is negative and 1 when its argument is positive. Thus if we consider the nonlinear evolution equation of the form

$$\frac{dy}{dt} = a + bH(y - y_c)$$ (EQ 4.29)

the solution will have two parts:

$$
\begin{aligned}
y(t) &= at & \text{for} \quad (t < t_c) \\
y(t) &= (a + b)(t - t_c) + y_c & \text{for} \quad (t \geq t_c)
\end{aligned}
$$ (EQ 4.30)

where we have assumed $y(0) = 0$. There is a discontinuous change of slope at the critical time defined by

$$t_c = \frac{y_c}{a}$$ (EQ 4.31)

Thus, sharp changes in behavior can be due to the natural evolution of a nonlinear system even when the external forcings are constant.

On the other hand, it may be that discontinuous changes are in fact caused by the influence of some other system, such as the hormonal system, on the learning system. How would we describe that? One way would be to introduce new dynamical variables, $x(t)$, with their own laws of evolution and allow them to play a role in the evolution of the learning system represented by $y(t)$. Thus we would write

$$\frac{dy}{dt} = f(x, y)$$

$$\frac{dx}{dt} = h(x)$$

(EQ 4.32)

where we continue to assume autonomous dynamics for simplicity. The system represented by x then can act as an internal clock, having its own evolution but not affected by the system represented by y, it can still influence y. The influence of x need not introduce discontinuous changes in $\frac{dy}{dt}$, but it might. An example of this would be the system

$$\frac{dy}{dt} = a + bH(x - c)$$

$$\frac{dx}{dt} = 1$$

(EQ 4.33)

With $x(0) = 0$, the solution for x is $x(t) = t$, that is, x is just time. Thus the rate of change of y for $t < c$ would be a, and for $t > c$ it would be $(a + b)$. If $c = t_c$, the result for $y(t)$ is precisely the same as the example given in Equation 4.33 above, although the conceptual models are very different. In the first case it is the growth of the variables describing the learning system that triggers the discontinuous change. No external influences are imposed, no set time is prearranged, and the change occurs when a certain level of learning is achieved. In the second, it is the interference of one system which has nothing inherently to do with learning (when growth is independent of learning) that can trigger the change at a prescribed age.

In many of those cases we know with some precision how and why a continuous shift in one parameter works to produce a discontinuous outcome. Take the continuous dimension of temperature: Ice turns to water, and water turns to gas, all with the turn of a single dial. Where does the "new stuff" come from? The answer lies in the interaction between temperature and the molecular properties of water, where the bonds between molecules are made and broken. The key insight here is that a continuous shift along a single parameter can put the entire system into a different problem space. In the second part of this chapter we discuss the notion of bifurcations, which provide one way in which small changes in a parameter which governs a dynamical system can lead to dramatically different behavioral regimes. Once the system is boot-strapped into this new problem space, a different set of forces take over to create the final outcome. In cases like this, we cannot even begin to count entries along the y axis until a certain threshold is reached along the x axis—a critical point t_c. To what extent is it meaningful to say that the phenomenon in question was caused at t_c? It appeared at t_c, but its causal history must include everything that led up to the critical value.

(6) Interacting patterns.   So far we have talked about the causal/temporal history of individual outcomes. In the last example (true discontinuity) we suggest that certain outcomes cannot occur until the system passes a threshold value, enough to permit entry into a new problem space. The new problem space, in turn, can be described (or so we hope) by principles and laws that dictate the eventual outcome. We have, therefore, introduced the notion of interacting patterns of growth. This brings us to a final point concerning the proper interpretation of change in two or more systems with different developmental trajectories, something we will call "co-development."

The simplest version of co-development is a linear correlation: two systems increase or decrease together over time, in a relationship that can be described by a line. It is quite common in developmental psychology to find claims about the causal relationship between two variables that change together in this way. A case in point is Lenneberg's (1967) argument for the connection between linguistic and motor milestones (e.g., children walk and talk around the same time). Of course no one is trying to suggest that language causes motor development, or vice-versa. The argument is usually cast in terms of a third force, i.e., a maturational clock of some kind that sets the pace for disparate domains. Similar arguments can be found in research on the cognitive bases of language, where investigators have pointed out analogies in the onset times and developmental trajectories of verbal and non-verbal events, e.g., parallels between language and symbolic play in the second year of life. Here too, no one wants to argue that first words are "caused by" symbolic play, or vice-versa. The claim is, instead, that these two representational domains depend upon the same underlying ability.

This brings us to the "layers of causality" issue that we introduced in our discussion of linear development. Let us suppose, for the moment, that many aspects of human development depend (at least in part) upon resources that grow continuously over time. These resources might include aspects of attention, memory, perceptual acuity, motor speed—for present purposes, the identity of this capacity parameter is not crucial. Assuming a continuous dimension of processing capacity, we can quantify the amount of capacity required for progress in any given task. In the simplest cast, there will be a linear relationship between capacity and output. Thus a 50% cut in capacity would result in a 50% deficit in behavior; and a 50% gain in capacity would result in a 50% jump in behavior. This relationship is illustrated by Function 1 in Figure 4.15 (from Bates, McDonald, MacWhinney & Appelbaum, 1991).

The linear relationship between performance and capacity in Figure 4.15 is, in fact, the assumption that underlies a great deal of research in neuropsychology (for a detailed discussion, see Bates et al. 1991, and Shallice, 1988). It is this assumption that gives force to arguments about the dissociation between cognitive domains. Say, for example, that we find a patient who performs normally on Task A (e.g., a spatial cognitive task) and abysmally on Task B (e.g., a language task). If language and space depended upon the same underlying resource, and there is a linear relationship between resource and capacity for all domains, then a dissociation of this kind could not happen. Since the dissociation does happen, we may conclude with some confidence that this patient has a selective deficit affecting only one domain. That is, language and spatial cognition must depend on two distinct sources that can be independently impaired. Suppose, however, that Task A and Task B are governed by different performance/capacity curves? For example, suppose that Task A follows a linear function (Function 1) while Task B involves a non-

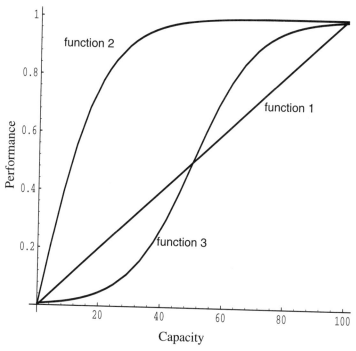

FIGURE 4.15 Function 1 illustrates a linear relationship between capacity and performance. Function 2 illustrates a nonlinear relationship in which rate of increase in performance drops rapidly after approximately a 30% threshold is reached. Function 3 illustrates a sigmoidal relationship in which performance changes rapidly in the region of a 50% threshold

linear relationship of the kind illustrated by Function 2 in Figure 4.15, i.e., a threshold model in which performance approaches ceiling after capacity exceeds 30%, and drops rapidly when capacity falls below 30%. If this is an accurate description of the situation, then a patient who has lost 50% of the resource in question will look very bad on Task A but close to normal on Task B. In other words, a single cause is responsible for both patterns.

Suppose, however, that we now find a patient who shows the opposite symptom profile: normal performance on Task B with serious deficits on Task A. Putting our two patients together, we have a classic double-dissociation, the sine qua non of cognitive neuropsychology. Surely, in this case, we can conclude that A and B are independent systems? In fact, as Shallice (1988) has pointed out, double dissociations can also arise from a single cause if the two tasks differ in the shape of their performance/capacity curves. Compare Function 1 (the linear function) in Figure 4.15 with Function 3 (the S-shaped curve). If our first patient has lost approximately 25% of the shared cognitive resource, he will show a performance profile in which Task B > Task A. If our second patient has lost up to 75% of the same shared resource, he will show a performance profile in which Task A > Task B. In short, claims about the dissociation between two cognitive systems must be evaluated with great caution, with detailed knowledge about the performance/capacity functions that underlie all the domains in question. If we run the arrow forward (capacity goes up, not down), then the same lesson applies to the study of normal development. Two domains can display markedly different developmental profiles, even though they both depend on the same underlying resource. Continuous shifts along that resource dimension can pull these domains apart and bring them back together again, in interesting patterns that tempt us to infer more independence and more discontinuity than we are entitled to claim.

Perhaps we can move our claims of independence to another level. Why, we might ask, do two domains obey such different laws? Why, for example, does Task A follow a linear pattern of growth while Task B reaches asymptote months or years before? Isn't this enough evidence, in and of itself, to stake out a modular claim? Yes, and no. We need to know, first of all, the extent to which these disparities in the shape of change derive from the ability in question (e.g., language vs. spatial cognition), or from the tasks we use to measure those abilities. We also need to know more than we do right now about the nature and availability of the input that drives learning in each of these domains. These are difficult questions, but they are the kinds of questions that are easier to answer if we are in a position to simulate learning under different parametric assumptions, to narrow down the critical range of possibilities for study in the human case. Indeed, if it is the case that cognitive change follows nonlinear laws, then simulation may become a crucial tool for developmental research.

The various examples of nonlinearity that we have described here constitute only a small subset of a huge class of possible nonlinear dynamic systems. This class includes an exotic and celebrated 𝑓-rm of nonlinearity called chaos, famous because it seems to elude our understanding altogether, being (by definition) completely deterministic but completely unpredictable. The behavior of nonlinear systems is difficult to predict because they have a property called *sensitivity to initial conditions*. In particular, because there is a nonlinear relationship between rate of change and the variable that is undergoing change, the outcomes that we ultimately observe are not proportional to the quantities that we start with. A very small difference in the starting points of two otherwise similar systems can lead (in some cases) to wildly different results. In principle, all these systems are deterministic; that is, the outcome could be determined if one knew the equations that govern growth together with all possible details regarding the input, out to an extremely large number of decimal places. In practice, it is hard to know how many decimal places to go out to before we can relax (and in a truly chaotic system, the universe does not contain enough decimal places). For this and other reasons, nonlinear dynamic systems constitute a new frontier in the natural sciences. For those of us who are interested in applying such systems to problems in behavioral development, this is both the good news and the bad news. The good news is that nonlinear dynamic systems are capable of a vast range of surprising behaviors. The bad news is that they are much harder to understand analytically than linear systems, and their limits remain largely unknown.

Whether or not we are comfortable with this state of affairs, it is likely that developmental psychology will be forced to abandon many of the linear assumptions that underlie current work. We have already shown how linear assumptions may have distorted our understanding of monotonic phenomena like the vocabulary burst, or the famous U-shaped curve in the development of grammatical morphology. It is hopefully clear by now why this assumption is unwarranted. Discontinuous outcomes can emerge from continuous change within a single system. Fortunately, connectionist models provide some of the tools that we will need to explore this possibility. Let's now take a look at how some of the concepts we have been discussing are implemented in connectionist networks.

Dynamical systems

Before proceeding to talk about how the above ideas may be implemented in connectionist networks, there is one additional set of concepts which we need to introduce which are central to notions of change: Dynamical systems.

A dynamical system is, very simply, any system whose behavior at one point in time depends in some way on its state at an earlier point in time. There are many systems for which this is not true. When a coin is tossed into the air, whether or not it lands heads up or tails up is totally independent from how it landed on previous tosses (the Gambler's Fallacy is to believe that after a run of heads, a coin is more likely to land tails up—in other words, to believe that this is a dynamical system, when it's not). Mathematically, we could express the outcome with the equation

$$P(\text{heads}) = 0.5 \qquad \textbf{(EQ 4.34)}$$

(where P denotes probability). The right-hand side of the equation does not in any way take into account prior outcomes.

But clearly there are many cases where behavior does depend on prior states. Suppose we catch a glimpse—so brief that we see only an instant—of a child on a swing, and see that the swing is right below the crossbar (that is, aligned perpendicular to the earth). If all we have to go on is this instantaneous view, we have no way of knowing what the position of the swing will be a second later. The swing might have been at rest, in which case we would expect it to remain there. But it might have been in motion when we took our brief look. In that case it will be in a different position a second later. Mathematically, the change in the position of the swing is given by the following pair of differential equations:

$$\frac{dp}{dt} = \frac{-g\sin\theta}{L} - rp$$

$$\frac{d\theta}{dt} = p$$

(EQ 4.35)

The length of the swing is denoted by the symbol L, the angular displacement of the swing from the vertical position is represented by θ, p represents speed, and r stands for the coefficient of friction. The second equation tells us the speed (in terms of change in angular displacement, θ), and the first equation tells us how, given that information, the speed will change over time.

The are various ways of classifying dynamical systems, but one useful way is in terms of the kind of behavior they exhibit over time. There are three kinds of behaviors one can distinguish: systems can exhibit *fixed points, limit cycles,* and *chaos.*

A system which has an attracting fixed point is one which over time eventually converges on a static behavior. A swing or pendulum, for instance, will ultimately come to rest (assuming friction). The position at rest is called the *attracting fixed point.* Let's look at a very simple equation which we will use to demonstrate a fixed point in a dynamical system.

Let us begin with the equation

$$f(x) = \frac{1}{(1+x)}$$

(EQ 4.36)

If we graph this for values of x ranging from 0 to 10 we get the curve shown in Figure 4.16.

Suppose we iterate this function. In other words, we begin with some value of x (we'll use 2) and calculate the result: $1/(1+2)$ or 0.33. Now let us take the output, 0.33, and use that as the new value for x. This time we get $1/(1+0.33)$ or 0.75. Continuing again eight more times gives us successive values of 0.571429, 0.636364, 0.611111, 0.62069, 0.617021, 0.618421, 0.617886, 0.61809. If we trace the pattern (using what's called a "cobweb graph") we see the trajectory shown in Figure 4.17.

It looks very much as if repeated iterations of this function lead to a single value. Indeed, if we calculate the result of iterating

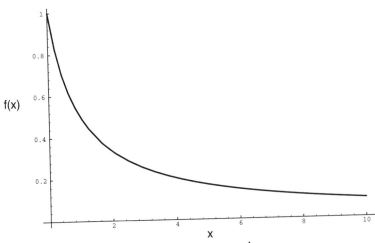

FIGURE 4.16 Graph of the equation $f(x) = \dfrac{1}{(1+x)}$ for values of x from 0 to 10.

Equation 4.36 on itself, starting with an initial value for x of 2, we find that after 50 iterations we converge on a value of 0.682328. Iterating another 25 or 100 or 1,000 times does not change this value. The system has settled into a fixed point.

Suppose we began with another initial value, say -10? Interestingly, on each iteration we get closer and closer to the value

$$x_\infty = \frac{1 + \sqrt{5}}{2}$$

which, for the precision available on the machine we used, is approximately = 0.682328. This is true no matter what value we start with. In a way, this equation is like a swing: No matter how hard our initial push, the swing will sooner or later always come to rest (unless, of course, we continue to push it; in this case we have what is called an external *forcing function*).

The fixed point for this function is said to be a global *attractor*, because no matter what initial value we start with, the function ultimately "attracts" us into a final value of 0.682328 (for the given precision available on our machine). (There are also repelling fixed points, which are regions where we remain if we are exactly at the

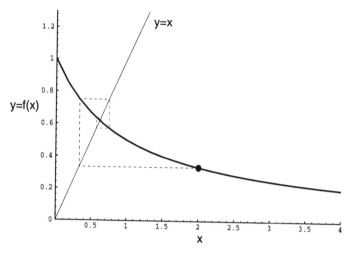

FIGURE 4.17 Graph showing result of iterating Equation 4.36 on itself 10 times, starting with an initial value of 2.0 for x (indicated by the dot). The dashed line traces the changing values of x. On the first iteration, the dotted line crosses the function at $x=2.0$ and $f(x)=0.33$. On the second iteration, the dotted line crosses the function at $x=0.33$ and $f(x)=0.571429$. After many iterations, the function settles at values of $x=0.682328$ and $f(x)=0.682328$, which is an attracting fixed point for this equation.

fixed point, but if we are anywhere else close by, we are pushed away from that value on subsequent iterations of the function.) We might wonder whether a function can only have one attracting fixed point (as is obviously true of the swing example). The answer is no; functions may have any number of fixed points, although it is typical for a function to have only a small number of fixed point attractors. For example, the function:

$$f(x) = \frac{1}{1 + e^{(3 + (-6x))}}$$ **(EQ 4.37)**

has three fixed points (two attracting fixed points and one repelling fixed point). Initial values of x which are greater than 0.5 converge toward 0.92928; initial values less than 0.5 converge toward

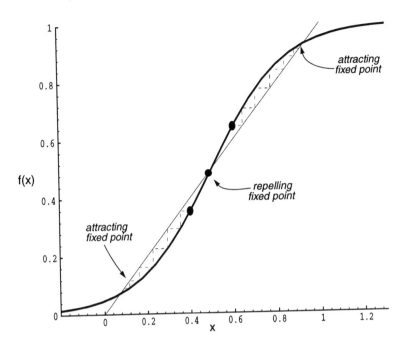

FIGURE 4.18 Graph showing result of iterating Equation 4.37 on itself 10 times, starting with three different initial values for x, shown by dots at 0.4, 0.5, and 0.6. Initial values greater than 0.5 converge toward a fixed point attractor of 0.92928; values less than 0.5 converge toward 0.0707202; and values of 0.5 remain at the fixed point 0.5. The first two are attracting fixed points and the third is a repelling fixed point.

0.0707202; and initial values of exactly 0.5 remain at 0.5 but slight deviations from 0.5 are forced away from that value. This is shown in Figure 4.18.

Fixed points are not the only kind of attractors one finds in dynamical systems. A second type of attractor is the *limit cycle*. In this case, the system exhibits a behavior which repeats periodically. A classic example of a function which can have limit cycles is the quadratic map, given in

$$f(x) = rx(1 - x)$$ **(EQ 4.38)**

where r is a constant. For example, if we let r be 3.2 and set the initial value of x to 0.5, then on successive iterations, this equation yields 0.8, 0.512, 0.799539, 0.512884, 0.799469, 0.513019, 0.799458, 0.51304, 0.799456, 0.513044, 0.799456, 0.513044, 0.799455, 0.513044, 0.799455, 0.513045, 0.799455, 0.513045, 0.799455, 0.513045. If we example these numbers closely, we see that it looks like we end up with a repeating sequence of two alternating numbers. If we graph the changing values, as is shown in Figure 4.19, it is easy to visualize this cyclic behavior.

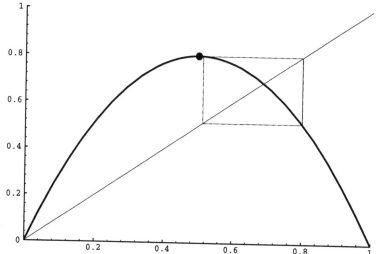

FIGURE 4.19 Graph showing result of iterating Equation 4.38 on itself 10 times, starting with an initial values of 0.5 for x and setting r=3.2. The function cycles through two different values (shown where the cobweb intersects the parabola).

More interesting is the fact that if we give x an initial value of 0.1, or 0.9, we end up with the same results. After many iterations, this equation (with r as we have set it) ends up cycling between the same two numbers: 0.513045 and 0.799455. This means that this limit cycle serves as an *attractor*, meaning that other initial values of the variable x end up being pulled into the same repeating pattern.

The quadratic equation given in Equation 4.38 has been well-studied because it has other interesting properties. If we change the value of the constant r (remember that in the above examples we kept it at 3.2), then the dynamics of this equation change dramatically. Figure 4.20 shows the behavior of the iterated quadratic equation with values of $r=3.45$ (left panel) and $r=4$, in both cases starting with x set to 0.1. In (a) the system soon settles on a repeat-

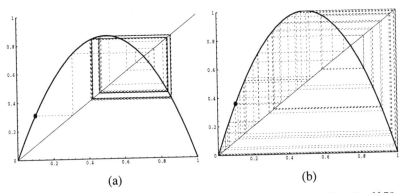

(a)   (b)

FIGURE 4.20 Graph showing result of iterating Equation 4.38 on itself 50 times, setting $r=3.45$ (a) and $r=4$ (b), in both cases starting with an initial values of 0.1 for x. (The dots indicate starting positions.) In (a) the system settles into a limit cycle with four possible values. In (b) the system enters a chaotic regime, never exactly repeating a prior state.

ing pattern of period four. In (b) the result is a *chaotic* attractor. Chaotic attractors are the third class of dynamical systems.

In this chaotic case, the system never returns to exactly the same state it was in before. For example, if we iterate Equation 4.38 (setting $r=4$) 50 times, starting with $x=0.1$, we end up on the last iteration with $x=0.54$. But if we start with a very slightly different initial value for x—say, 0.1001—after 50 iterations we have $x=0.08$. So not only does the attractor not repeat itself, but its behavior varies dramatically depending on the initial values for variables. This property of chaotic systems is called *sensitivity to initial conditions*. Note also that chaotic behavior is not the same as random behavior; if we start with the same initial conditions, we repeat the whole sequence exactly, unlike what happens when we toss a coin. One of the best known chaotic attractors is the Lorenz attractor (which was developed as a model for changes in weather), shown in Figure 4.21. The

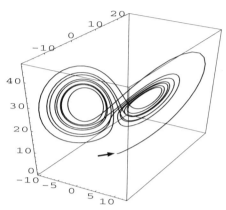

FIGURE 4.21 The Lorenz attractor. This system is an example of a deterministic chaotic system; the behavior never repeats but is not random.

system displays patterned movement through the space of the three variables in the equations which define it, but never returns to exactly the same position in state space as it was previously.[2]

There is a final point we want to make with regard to the behavior illustrated in Figure 4.20. We find that the same equation yields qualitatively different dynamical behavior as a function of a change in one of the constant parameters of the equation (in this case, we changed r from 3.45 to 4). Let's explore the effect of changing this parameter in a more systematic fashion. We'll begin by setting r to 2, feed in an initial value for x (we'll use 0.5 in this example), and then examine the behavior of the system when we iterate the equation. The question we're interested in is how many attractors the

---

2. Here the variations are continuous in time, as opposed to the examples we have given so far, which involve discrete time steps. Although there are differences between discrete and continuous dynamical systems, all the attractors we have discussed with discrete systems have analogues in continuous time systems.

equation has.

If we carry this experiment out with Equation 4.38, we discover that when r is 2 and x is 0.5, the system converges immediately on a single fixed point (0.5). If we run this experiment again many times, each time increasing the value of r in very small increments, we discover that all versions of this equation have a single fixed point attractor until we get to the case where r is 3.0. Suddenly the behavior of the system shifts and we find a limit cycle with two attractor fixed points. Increasing the value of r even more reveals successive changes in behavior as the system switches abruptly at $r=3.44949$ from a 2-attractor limit cycle to a 4-attractor cycle, then with $r=3.54409$ to an 8-cycle, and so on. These shifts always result in increasing the number of cycles by a power of two. Then we find that when $r=3.569945$ is reached we enter a chaotic regime in which the dynamical behavior never repeats. But if r is increased even more, we revert to a limit cycle regime for a while, then return to chaos. And so on.

The results of this experiment have been graphed in Figure 4.22. The vertical axis shows us the different output values yielded by the equation for the different values of r, shown along the horizontal axes. The changes in behavior are called *bifurcations* (and the graph in Figure 4.22 is therefore called a bifurcation plot). This is because the behavior of the system splits (bifurcates) at the change points.

The relevance of bifurcations for our purposes should be obvious. Children often go through distinct stages in development, exhibiting qualitatively different behavior in each stage. It is tempting to assume that these stages reflect equally dramatic changes and reorganization within the child. Bifurcations illustrate that this need not be the case; the same physical system may act in different ways over time merely as a result of small changes in a single parameter. Later in this chapter we shall present a neural network model of this phenomenon. We'll see how a network which appears to appears to go through different stages of learning and abstraction does so as a result of bifurcations.

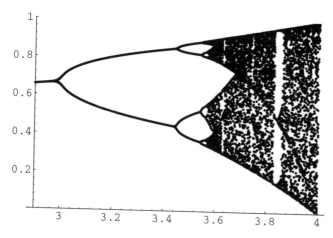

FIGURE 4.22 Bifurcation diagram.

Dynamics and nonlinearity in neural networks

The central topic of this chapter is change, and that means that we are ultimately interested in the dynamical properties of children and networks. One of the things which makes network dynamics so interesting is the presence of nonlinearities in learning and processing. We therefore begin by discussing where nonlinearity arises and what its effects are. Then we shall see how this affects the shape of change.

Nonlinearity in networks

It is possible to have a connectionist network that is fully linear, and some very interesting things can be done with such networks. But today most people are interested in networks that have nonlinear characteristics, as discussed in Chapter 2. The major locus of the nonlinearity appears in the way in which a unit responds to its input.

The basic job carried out by the units (or nodes) in a network is simple. A node collects input from other nodes (excitatory or inhibitory) and also sends output to other nodes. Viewed this way, a node merely serves as a way-station shuttling activation values around the network. But each node does one small thing in addition, and this has serious consequences. Before passing its input on to other nodes, a node performs a relatively simple transformation on it, so that what the node puts out is not exactly what it took in. This input/output function (what is called the activation function) is typically nonlinear. One common function is graphed in Figure 4.23.

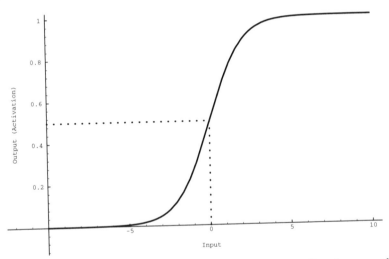

FIGURE 4.23 The activation function used most often in neural networks. The function is $1/1 + \exp^{(-net)}$, where *net* is the net input to a node and *exp* is the exponential function. The net input is potentially unbounded, whereas the resulting output is limited to (in this case) a range of activation values from 0.0 to 1.0. When net input is 0.0, the output is 0.5, which is in the middle of the unit's range of possible activations.

The shape of this function shows how the node's output (measured along the y-axis) changes in relation to the node's total input (shown along the x-axis). Notice that the range of possible values along the two axes differs. The range of possible inputs is unrestricted, since a node might receive input a very large number of other nodes. The range of outputs, on the other hand, is restricted (with this particular activation function) to the interval {0, 1}. The activation function thus squashes the nodes net input first before sending it off to the other nodes it excites or inhibits.

Why might it be useful for the activation function to take such a form? One of the obvious benefits is that the sigmoid function keeps a node's activation under control. Should the input grow too large, the node will simply saturate rather than growing out of bounds. Unconstrained systems have a habit of "blowing up," which can be nasty if one wants stability.

There are other important benefits to this nonlinearity. When a node receives a great deal of inhibitory (negative) or excitatory (positive) input, its response will be binary; the resulting activation will be asymptotically close to 0 or 1, respectively. Furthermore, once the magnitude of the input has reached a certain value, it doesn't matter how much more input is received. The output will still be either 0 or 1. So when it has to, a node can generate an all-or-nothing decision.

At the same time, there is a large gray area of input conditions within which a node is capable of graded responses. When the net input is close to 0 (the dashed line in Figure 4.23), the output is 0.50, which is in the middle of the node's activation range. Notice that small differences in this region of input produce significant differences in output. An input of 2 gives an output of 0.88, whereas inputs of 10 and 12 both give outputs of 0.99. Thus, nodes can make finely tuned responses which vary subtly as a function of input. Importantly, the same unit can make both types of decisions along the same continuum. In certain regions along an input dimension (i.e., near the ends of the continuum) a node may be relatively insensitive to stimulus differences and output nearly the same response to inputs which may be very different. In other regions along the same dimension, the same node may be exquisitely sensitive and respond in very different ways to inputs which are very similar.

Such responses are often found in human psychophysics. The perception of many speech distinctions, for instance, is characterized by an apparent insensitivity to acoustic differences which fall within a phonetic category, accompanied by great sensitivity to small differences which straddle a category boundary. Humans may perceive a difference of 5/1000 of a second as signalling a category distinction (e.g., the sound [b] vs. [p]) in some cases; whereas in other regions of the input dimension a 5/1000 goes undetected.

Finally, the nonlinearity extends the computational power of networks in an important way. Recall from Chapter 2 that there are problems which require that a network have additional layers (e.g., those which are nonlinearly separable). Why not use linear units? The reason is that it is the nature of linear networks that for any multi-layer network composed of linear units, one can devise a simpler linear network composed of just two layers. So the additional layers do no work for us. Since we need additional layers ("real" layers), the only way to get them is to have the units be nonlinear. That way we can have multi-layer networks which cannot be reduced to two-layer systems.

Interactions: A case study in nonlinearity

The simple nonlinearity introduced in node activation functions turns out to have even more interesting and useful properties when we consider how it manifests itself in networks which are learning complicated tasks. What we find is that behaviors which at first glance might appear to arise because of multiple mechanisms can actually be accounted for as well (or better) by single but nonlinear networks.

Consider the question of stimulus frequency and consistency. To make this example we turn to a model developed by Plaut, McClelland, Seidenberg, and Patterson (1994). Plaut and his colleagues have been interested in the question of what mechanisms underlie reading. One proposal that has been made is that skilled readers use two independent processing routes when going from written form to pronunciation. One route involves the application of grapheme-phoneme correspondence rules (e.g., "c followed by i or e is soft"). A second route is also hypothesized. This second pathway allows readers to look words up in their mental lexicons and retrieve pronunciations directly. Both routes seem to be needed. The first route is useful in reading novel words which can't be found in the reader's mental dictionary; the second route is needed to read words with exceptional pronunciations which do not accord with the rules.

Certain empirical facts would seem to lend support to this analysis of reading as involving two routes. One important fact is that the speed with which subjects can name words (presented briefly on a visual display) is highly correlated with word frequency. Overall, frequent words are read more rapidly than low frequency words. However, this effect interacts with the regularity of a word's pronunciation. The pronunciation of regular word is more or less independent of frequency, whereas pronunciation latencies are much faster for frequent irregular words than infrequent irregular words.

The apparent insensitivity of regular forms to frequency has been hypothesized to occur because the regulars are produced by rule, whereas the irregulars are stored directly and their retrieval time is facilitated by frequent access. (A similar conclusion has been drawn in the case of English verbs, since a frequency/regularity similar to that found in reading has been observed: subjects are faster at producing the past tense of regular English verbs ("walked" given the prompt "walk") compared with irregular verbs ("went", given "go").) Thus, there are two reasons why one might believe that reading involves two distinct mechanisms: (1) the ability to read non-words (which cannot be stored in the lexicon), and (2) the effect of word frequency on the reading of irregulars, coupled with the lack of such an effect on regulars.

In fact, such interactions can also be produced by a single mechanism which has nonlinear characteristics. Plaut et al., building on earlier work by Seidenberg and McClelland (1989) trained several networks to read. (That is, the networks produced the correct phonological output for graphemic input.) Their network, shown in

Figure 4.24, was trained on 2998 monosyllabic words, with the training of each word weighted by its frequency of occurrence. After training, the network was tested on several sets of nonwords. Performance closely resembled that reported for human subjects reading the same words.

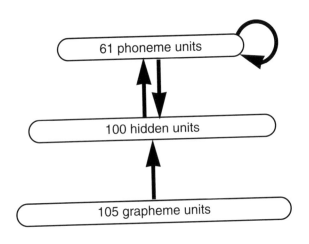

FIGURE 4.24 The architecture of the Plaut et al. (1994) reading network. (After Plaut et al., Figure 12.)

The network also showed systematic differences in the speed with which it processed words, depending on the frequency of a word's occurrence. However, this frequency advantage disappeared in the case of regular words. More interestingly, there was also an effect for consistency. As Glushko (1979) first noted, a word can be regular in its pronunciation but resemble another word which is irregular. "have" and "mint" are pronounced in a way which follows the normal rules, so they are regular. However, there are orthographically similar words "gave" and "pint" which are irregular. The presence of these irregulars in the neighborhood of the "save" and "mint" mean that the latter are regular but inconsistent. "slap," on the other hand, is not only regular but has no irregular neighbors. So it is regular and consistent.

The network, like skilled readers, also shows differences in processing latency for these two classes of words. Regular but inconsistent words are processed more slowly than regular consistent words. Furthermore, word frequency has no effect on regular consistent words, but regular inconsistent words show a pattern which is an attenuated version of what happens with irregulars.

As Plaut et al. point out, all three effects (frequency, regularity, and consistency) have a common source in the network. The weight changes which occur when a network is trained are in proportion to the sources of error. Because more frequent items are by definition encountered more often during training, these items contribute a larger source of error. So the network's weights are altered to accommodate them more than infrequent words. Furthermore, regular words all induce similar changes. This means that the frequency of a given regular form is of less consequence to regulars. The network will learn a rare regular word anyway since it resembles all the other regulars and benefits from the learning they induce. Consistency is simply a milder version of this. A word may be regular, but if there are not many other regular forms which resemble it (or, worse, there are high frequency irregulars which do), then will be learned less well, and learning will depend more on its individual frequency.

These factors are additive, but because the network is nonlinear, the cumulative consequence of their contribution asymptotically diminishes. This follows directly from the sigmoidal shape of the activation function of nodes. A schematic representation of this is shown in Figure 4.25.

Imagine that a word receives 1 or 3 units of activation, depending on its frequency of occurrence, and that regularity contributes an additional 1 units of activation. So we might imagine that High Frequency Regulars would receive the most (4 units of input), Low Frequency Exceptions the least (1 unit of input), and High Frequency Exceptions and Low Frequency Regulars intermediate amounts (3 and 2 units of input respectively). We see in the top panel of Figure 4.25 that although the contribution of frequency and regularity is additive, because the output of nodes in the network is a nonlinear function of their input, the difference between high and

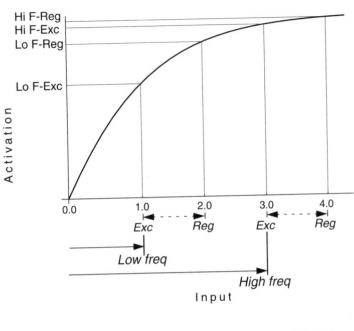

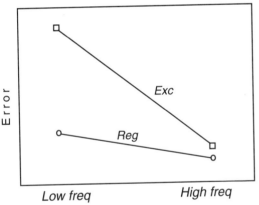

FIGURE 4.25 The nonlinear outcome of additive effects of word frequency and regularity of pronunciation. Top panel: Frequency and regularity each contribute the same amount to processing of both regular and irregular words, but the effect of frequency is less in the case of regular words because of the nonlinear activation function. Bottom panel: The differences in activation translate into error differences in performance. (After Plaut et al., 1994, Figures 7 and 9.)

low frequency regulars is much less than between high and low frequency irregulars. This translates into performance differences shown in the bottom panel of the figure; word frequency results in relatively minor differences in performance in the case of regulars, but large differences in the case of exceptions.

Dynamics

The previous example shows how a single mechanism with nonlinear characteristics can produce behaviors which closely resemble those expected from multiple mechanisms. What about change? This is, after all, the critical issue of this chapter. Let us now turn to dynamics.

There are two contexts in which one finds dynamics in networks. The first has to do with the pattern of change which occurs as a network learns. This comes closest to the sense in which we talked about dynamics in the first part of this chapter. Secondly, in certain kinds of networks (those with recurrent connections), we can also find behaviors (after learning) which exhibit dynamics.

Dynamics in learning

At first it might seem odd to talk about dynamics in feedforward networks. These are networks in which the output is a direct function of the input. When we look at their instantaneous behavior, these are "one-shot" networks. Indeed, in this sense, they have no dynamics.

The dynamics arise when these systems are changed in response to training. The dynamics now has to do with the changes in weight space, which are (in a back-propagation network) governed by the equation given in Chapter 2 as Equation 2.5, reproduced below.

$$\Delta w_{ij} = -\eta \frac{\partial E}{\partial w_{ij}} \qquad \textbf{(EQ 4.39)}$$

This equation tells us how to change the weights in order to minimize error. It says that the change in the weight between units i and j should be changed in proportion to the product of a small learning rate (η) times the derivative of the network's error with respect to the weight connecting units i and j. The derivative tells us how changes in the error are related to changes in the weights; if we know this then we know how to learn: Change the weights in such a way as to reduce error. We also see from this equation that because the change in weight is depending on the existing weight, this is a system which has the potential for interesting dynamics.

We might have supposed, for example, that because network learning is incremental (in the sense that we employ the same learning rule at all points in time, making small adjustments in accord with current experience) the change in performance will be incremental in some straightforward manner. But consider what happens when a network learns the logical XOR function. If we measure the performance error over time, sampling the mean error every 100 patterns, we see the expected monotonic drop in performance. This is illustrated with the heavy line in Figure 4.26.

However, if we examine error more precisely, looking at the performance on each of the four patterns which comprise XOR (the lighter lines in Figure 4.26), we find that some of the patterns exhibit nonmonotonic changes in performance. The most dramatic instance of this is the error associated with the pattern 11. Initially, error on this pattern actually increases while the error on the other patterns decreases. During this phase, the network has found a temporary solution which assumes the function underlying the data is logical OR. Because this solution works for three of the four patterns, the overall drop in error is sufficient that the network tolerates increasing error on the one pattern for which OR does not fit (an input of 11 produces an output of 1 for OR rather than the output of 0 which is correct for XOR). It is only when the price paid by this outlier is sufficiently great that the network backs off from this solution and finds another. There is a small trade-off during the transition period; as the error for 11 drops, the error for another pattern (01) temporarily increases. But eventually the network finds a setting of weights which accommodates all the patterns.

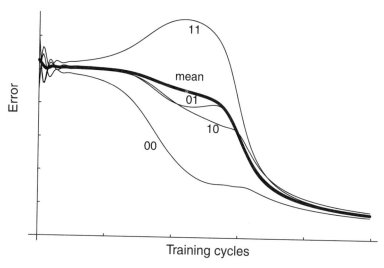

Training cycles

FIGURE 4.26 Error plot during learning the XOR function. The heavy line shows the mean error calculated over successive batches of 100 training cycles. The lighter lines show the error produced by each of the four patterns which comprise XOR. Two of the patterns (00 and 10) have monotonically decreasing error; the other two (01 and 11) temporarily increase error before a solution is found which accommodates all four patterns.

This sort of interaction, in which performance on different components of a complex task varies over time, has obvious parallels with phenomena in child development. For example, many children display nonmonotonic changes in performance in the course of learning the morphology of the English past tense. The classic story, as detailed in Chapter 3 (Berko, 1958; Ervin, 1964) is that children begin by producing both regular ("talked") and irregular forms ("came") correctly. As time passes, children continue to produce the regulars correctly but begin to make errors on the irregulars, regularizing forms which they used to produce correctly as irregulars (e.g., "comed"). A reasonable interpretation is that the first phase reflects the rote memorization of verbs, whereas the second phase occurs when children begin the process of rule abstraction. This rule should apply only to the regulars (by definition); over-extension occurs because the children have not yet fine-tuned the conditions of the rule's application.

As we pointed out in Chapter 3, both the data and the interpretation have been called into question (Bybee & Slobin, 1982; Marchman, 1988; Marcus et al., 1992; Plunkett & Marchman, 1991, 1993). The U-shape is in fact composed of many "micro-Us", and correct irregulars often co-exist in free variation with the incorrect regularized form. Furthermore, regulars themselves are sometimes "irregularized." Plunkett and Marchman's simulation, which we described in Chapter 3, shows similar patterns of nonmonotonic change and irregularizations. This nonmonotonicity arises for essentially the same reasons it does in the XOR simulation, not because the network has suddenly moved from a strategy of memorization to rule abstraction.

Readiness, stages of learning, and bifurcations

Another phenomenon which is often observed with children is the sudden transition from what appears to be insensitivity to input a stage where the child seems to be extraordinarily sensitive to new data. The abruptness with which this occurs suggests that some internal event has occurred which has moved the child from a prior stage in which she was not able to make use of information, to a new stage in which the information can now have an effect. Until the child is ready for it, the information is simply ignored.

One example of stages of learning, discussed in Chapter 3, is the sequence of phases children go through in the course of learning to balance weights on a beam. As the McClelland and Jenkins simulation demonstrates, such stages of learning can be readily reproduced in a network which uses an incremental learning strategy. But of course simulating one black box with another does us little good. What we really want is to understand exactly what the underlying mechanism is in the network which gives rise to such behavior. We outlined the rudiments of an explanation in Chapter 3. We are now in a position to flesh the explanation out in greater deal.

To do this we shall use an example of a behavior in child development which in some ways resembles the balance beam experiment. When children first learn the principle of commutativity (i.e., the principle that $x + y \equiv y + x$), they demonstrate a limited ability to generalize beyond the sums that were used to teach them the principle. Children thus can reliably report that 11+12 and 12+11 yield the same sum, even though they may never have seen this specific example. The ability to generalize is fairly limited, however. Children at this stage will fail to recognize that sums such as 645,239,219+14,345 (which differ greatly in magnitude from those they encounter as examples of the principle) are the same as 14,345+645,239,219. In Karmiloff-Smith's Representational Redescription model (Karmiloff-Smith, 1992a), children are in the Implicit phase. At a later stage, children do achieve complete mastery of the principle and are able to apply it to any arbitrary sum, no matter what the magnitude. In Karmiloff-Smith's terms, children have now moved to the stage of Explicit-1 (and subsequently move to Explicit-2 and then to Explicit-3, when they can verbally articulate their knowledge).

The question is, what sort of mechanisms might be responsible for what seems to be qualitatively different sorts of knowledge, and how can we move from one phase to the next? The character of the knowledge seems quite different in the two phases, and it is not clear how a network might simulate such drastic transformations. What we shall see, however, is that there is a far simpler explanation for the change in network terms than we might have imagined.

We shall build a model of a somewhat different mathematical task. Instead of commutativity, we chose the problem of detecting whether a sequence of 1s and 0s is odd or even (counting the number of 1s in the string). If the sequence is odd, we shall ask a network to output a 1. If the sequence is even, the correct output is 0. Furthermore, we shall ask the network to give us a running report, so that if the network sees a sequence of five inputs, it should tell us after each input whether the sequence to that point is odd or even. (This is a temporal version of the *parity* task.) Given the input string

    1 1 0 1 0

the correct outputs would be

    1 0 0 1 1 (*odd, even, even, odd, odd*).

Since this task involves a temporal sequence, we shall use the simple recurrent network architecture shown in Figure 4.27

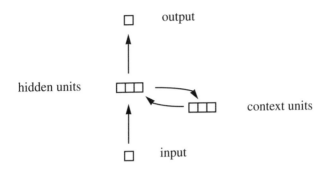

FIGURE 4.27 Simple recurrent network architecture used to learn the odd/even task. Context units store the state of the hidden units from the prior time step.

We train the network on a series of short sequences, two to five bits in length. At each stage during a sequence the network outputs a 1 just in case the sequence up to that point contains an odd number of 1s; otherwise, it should output a 0. After a sequence is done, the network context units are reset to 0.5 and the next sequence is input.

To evaluate the network's performance during the course of training, we ask it to do the odd/even task on a string of 100 1s. The length of this string is an order of magnitude greater than any of the sequences encountered in training. What we are interested in is just how far into the sequence the network can go before it fails. If the network generalizes perfectly, then the output will be an alternating sequence of 1010... (that is, *odd, even, odd, even...*). Graphically, such output would be a zig-zag pattern with 100 peaks and 100 valleys.

In Figure 4.28 we see the results of tests performed at several stages in learning. The top panel shows the network's outputs to the test string after 15,000 inputs. The network shows the correct pattern of alternating *odd, even* responses for four successive inputs. Since the training data contained sequences which varied randomly

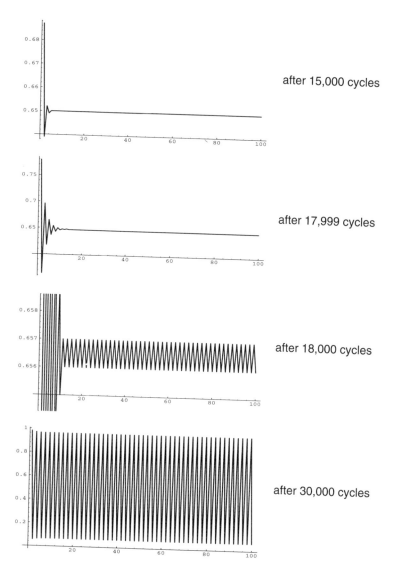

FIGURE 4.28 Performance of the simple recurrent network while learning the odd/even task at several points in time (measured in cycles). Generalization occurs at 17,999 cycles but is limited. One learning cycle later, however, the network is able to extend the generalization indefinitely.

between two and six inputs in length, the network has not yet mastered the task even for lengths it has seen.

After additional training (the panel marked "after 17,999 inputs"), we find that the network generalizes to the extent that it can give the correct answer for the first 13 inputs from the test sequence. Thus the network has generalized beyond its experience (Karmiloff-Smith's Implicit I phase). The network has not yet learned the principle of odd/even in a fully abstract manner, because the network is unable to give the appropriate response beyond 13 inputs.

At least, this is the state of affairs after 17,999 training cycles. What is remarkable is the change which occurs on the very next training cycle. After one additional input, the network's performance changes dramatically. The network now is able to discriminate odd from even. The magnitude of the output is not as great for longer sequences, so we might imagine that the network is not fully certain of its answer. But the answer is clearly correct and in fact can be produced for sequence of indefinite length. The network seems to have progressed to Karmiloff-Smith's Explicit-1—in a single trial! Subsequent training simply makes the outputs more robust.

We might say therefore that early in training, the network was not able to learn the odd/even distinction; and that at 17,999 cycles the network has moved into a stage of readiness such that all that was needed was one further example for it to learn the distinction fully. Yet we know that there are no abrupt endogenous maturational changes which occur in the network. And we know that the learning procedure is the same throughout, namely, gradient descent in error space through small incremental changes in weights. So why does performance change so dramatically?

To explain why this occurs, it will be useful to simplify matters even further. Let's consider the case where we have a single node, as shown in Figure 4.29. We will give it a constant input, labeled the bias, and a recurrent weight, labeled w_r. Finally, we inject some initial input to the node, and then we let the node activate itself for some number of iterations before we look at its final output.

The point of this example will be to look at what kinds of dynamics develop as the node recycles activation back to itself over

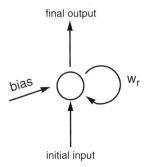

FIGURE 4.29 A one-node network which receives an initial input; the input is then removed, and processing consists of allowing the network to fold its activation back on itself through the recurrent weight. After some number of iterations, we examine the output.

the recurrent weight. We can see that for the odd/even task it is important for the network to remember what an initial input was, since this can make the difference between a string having an odd vs. an even number of 1s. The recurrence in the network allows nodes to "remember" their prior activations. When we look at the dynamics of this simple system, we are therefore looking at the network's memory.

One problem we notice immediately is that the sigmoidal activation function will continuously map the node's activation back to the interval {0.0, 1.0}, squashing it in the process. For example, suppose the recurrent weight has a value of $w_r = 1.0$ and there is a constant bias of $b = -0.5$ Then if we start with the node having an initial activation of 1.0, on the next cycle the activation will be as given in Equation 4.40:

$$a(t+1) = \frac{1}{1 + \exp(-(a(t) - 0.5))} \qquad \textbf{(EQ 4.40)}$$

or 0.62. If this diminished value is then fed back a second time, the next activation will be 0.53. After 10 iterations, the value is 0.50—and it remains at that level forever. This is the mid-range of the node's activation. It would appear that the network has rapidly lost the initial information that a 1.0 was presented.

This behavior, in which a dynamical system settles into a resting state from which it cannot be moved (absent additional external input) is called a *fixed point*. In this example, we find a fixed point in the middle of the node's activation range. What happens if we change parameters in this one-node network? Does the fixed point go away? Do we have other fixed points?

Let's give the same network a recurrent weight $w_r = 10.0$ and a bias $b = -5.0$. Beginning again with an initial activation of 1.0, we find that now the activation stays close to 1.0, no matter how long we iterate. This makes sense, because we have much larger recurrent weight and so the input to the node is multiplied by a large enough number to counteract the damping of the sigmoidal activation function. This network has a fixed point at 1.0. Interestingly, if we begin with an initial activation of 0.0, we see that also is a fixed point. So too is an initial value of 0.5. If we start with initial node activations at any of these three values, the network will retain those values forever.

What happens if we begin with activations at other values? As we see in Figure 4.30, starting with an initial value of 0.6 results over the next successive iterations in an increase in activation (it looks as if the node is "climbing" to its maximum activation value of 1.0). If we had started with a value of 0.4, we would have found successive decreases in activation until the node reached its fixed point close 0.0. Configured in this way, our simple one-node network has three stable fixed points which act as basins of attraction. No matter where the node begins in activation space, it will eventually converge on one of these three activation values.

The critical parameter in this scheme is the recurrent weight (actually, the bias plays a role here as well, although we shall not pursue that here). Weights which are too small will fail to preserve a desired value. Weights which are too large might cause the network to move too quickly toward a fixed point. What are good weights?

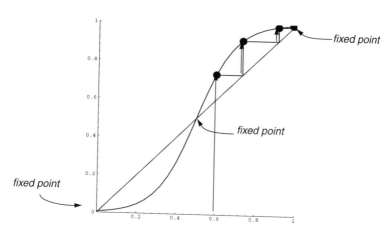

FIGURE 4.30 If a recurrent unit's initial activation value is set to 0.6, after successive iterations the activation will saturate close to 1.0. An initial value of 0.5 will remain constant; an initial value of less than 0.5 will tend to 0.0 (assumes a bias of -5.0 and recurrent weight of 10.0).

Working with a network similar to the one shown in Figure 4.29, we can systematically explore the effects of different recurrent weights. We will look to see what happens when a network begins with different initial activation states and is allowed to iterate for 21 cycles, and across a range of different recurrent weights. (This time we'll use negative weights to produce oscillation; but the principle is the same.) Figure 4.31 shows the result of our experiment.

Along the base of the plot we have a range of possible recurrent weights, from 0.0 to -10.0. Across the width of the plot we have different initial activations, ranging from 0.0 to 1.0. And along the vertical axis, we plot the final activation after 21 iterations.

This figure shows us that when we have small recurrent weights (below about -5.0), no matter what the initial activation is (along the width of the plot), we end up in the middle of the vertical axis with a resting activation of 0.5. With very large values of weights, however, when our initial activation is greater than 0.5 (the portion of the surface closer to the front), after 21 iterations the final

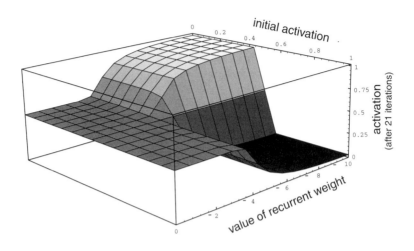

FIGURE 4.31 The surface shows the final activation of the node from the network shown in Figure 4.29 after 21 iterations. Final activations vary, depending on the initial activation (graphed along the width of the plot) and the value of the recurrent weight (graphed along the length of the plot). For weights smaller than approximately -5.0, the final activation is 0.5, regardless of what the initial activation is. For weights greater than -5.0, the final activation is close to 0.0 when the initial activation is above the node's mid-range (0.5); when the initial activation is below 0.5, the final activation is close to 1.0.

value is 0.0 (because the weight is negative and we iterate an odd number of times, the result is to be switched off; with 22 iterations we'd be back on again). If the initial activation is less than 0.5, after 21 iterations we've reached the 1.0 fixed point state. The important thing to note in the plot, however, is that the transition from the state of affairs where we have a weight which is too small to preserve information to the state where we hold on (and in fact amplify the initial starting activation) is relatively steep. Indeed, there is a very precise weight value which delineates these two regimes. The abruptness is the effect of taking the nonlinear activation function and folding it back on itself many times through network recurrence. This phenomenon (termed a bifurcation) occurs frequently in nonlinear dynamical systems.

CHAPTER 5

Brain development

In previous chapters we have repeatedly stressed the importance of biological factors within our connectionist approach to development. However, the connectionist models we have discussed so far have not attempted to integrate information from brain structure, function, or development. In Chapter 6 we will present some initial attempts to generate such models. Before doing this, however, we need to provide some background information on vertebrate and human brain development. The goal of this chapter is to review some facts and hypotheses about brain development, and to see what light these processes can shed on connectionist models of behavioral change.

While the primary function of this chapter is to serve as an overview of neural development with special emphasis on the facts that can inform a connectionist theory, the reader should be aware that what follows is a selective review, and contains some conclusions and theoretical statements which remain controversial. There are still many aspects of brain development that are poorly understood.

The chapter is divided into three main parts. In the first of these parts, *Basic questions*, we raise a number of issues about the relation between cognition and behavior. The second part of the chapter, *Building a vertebrate brain*, is a sketchy overview of vertebrate brain development from embryogenesis to the effects of postnatal experience. We focus particularly on the origins of specificity in the cerebral cortex, since this is relevant for our claim that there is no compelling evidence for innate representations in the cortex.

While this discussion of development in the mammalian cerebral cortex is of clear relevance to the human brain, it is only the last part of the chapter, *And finally...the human brain*, that we specifically focus on brain development in our own species. We find no evidence that our species possesses unique brain cells or circuitry. Rather, humans simply seem to have a greater extent of cerebral cortex, with a delayed timetable of postnatal development.

Basic questions

Before we go into the facts about brain development, it may be useful to grapple a little with some conceptual issues. How could information about brain development constrain a cognitive theory? In particular, how does this information affect our understanding of major issues like innateness and domain specificity? We will attempt to unpack these issues into questions of *what* and *where*, *how* and *when*, and also *who* (i.e., is the course of development the same for everyone? How does the brain respond if default conditions do not hold?).

What and where: The issue of localization

When presented with a correlation between a psychological phenomenon and a particular change in brain structure or function, certain theorists have assumed that the aspect of cognition in question must be innate, or, (at the very least) subject to powerful biological constraints that do not hold for other forms of thought or behavior. For example, some might argue that the brain honors the contrast between grammar and semantics, a statement supported by neuroimaging studies showing that different linguistic structures are correlated with different patterns of brain activity. This view implies that the contrast between grammar and semantics differs in an important way from (for example) the contrast between politics and religion, or the difference between subtraction and long division. Of course such statements are certainly right at one level of analysis: Grammar and semantics are natural kinds, mental/behavioral systems that emerge in all normal humans at some point in their lives. Such pervasive and important human activities must be adequately served and represented somewhere in the human brain, and because of their universal nature, it may be that these functions have a particularly high priority in the allocation of neural resources. But politics, religion and long division have to be somewhere too.

In fact, it is fair to say that everything humans know and do is served by and represented in the human brain (although we also make use of external memory systems to store portions of our knowledge, e.g., books, objects, and other human beings—Hutchins, 1994). Our best friend's phone number and our spouse's shoe size must be stored in the brain, and presumably they are stored in nonidentical ways, which could—at least in principle—show up someday on someone's future brain-imaging machine. In fact, every time we learn something new (the outcome of a presidential race, the announcement of a friend's engagement, a change in the local area code), we have a slightly different brain than we had before that learning took place. The existence of a correlation between psychological and neural facts says nothing in and of itself about innateness, domain specificity or any other contentious division of the epistemological landscape. It merely helps us make some progress in understanding the links between brain and behavior.

The *what* question has an obvious answer, once we stop to think about it. The *where* question is a little more subtle, and more controversial. Neuropsychologists often make two related assumptions (implicitly or explicitly) about localization in the brain:

- *Compact coding*: Innate and domain-specific behavioral functions are localized to specific regions of the brain, while learned functions are distributed more broadly across neural systems.

- *Universal coding*: Innate and domain-specific behavioral functions are assigned to the same locations in all normal human beings, while the distribution of learned functions can vary markedly from one individual to another.

These are both reasonable claims, and they can be tested. But for reasons that we have already discussed in some detail in other chapters, neither claim is logically necessary. In fact, all logical combinations of innateness, domain specificity and localization may be found in the minds and brains of higher organisms (see Table 5.1). Here are a few possible examples.

1. Well-defined regions of the brain may become specialized for a particular function as a result of experience. In other words, learning itself may serve to set up neural systems that are localized *and* domain specific, but *not* innate. A good example comes from positron emission tomography studies of brain activity showing a region of visual cortex that is specialized for words that follow the spelling rules of English (Petersen, Fiez, & Corbetta, 1992). Surely we would all agree that English spelling is not part of our biological heritage. The ultimate location of a "spelling module" must be based on general facts about the organization of visual cortex, and its connections to the auditory system (in particular, the areas with primary responsibility for language—see below). In the same vein, Nichelli et al. (1994) have used positron emission tomography to study the areas of cortex that are most active in skilled chess players at different points across the course of the game. At it turns out, there are indeed regions that are uniquely active at crucial points in the game, but no one would want to argue for the evolution of an innate "checkmate center."

2. There may be a strong predisposition to set up domain-specific functions in a form that is broadly distributed across many different cortical regions, and in patterns that vary widely from one individual brain to another. In other words, these systems may be innate *and* domain specific, but *not* strongly localized. A possible example comes from cortical stimulation showing that many different regions of the left hemisphere can interrupt naming, although some sites are more vulnerable than others (Lüders, Lesser, Dinner et al., 1991; Lüders et al., 1986; Lüders, Lesser, Hahn et al., 1991; Ojemann, 1991).

3. There may be systems that are innate *and* highly localized, but not domain specific. Instead, they are used to process many different kinds of information. An example in this category might be the attentional systems described by Posner and his colleagues (e.g., Posner & Petersen, 1990). These systems are postulated to play a role in the disengaging and shifting of attention, but they can operate on a variety of sensory domains. Posner's posterior attention system is supported by a network of structures such as the posterior parietal lobe and the superior colliculus. This system is capable of increasing the sensitivity of particular parts of early visual or auditory processing, such as color, shape, location and frequency. Thus, while this system may be innate and localized to a particular group of structures, it is not domain specific.

TABLE 5.1

Example no.	Domain specific?	Localized?	Innate?
1	Yes	Yes	No
2	Yes	No	Yes
3	No	Yes	Yes

In short, answers to the *where* question are extremely important for our understanding of brain function, but they do not (in and of themselves) justify conclusions about innateness or domain specificity. Those issues are better addressed by inquiries into the nature (how?) and timing (when?) of brain development.

How and when do things develop in the brain?

To tackle the *how* and *when* questions and apply them to classic issues in cognitive development, many theorists have found it useful to draw a distinction between maturation and learning (see Chapter 3 for a further discussion of this point). But as we shall see, it is no longer clear where to draw that line, or whether it can be drawn at all. Let us consider two proposed dichotomies: *experience-independent vs. experience-dependent change*, and *early* (time-limited) vs. *late* (open-ended) *change*.

We can define an experience-independent change in brain structure as an event that takes place in the absence of particular inputs from the environment. Conversely, an experience-dependent change only comes about in the presence of specific types of environmental inputs. At first glance, this seems like a straightforward way to operationalize the nature-nurture issue in brain development. However, as it turns out, it isn't clear where experience or environment begin in the developing organism. Environment is difficult to define because every cell and every region in the brain serves as an environment for its neighbors, and experience is difficult to separate from an intra-organismic notion called *activity dependence*.

Activity dependence refers to the causal effects of inputs from one part of the system to another. As we shall see in more detail later, some major features of brain organization in vertebrates (e.g., the formation of cortical columns) are caused by a competition among inputs to a target region (e.g., a competition between inputs from the two eyes during the development of visual cortex). If that competition is blocked (e.g., regions of visual cortex only receive input from one eye), then the familiar cortical columns do not form. Notice that activity-dependent processes can take place in the absence of input from the outer world (i.e., from events outside the skull). For example, the formation of cortical columns can be driven by spontaneous firing from peripheral neural structures (Miller, Keller, & Stryker, 1989; see also Chapter 2). If these peripheral systems are active and intact, then cortical columns will emerge even though the animal's two eyes have been closed shut at birth; but if spontaneous firing is blocked from one or both of the peripheral ocular systems, the formation of cortical columns is abnormal or altogether absent. These particular examples of activity-dependent growth take place after birth, when the line between internally generated and externally generated activity is especially difficult to draw. But other forms of activity-dependent growth take place in utero, when the animal still has virtually no inputs from the outside world. Should we classify these forms of internal experience as "learning" or "maturation?" Does the classification serve any purpose? It is not clear to us that it does, particularly since the effects of external vs. internal stimulation often look quite similar in the structures that receive those inputs.

In the same vein, the notion of environment is perhaps best conceived as a Russian doll, a nested series of structures organized from "outside" to "inside." Mothers serve as the environment for the fetus. Organs serve as environments for one another—scaffolding, supporting, blocking and shaping one another into a final configuration. And individual cells are powerfully influenced by their neighbors. For example, neuronal migration in the cortex is guided by glial fibers that take them from point to point (see cell migration, below), and their ultimate form and function is strongly determined by the cells that live around them when they finally come to a stop, and the work that they must do from that point on (see below for experiments on plugs of visual cortex transplanted in a somatosensory region).

A second approach to the distinction between maturation and learning is based not on spatial dimensions (i.e., inside or outside the skull), but on the dimension of time. Specifically, it has been argued that maturational effects are early and limited in time, while learning is a later process that continues throughout life. This temporal distinction would be easier to defend (although the line is difficult to draw) if it could be demonstrated that early and late processes are also qualitatively different. To some extent, this is true: The principles that govern neural development do change from conception to death, along a number of dimensions. But many of these changes are a matter of scale, not an absolute difference in kind.

To make this point, we need to introduce an important distinction between *progressive events,* and *regressive events.* Progressive events (a.k.a. "additive" or "constructive events") are changes in neural organization that come about by adding new structure. These would include the birth and proliferation of neurons, the migration of neurons from the proliferative zone to their final destination, the sprouting and extension of neural processes (long-range axonal connections, and local dendritic branching), and those changes that take place at the synaptic junction as a result of learning. Regressive events (a.k.a. "subtractive" or "destructive events") are changes in neural organization that come about through the reduction or elimination of existing structures. Examples include cell death, axonal retraction, and synaptic pruning. The existence of progressive events has been known for some time and, until recently, most speculations concerning the maturational basis of behavioral change have concentrated on these factors. The regressive events are a much more recent discovery (for reviews see Janowsky & Finlay, 1986; Oppenheim, 1981). We will talk about these phenomena in more detail below. For present purposes, the point is that brain development involves a combination of additive and subtractive events. This provides a useful framework for thinking about maturation, learning and the dimension of time.

The first additive events are easy to classify: The birth and migration of cells takes place only once (at least in higher species, for most areas of the nervous system). These are (by definition) early events, limited in time, and it is widely assumed that they are under close genetic control, a property of interactions at the molecular and cellular level. It seems fair to conclude that these are truly maturational phenomena that cannot be confused with learning. From this point on, however, things get murky. The sprouting and extension of neural processes that take place after migration are certainly part of the genetic program (although that program may operate indirectly, e.g., through release of nerve growth factors). However, we now know that some aspects of sprouting and growth can occur throughout the animal's lifetime (although the first blush of youth is never seen again), and we also know that sprouting and growth of dendrites and synapses can be stimulated by experiential factors, even in an aging adult (Greenough et al., 1986). So should we classify these events as "maturation" or "learning?" Regardless of their source, the changes that occur through additive events (sprouting of new connections, and/or the strengthening of existing connections) look a lot alike at the local level. The line is even harder to draw for subtractive events. We know that genetic factors can play a powerful role in early cell death, and in the large-scale elimination of axons and dendrites that occurs in the first months of life. On the other hand, we also know that "synaptic pruning" is heavily influenced by experience. Indeed, it has been argued that pruning is the direct result of activity-dependent competition, a process by which experience sculpts the brain into its final form (i.e., Changeux, Courrège, & Danchin, 1973; Changeux & Danchin, 1976; Edelman, 1987).

Thus, while in general subtractive events are more commonly associated with the effects of experience than constructive ones, there is no clear, neat division between them. Answers to the questions "how" and "when" do not support a strict dichotomy between maturation and learning. The same additive and subtractive processes occur again and again, at different scales, in response to chemical and electrical inputs that can be generated inside or outside of the organism.

The question of "who": Is it the same for everyone?

Another approach to innateness and domain specificity in brain development revolves around the problem of variability (see "universal coding," above). Specifically, it has been argued that innate behavioral systems involve a universal set of structures that unfold on a universal schedule of development, displayed by every normal member of the species. In other words, the answer to the question "who" is "almost everybody." By contrast, it was once assumed that variations in structure and timing are the hallmark of learned behaviors, reflecting those variations that are known to occur in the environment. For those who find this dichotomy convenient, recent findings that reveal a substantial degree of plasticity in brain development present a logical and an empirical challenge. We will review those findings in some detail below. For present purposes, the point is that the same behavioral outcome (or one that is good enough) can be achieved in a number of ways, i.e., with different forms of cortical representation, and with the collaboration of several different brain regions, in several different working coalitions.

Plasticity was once viewed as a rather limited and "special" process, a kind of Civil Defense system[1] that is only called into play in response to pathological conditions (i.e., when injuries to the brain prevent the establishment of a default pattern). Examples might include massive internal injuries (e.g., focal brain lesions following a perinatal stroke) or massive variations in the input from peripheral systems (e.g., deafness, blindness, amputation). Recently, more and more researchers have come to view these cases simply as the outer extremes of yet another developmental continuum. There are two reasons for this shift in orientation. First, it is increasingly difficult to distinguish the mechanisms that produce "special" modifications from those that produce "default" patterns of brain organization. Many of the same additive and subtractive processes appear to be at work in both cases (i.e., sprouting and growth of new connections, elimination of unsuccessful links, strengthening or weakening of local connections, competition and lateral inhibition). Second, new techniques for structural and functional imaging have revealed a surprising degree of variation across *normal* brains (e.g., variations in shape, size and degree of symmetry in structural-imaging studies; variations in the size and location of the "hot spots" associated with specific tasks in functional-imaging studies). For example, Damasio (Damasio, 1989; Damasio & Damasio, 1992) has suggested that brain organization for language and other higher cognitive functions may vary markedly across normal individuals, in idiosyncratic patterns that are as unique as their finger prints (see also Goodglass, 1993). Findings of this sort complicate our interpretation of neuropsychological data with patient populations. Because we have no access to "the man that was" or "the child that might have been," it is difficult to know how much of the variation we find in studies of a brain-damaged patient are due to the disease process itself, or to the individual differences that were established under normal conditions (see Bates, Appelbaum & Allard, 1991, for a discussion of this point).

The additive and subtractive mechanisms that are available for both normal and abnormal variations in brain development include the following.

---

1. We thank Joan Stiles for this term.

1. **Growth of new connections.** With a few exceptions (e.g., the olfactory lobes), it is commonly supposed that no new neurons are formed in the brains of higher organisms after the initial phase of cell formation and migration. The neurons that we have in place a few weeks before birth are the only neurons we are ever going to get, no matter how much bad luck we have to face. With the exception of neurochemical changes, synaptogenesis is the only major *additive* mechanism that we have available to explain alternative forms of neural organization in response to brain injury. As we shall see, synaptic sprouting has been demonstrated in response to a marked change in environmental conditions, in young and aging rats (Greenough et al., 1986). This is a particularly clear example of a mechanism for plasticity that operates in normal development and in response to abnormal conditions.

2. **Elimination of normal connections.** As we will see later, subtractive processes play a major role in normal brain development. Experience contributes to "sculpting" the brain into its mature form, through a competitive process in which unsuccessful connections are eliminated while successful connections are retained. This normal subtractive process is also available for the creation of alternative forms of organization following focal brain injury (Stiles, 1995).

3. **Retention of exuberant connections.** If insults occur early enough in neural development, it is possible to retain exuberant neurons and connections that would be eliminated under normal (default) conditions (Bhide & Frost, 1992; Frost, 1990; Langdon & Frost, 1991). The axons from these "salvaged" units can go on to form connections that would not otherwise occur, and as a result, displaced functions can be rerouted to a different cortical site. Some research suggests that this process can interact with #2 above, i.e., retention of exuberant connections is sometimes accompanied by elimination of normal connections, so that the gross "quantity" or ratio of connectivity is preserved (see Johnson, 1988; Johnson & Karmiloff-Smith, 1992).

4. **Reprogramming of existing connections.** Normal and abnormal learning and development can also take place (at least in principle) without adding connections or taking them away, through a strengthening or weakening of existing synaptic contacts (e.g., Hebbian learning). This form of reprogramming is thought to occur in normal learning, and it is generally assumed to play a role in the alternative forms of organization that emerge after brain injury (Churchland & Sejnowski, 1992; Hebb, 1949; Linsker, 1986, 1990; Pons, Garraghty & Mishkin, 1992; Pons et al., 1991).

5. **Compensation/redistribution of function.** Under this mechanism of plasticity, intact regions take over a displaced function by processing that functions in the usual manner (i.e., by applying the operations that normally take place in that region). Under some interpretations of this form of plasticity, nothing has to change at the level of brain structure. For example, if the right hemisphere has to take over the language function, it will do so by applying strategies or modes of processing that are normally used for aspects of spatial analysis. This is an interesting possibility, but it is problematic on both empirical and conceptual grounds. Within the connectionist view of development outlined in this book, a large-scale compensatory reorganization of this kind could not take place without some kind of local structural change. That is, because all learning involves some kind of change in patterns of connectivity, a region that is forced to take over a new function will have to change its structure in some way (i.e., through addition, subtraction, retention and/or reprogramming, 1-4 above). We are presenting compensatory redistribution of function under a separate category because it involves reorganization on a very large scale, in which regions that are located far away from the damaged area are recruited to solve a computational problem that cannot be solved in the usual way. A particularly clear example of reorganization on this scale will be offered later, from research by Webster, Bachevalier, and Ungerleider on object recognition in lesioned monkeys.

To summarize: Current information about the structure and development of the human brain offers no easy solutions to the nature-nurture problem. All of our Wh-questions (what, where, when, how and who) receive a similar answer: It's interaction all the way down, from top to bottom, and apparent universals often mask a range of alternative outcomes that can be demonstrated when default input conditions do not hold. With these lessons in mind, let us turn to an examination of brain development in vertebrates in search of useful constraints for a connectionist theory of development. Following this, in the final part of the chapter, we will turn to the question of the uniqueness or otherwise of human brain development.

Building a vertebrate brain

Embryological development of the brain

Shortly after conception (in a matter of moments), the fertilized cell undergoes a rapid process of cell division, resulting in a cluster of proliferating cells called the *blastocyst*—a structure that looks, to the uninformed eye, like nothing more differentiated than a clump of grapes. Within a matter of days, the blastocyst differentiates into a three-layered structure called the *embryonic disk*. The three layers of this disk have been termed the *ectoderm* (outer layer), the *mesoderm* (middle layer) and the *endoderm* (the inner layer which, in placental species, hugs the wall between the budding organism and its host). Despite their rather uninformative names, each of these layers will differentiate into a major system. The endoderm will differentiate into the set of internal organs (digestive, respiratory, etc.), the mesoderm turns into the skeletal and muscular structures, and the ectoderm gives rise to the skin surface and the nervous system (including the perceptual organs).

The nervous system itself begins with a process known as *neurolation*. A portion of the ectoderm begins to fold in on itself to form a hollow cylinder called the *neural tube*. The neural tube differentiates along three dimensions: the length, the circumference, and the radius (see Figure 5.1). To visualize how this deceptively simple

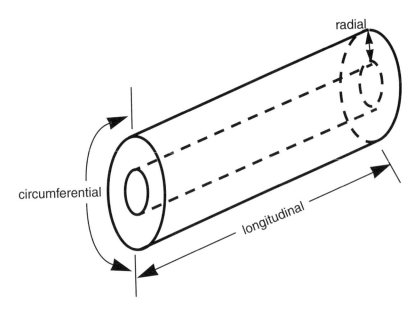

FIGURE 5.1 Stylized representation of the neural tube (after Nowa-kowski, 1993).

structure gives rise to a more complex brain, consider your index finger as we review these three dimensions of differentiation.

The length dimension gives rise to the major subdivisions of the central nervous system, with the forebrain and midbrain arising at one end (the finger tip in your visual image), and the spinal cord at the other (from the second to the third knuckle). The end which will become the spinal cord differentiates into a series of repeated units or segments (think of them as a series of repeated knuckles). However, the front end of the neural tube organizes in a different way, with a series of bulges and convolutions. By 5 weeks after concep-tion these bulges can be identified as protoforms for major components of the mammalian brain. Proceeding from front to back (on your mental index finger): the first bulge gives rise to the cortex (telencephalon), the second will give rise to the thalamus and hypothalamus (diencephalon), the third turns into the midbrain (mesencephalon), and others to the cerebellum (metencephalon), and to the medulla (myelencephalon).

To visualize how these linearly arranged structures turn into something recognizably like a vertebrate brain, fold your index finger tightly into a configuration with the finger tip on top (the cortex), and the rest of the finger wrapping around down to your bottom knuckle (the medulla, followed by the spinal cord). This approximates (crudely) the differentiation and folding that characterizes development along the length dimension.

To understand differentiation along the circumferential dimension (tangential to its surface), orient your index finger with the fingernail facing up, and palm side down. The upper side corresponds roughly to the dorsal (back) structures in the developing brain (the alar plates in the developing brain stem), while the lower side corresponds to the ventral (front) structures (basal plates in the lower brainstem).

Why should up-down turn into front-back? To understand this dimension, remember that humans are bipedal vertebrates, but most other mammals are not. Consider a hunting dog standing still, pointing towards his prey. We can draw a line from the animal's nose, over his head and down his back (the dorsal/alar line), and we can also draw a line from the animal's nose, down his chin and along his stomach (ventral/basal line). In other words, the dog's alignment corresponds more directly to the circumferential dimension along the neural tube. This correspondence is less obvious in the human brain because of the way that we have folded things up to achieve bipedalism (fold your index finger again, as instructed above, and see where the ventral and dorsal surfaces end up). The circumferential dimension in the brain is critical, because the distinction between sensory and motor systems develops along this dimension: Dorsal corresponds roughly to sensory cortex, ventral corresponds to motor cortex, with the various association cortices and "higher" sensory and motor cortices aligned somewhere in between. Within the brain stem and the spinal cord, the corresponding alar (dorsal) and basal (ventral) plates play a major role in the organization of nerve pathways into the rest of the body. Differentiation along the radial dimension gives rise to the complex layering patterns and cell types found in the adult brain. Across the radial dimension of the neural tube the bulges grow larger and become further differentiated. Within these bulges cells *proliferate, migrate,* and *differentiate into particular types.* The vast majority of the cells that will compose the brain are born in the so-called *proliferative zones.* These zones are close to the hollow of the neural tube (which subsequently become the ventricles of the brain). The first of these proliferation sites, the *ventricular zone,* may be phylogenetically older (Nowakowski, 1993). The second, the *subventricular zone,* only contributes significantly to phylogenetically recent brain structures such as the neocortex (i.e., "new" cortex). These two zones yield separate glial and neuron cell lines and give rise to different forms of migration. But first we must consider how young neurons are formed within these zones.

Neurons and glial cells are produced by division of proliferating cells within the proliferation zone to produce clones (a clone is a group of cells which are produced by division of a single precursor cell—such a precursor cell is said to give rise to a lineage). *Neuroblasts* produce neurons, and *glioblasts* produce glial cells. Each of the neuroblasts gives rise to a definite and limited number of neurons (we will return to this point later). In at least some cases particular neuroblasts also give rise to particular types of cell. For example, less than a dozen proliferating cells produce all the Purkinje cells of the cerebellar cortex, with each producing about 10,000 cells (Nowakowski, 1993).

After young neurons are born, they have to travel or *migrate* from the proliferative zone to the particular region where they will be employed in the mature brain. There are two forms of migration observed during brain development. The first, and more common, is passive cell displacement. This occurs when cells that have been generated are simply pushed further away from the proliferative zone by more recently born cells. This form of migration gives rise to an outside-to-inside spatiotemporal gradient. That is, the oldest cells are pushed toward the surface of the brain, while the most recently produced cells are toward the inside. Passive migration gives rise to brain structures such as the thalamus, the dentate gyrus of the hippocampus, and many regions of the brain stem.

The second form of migration is more active and involves the young cell moving past previously generated cells to create an "inside-out" gradient. This pattern is found in the cerebral cortex and in some subcortical areas that have a laminar structure.[2] In the next section we will examine how the laminar and radial structure of the cerebral cortex arise from an "inside-out" gradient of growth.

The basic vertebrate brain plan

It is important to note that the mammalian brain closely follows the basic vertebrate brain plan found in lower species such as salamanders, frogs, and birds. The major differences between these species and higher primates is in the dramatic expansion of the cerebral cortex together with associated internal structures such as the basal ganglia. The rapid expansion of the area of cortex has resulted in it becoming increasingly convoluted. For example, the area of the cortex in the cat is about 100cm, whereas that of the human is about 2400cm. The extra cortex possessed by primates, and especially humans, is presumably related to the higher cognitive functions they possess. However, the basic relations between principal structures remain similar from toad to man. The cerebral cortex of mammals is basically a thin (about 3-4mm) flat sheet, and it has a fairly constant structure throughout its extent. This does not mean, however, that this structure is simple. It is not.

The basic structure of the human brain is shown in Figure 5.2. This simple sketch serves to illustrate some of the basic relationships of the cortex to other parts of the brain. Most of the sensory inputs to the cortex pass through a structure known as the thalamus. Each type of sensory input has its own particular nuclei

---

2. By lamination we mean that they are divided up into layers that run parallel with the surface of the proliferative zone, and, in the case of the cerebral cortex, with the surface of the structure.

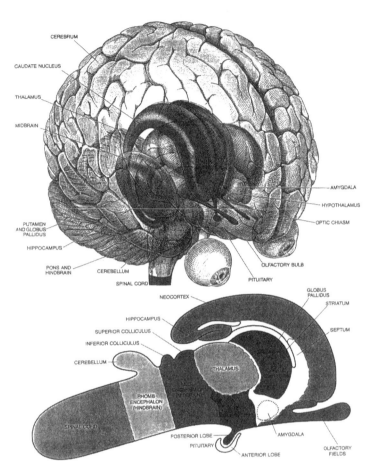

FIGURE 5.2 Basic structure of the human brain.

within this region. For example, the lateral geniculate nucleus carries visual input to the cortex, while the medial geniculate nucleus carries information from the auditory modality. Because of the crucial role of the thalamus in mediating inputs to the cortex, some have hypothesized that it also plays a crucial role in cortical development—an idea which we will discuss at greater length later. We should point out, however, that the flow of information between thalamus and cortex is not unidirectional. Most of the projections from lower regions into the cortex (i.e., input streams) are matched by projections from the cortex back down (i.e., output streams). Some output projections pass to regions that are believed to be involved in motor control, such as the basal ganglia and the cerebellum. However, most of the projections from the cortex to other brain regions terminate in roughly the same regions from which projections arrived (such as the thalamus). In other words, the flow of information in the cortex is largely bidirectional. For this reason, it is important not to confuse the terms "input" and "output" with "sensory" and "motor." All sensory and motor systems make extensive use of both input and output fibers, with information passing rapidly in both directions along collateral pathways (like highways with traffic running east and west, along separate but nearby tracks). The fact that sensory cortex "talks back" to the sources that feed it information suggests that cortex is not a passive registry of input from the environment; instead, inputs are modulated actively and constantly via these output fibers.

The cortex is composed of two general types of cells, neurons and glial cells. Glial cells are more common than neurons, but it is generally assumed that glial cells play no direct role in the computations that the cortex subserves. However, as we shall see, they probably play a very important role in the development of the cortex. Neurons come in many shapes, sizes and types, each of which presumably reflects their particular computational function or "style." By current count, there appear to be approximately 25 different neuronal types within the cortex, although most of these types are relatively rare, and some are restricted to particular layers. About 80% of neurons found in the cortex are pyramidal cells, so-called because of the distinctive pyramid shape of the soma (cell body) produced by the very large apical dendrite (input process) which is always tangential to the surface of the cortex. These are the neurons whose long axons (output processes) are so often found in the fibers feeding into other cortical and subcortical regions. While pyramidal cells are found in many of the layers of the cortex (generally they are larger in the lower layers and smaller in the upper layers), their apical dendrites often reach into the most superficial layers of the cortex, layer I (see below). This long apical dendrite allows the cell to be influenced by large numbers of cells from other (more superficial) layers and regions. This may be computationally important if the pyramidal cell is a very stable and inflexible class of cell whose output is modulated by groups of plastic and flexible inhibitory regulatory neurons.

Aside from pyramidal cells, the cortex possesses a variety of other cell types. While classifications of these cell types can vary, two of the most common are stellate cells and Martinotti cells. Stellate cells are so called because the arrangement of dendrites around the cell body is stellate, i.e., without any preferential orientation, as the dendrites are equally likely to point in any direction. In Martinotti cells the pattern of dendrites around the cell body varies, being sometimes confined to a sphere, and other times elongated in one direction or another. In contrast to pyramidal cells, the axon leaves the cell body in a vertical direction. Non-pyramidal cells together comprise about 20% of cells in the cortex, though estimates of the exact frequencies of each of the cell types varies between species and laboratories.

Figure 5.3 shows a schematic section through an area of primate cortex, the primary visual cortex of the macaque monkey. This section is cut at right angles to the surface of the cortex, and reveals the layered structure. We will refer to it as the *laminar* structure of the cortex. As we have just noted, each of the laminae has particular cell types within it, and each layer has typical patterns of inputs and outputs.

All cortical areas (in all mammals) are made up of six layers. The basic characteristics that define each layer appear to hold in every region of the cortical sheet (although there are some interesting variations between sensory and motor cortex to which we shall turn shortly). Let us assume for the moment that we are looking at the side of a cortical sheet that has been ironed out flat, lying in front of us on a table. Layer 1 is on the top, and Layer 6 is on the bottom. Layer 1 has relatively few cell bodies. It is made up primarily of long white fibers running along the horizontal surface, linking one area of cortex to others some distance away. Layers 2 and 3 also contain horizontal connections, projecting forward from small pyramidal cells to neighboring areas of cortex. Layer 4 is the pri-

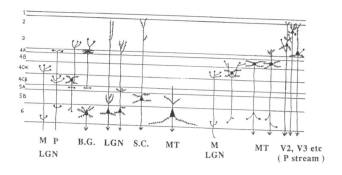

FIGURE 5.3 Schematic section through an area of primate cortex. The section is cut at right angles to the surface of the cortex and reveals the layered structure.

mary input layer (i.e., it is the layer where input fibers terminate), and it contains a high proportion of spiny stellate cells which are (presumably) well suited to this purpose. Layers 5 and 6 have the major outputs to subcortical regions of the brain. These layers contain a particularly high proportion of large pyramidal cells, with long axons for the long journey down.

Although this basic laminar structure holds throughout the cortex, there are important regional variations. For example, the input layer (Layer 4) is particularly thick and well developed in sensory cortex. Indeed, in the visual system, it is possible to distinguish at least four sublayers within Layer 4. Conversely, Layer 5 (one of the two critical output layers) is particularly well developed in motor cortex, presumably due to its importance in sending output signals from the cortex. It is also clear that different parts of the cortex have different cortico-cortico projection patterns. While there may only be a small number of these characteristic projection patterns, there is no single pattern that can be said to be characteristic of all cortical regions. Hence this is another dimension that contributes to regional specialization (see Figure 5.4 for a typical example of feedforward and feedback pathways). There are also regional variations in the absolute number of cells within and across layers. For example, the cell population in primary visual cortex is exceptionally dense and tightly packed in primates. Indeed, Area 17 in primates contains twice as many cells as any other cortical area. Other possible dimensions that could be responsible for regional specialization include the manufacture of neurotransmitters within cells, and the relative contribution of excitatory vs. inhibitory neurotransmitters. Finally (an important point for our purposes here), regions may vary in the timing of key developmental events, such as the postnatal reduction in the number of synapses (Huttenlocher, 1990).

This brings us to the controversial issue of regional specialization, i.e., development along the dimension tangential to the layers. The tangential dimension can be described at a number of different scales. For most of this chapter (and much of this book), we will focus on the large-scale areas of cortex, the largest component. However, in the next section we will discuss evidence for the smallest level of tangential structure, the radial unit initially proposed by Pasko Rakic (1988). We will see how differentiation into areas may arise as a product of constructive developmental processes. Some, but not all, of these units extend through most of the cortical layers described above.

To summarize so far, the cerebral cortex is highly structured in both laminar and tangential dimensions. How does this structure arise during development?

Cerebral cortex and functional areas: Protomap or protocortex?

An ongoing debate among those who study the developmental neurobiology of the cortex concerns the extent to which its structure and function are prespecified. Returning to the themes outlined in Chapter 1, we can enquire which of the ways to be innate discussed earlier might be applicable to aspects of the structure and function of the cortex. For example, are cortical representations specified prior to experience (i.e., representational nativism)? How do the specific computational properties of different cortical regions arise (i.e., architectural and/or representational nativism)? As we will see, one view is that the answer to the latter question is different for tangential (area) vs. radial (layered) structure of cortex (e.g., Shatz, 1992a,b,c). While there is little dispute that the laminar structure of cortex is the product of interactions within the cortex, controversy rages about the extent to which the tangential (area) structure of the cortex is the result of a pre-existing protomap, or the result of self-organizing and experiential factors. We begin by discussing arguably the most complete theory of the development of cortical structure, the radial unit model proposed by Pasco Rakic, before describing some alternative viewpoints.

Most cortical neurons in humans are generated outside the cortex itself in a region just underneath what becomes the cortex, the "proliferative zone." As mentioned earlier, this means that these cells must migrate to take up their locations within the cortex. How is this migration accomplished? Pasko Rakic has proposed a "radial unit model" of neocortical parcellation which gives an account of how both the tangential (area) and the radial (laminar) structure of the mammalian cerebral cortex arise (Rakic, 1988). This model gives a crucial role to glial cells. According to the model, the tangential organization of the cerebral cortex is determined by the fact that each proliferative unit (in the subventricular zone) gives rise to about 100 neurons. The progeny from each proliferative unit all migrate up the same radial glial fiber, with the latest to be born travelling past their older relatives. Thus, radial glial fibers act like a climbing rope to ensure that cells produced by one proliferative unit all contribute to one radial column within the cortex. Rakic's proposed method of migration is illustrated in Figure 5.4.

There are some consequences of the radial unit model for the role of genetic regulation in species differences. For example, Rakic (1988) has pointed out that a single round of additional symmetric cell division at the proliferative unit formation stage would double the number of ontogenetic columns, and hence the area of cortex. In contrast, an additional single round of division at a later stage, from the proliferative zone, would only increase the size of a column by one cell (about 1 per cent). As we pointed out earlier, although there is very little variation between mammalian species in the layered structure of the cortex, the total surface area of the cortex can vary by a factor of 100 or more between different species of mammal. It

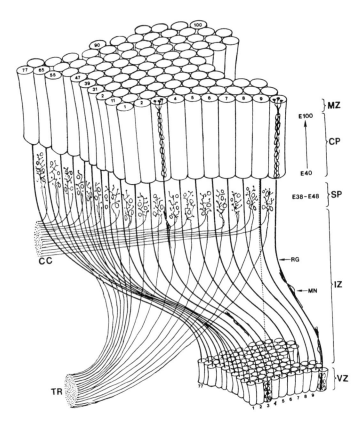

FIGURE 5.4 Schematic of radial unit hypothesis. Glial cells serve as "guide wires" along which developing neurons travel as they migrate from the proliferative zone (bottom) to their final cortical destination (top).

seems likely, therefore, that species differences originate (at least in part) in the timing of cell development (i.e., the number of "rounds" of cell division that are allowed to take place within and across regions of the proliferative zone).

How does the laminar structure of the cerebral cortex emerge? While we are far from being able to answer this question definitively at this point, one view is that differentiation into a particular cell type occurs *before* a neuron reaches its final location. That is, a cell "knows" what type of neuron it will become (pyramidal, spiny stellate, etc.) before it reaches its adult location within the cortex. Some recent evidence suggests that cells do indeed begin to differentiate before they reach their final vertical location. For example, in genetic mutant "reeler" mice, cells that acquire inappropriate laminar positions within the cortex still differentiate into neuronal types according to their time of origin, rather than the types normally found at their new location. This implies that the information required for differentiation is present at the cell's birth in the proliferative zone; it is not dependent upon their distance from the proliferative zone, nor on the characteristics of the neighborhood in which that cell ends up. That is, in the proliferative zone for the neocortex, some cell types may be determined at the stage of division.

Although in many cases a cell's identity may be determined before it leaves the proliferative zone, some of the properties that distinguish among cell types may form later on. Marin-Padilla (1970) has proposed that the distinctive apical dendrite of pyramidal cells, which often reaches into layer 1, is a result of the increasing distance between this layer and other layers resulting from the inside-out pattern of growth. Specifically, the increasing separation between layer 1 and the subplate zone which results from young neurons moving into what will become layers 2 to 6, means that cells which have their processes attached to layer 1 will become increasingly "stretched"—that is, their leading dendrite will become stretched tangential to the surface of the cortex resulting in the elongated apical dendrite so typical of cortical pyramidal cells. As mentioned earlier, this long apical dendrite allows the cell to be influenced by large numbers of cells from other (more superficial) layers.

Another aspect of the radial (layered) structure of cortex that appears to be regulated by intrinsic cellular and molecular interactions concerns the connections between cells, in particular the inputs from the thalamus. Recall from the earlier sections that the main input layer in the cortex is layer 4. A series of experiments by Blakemore and colleagues established that the termination of projections from thalamus in layer 4 is governed by molecular markers.

Slices of brain tissue are able to survive and grow in a petri dish for several days under the appropriate conditions. Indeed, pieces of thalamus of the appropriate age will actually innervate other pieces of brain placed nearby. Molnar and Blakemore (1991) investigated if and how a piece of visual thalamus (LGN) would innervate various types of cortical and noncortical brain tissue.

In initial experiments they established that when a piece of thalamus (LGN) and a piece of visual cortex were placed close together in the dish, afferents from the LGN not only invaded a piece of visual cortex of the appropriate age, but also terminated in the appropriate layer 4. Thus, layer 4 appears to contain some molecular stop signal that tells the afferents to stop growing and form connections at that location. Next, they conducted a series of choice experiments in which the visual thalamus had a piece of visual cortex and some other piece of brain placed nearby. The thalamic afferents turned out to dislike cerebellum, rarely penetrating it, but the afferents did grow into hippocampus. However, the growth into hippocampus (a piece of brain that is closely related to cortex) was not spatially confined the way it had been for visual cortex, suggesting that it was just a "growth-permitting substrate."

The evidence discussed so far indicates that the laminar structure of cortex probably arises from local cellular and molecular interactions; it is not shaped as a result of thalamic and sensory input. That is, neurons "know" at birth or shortly thereafter what kind of cell they have to be, and they are biased to take up residence in particular cortical layers. Similarly, incoming fibers are told in which layer to stop and make synaptic contacts. This bring us to a crucial and more controversial set of questions along the radial dimension. Do cells "know" that they are destined to participate in a particular cortical area, and are they specialized at birth for a special role within that area (e.g., "I am going to work for Area MT when I grow up")? There are a number of possibilities that have been put forward to account for the parcellation of cortex into areas, including the following;

1. Rakic (1988) has proposed that the areal differentiation of the cortex is due to a *protomap*. The protomap either involves pre-specification of the proliferative zone or intrinsic molecular

markers that guide the division of cortex into particular areas. One mechanism for this would be through the radial glial fiber path from the proliferative zone to the cortex.

2. The area differentiation of cortex arises out of an undifferentiated *protocortex* that is divided up into specialized areas as a result of input through projections from the thalamus (which result, in turn, from stimulation by the external world).

While the first of these options may initially seem attractive, there is surprisingly little evidence for it. The idea that there is a detailed protomap in the proliferative zone is difficult to defend against experimental evidence showing that the cortex is relatively equipotential and plastic early in life (cf. the evidence for plasticity that we will review in the next section). Nor is there any evidence to date for specific molecular markers or gradients that map onto the cortical areas that compose the tangential structure of cortex (e.g., Shatz, 1992a,b,c), in contrast with the evidence that does exist for molecular determination of laminar structure. Further, there is recent evidence that neurons born from a particular proliferative unit can wander quite far from their original locations (Goldowitz, 1987; Johnston & van der Kooy, 1989; O'Leary, 1993). Thus there seems to be no reliable mechanism for imposing a protomap *onto* the cortex from the proliferative zone.

An alternative way that a protomap could be manifest is by means of molecular markers within the cortex that guide particular inputs (from the thalamus) to particular regions. Thus, for example, the area which normally becomes auditory cortex could specifically attract inputs from the auditory part of the thalamus. When discussing the radial (layered) structure of cortex we described some experiments by Blakemore and colleagues in which evidence for specific molecular markers in layer 4 that indicate the termination location for projections from the thalamus was gathered. While these initial petri dish experiments provided evidence for radial (layered) structure prespecification, and for thalamic afferents being attracted to cortical tissue, further experiments of a similar kind indicated that thalamic projections do not innervate particular tangential areas of cortex preferentially. In experiments in which a piece of visual thalamus (LGN) was placed close to a variety of pieces of cortex from different areas, the visual thalamus did *not* preferentially innervate visual cortex: all cortical targets were equally acceptable to all parts of the thalamus. Molnar and Blakemore tried several different combinations of thalamic areas and cortical ones and found that regardless of the origin of the piece of thalamus or cortex, equivalent innervation of layer 4 occurs. If these results are extended to the developing brain *in situ* (i.e., inside the animal, as opposed to a dish), they imply that while the cortex provides an attractive growth substrate for the thalamus, and a stop signal at layer 4, there is no area-specific targeting. Thus, for example, the visual thalamus is not specifically attracted to the visual cortex by some molecular marker, and auditory thalamus is not attracted to auditory cortex etc.

Another alternative version of the protomap theory might be that the cortex sends connections to specific regions of thalamus that then guide the afferents from the thalamus. In other words, the developing cortex may send guidewires back down that say "Come and get me" (we might refer to this as the "Rapunzel hypothesis"). Unfortunately for this hypothesis, however, Molnar and Blakemore found that cortical projections to thalamus lacked regional specificity in the same way as the reciprocal connections had. Any zone of embryonic cortex would innervate any zone of thalamus. These experiments indicate that there are probably no region-specific molecular signals available to guide thalamocortical projections. (See below for one possible exception to this conclusion, involving effects of visual thalamus on the extent of cell proliferation in the ventricular zone—Kennedy & Dehay, 1993.)

Given that there is little evidence for a cortical protomap that provides different "targets" for different thalamic projections, and little evidence that thalamic afferents are prespecified to project a thalamic map to the cortex, how are we to account for thalamic specificity in which particular thalamic afferents normally innervate particular regions of cortex, but do not appear to be specifically or rigidly targeted? Molnar and Blakemore (1991) propose an account of thalamic innervation in which the regional specificity of projections between thalamus and cortex arises from a combination of timing and spatial constraints. Briefly, this account states that afferents grow from regions of the thalamus according to a particular spatiotemporal gradient. Different regions of thalamus grow projections at slightly different times, and, as a new afferent grows, it always grows on top of, or beside, existing afferents. By following these existing afferents, the newly grown axons terminate in a location adjacent to those that grew just before. Thus, there is a chronotopic innervation of cortex by thalamus which uses the physical presence of other afferents as a spatiotemporal constraint on the pattern of innervation. The proposal is still controversial (O'Leary, personal communication, September, 1994), but in the absence of convincing evidence of molecular markers that connect specific areas of the thalamus to specific regions of cortex, it remains a viable option. (For evidence that thalamus may know more about its preferred cortical targets when tested *in vivo* rather than *in vitro*, see Niederer, Maimon, & Finlay, 1995.)

To summarize so far, the general structure of the cerebral cortex in mammals appears to be almost universal. Cellular- and molecular-level interactions determine many aspects of the layered structure of cortex and its patterns of connectivity. Cortical neurons are often differentiated into specific computational types before they reach their destination (although some of the characteristic features of cell types are shaped by their journey to that site, e.g., the long apical dendrite that typifies pyramidal cells reflects a literal "stretching" of processes during the migration process). However, this does not mean that cells "know" what kind of information they will have to compute. The division of cortex into distinct areas takes place along the tangential dimension. This area differentiation is heavily influenced by the thalamic inputs to each region. There is at present little direct evidence for the claim that cortex emerges from a protomap in the proliferative zone. In other words, neurons do not appear to be "born" destined to carry visual or auditory information. There is also little evidence to support the notion of an internal molecular marker to guide inputs from the thalamus (i.e., the way station that conveys sensory input to each cortical region). Indeed, as Blakemore and his colleagues have shown, axons from an occipital thalamic source will happily invade and innervate slabs of frontal or auditory cortex, in patterns that are indistinguishable from those that the same axons produce in slabs of occipital cortex. Current evidence suggests that there are no intrinsic, predetermined areal maps in either the cortex or the thalamus; instead, both develop their area specializations as a consequence of their inputs and the temporal ("chronotopic") dynamics of neural growth. This means that normal cortical development (i.e., the basic vertebrate brain plan) permits a considerable degree of cortical plasticity. That is, cortical regions could support a number of different types of representations depending on their input.

A good example of how a region of cortex can become differentiated comes from work on the so-called "barrel fields" that develop in the somatosensory cortex of rodents. Each barrel field is an anatomically definable functional grouping of cells that responds to a particular whisker on the animal's snout (see Figure 5.5). Barrel

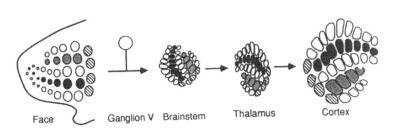

Face   Ganglion V   Brainstem   Thalamus   Cortex

FIGURE 5.5 Barrel fields (right) are collections of cells which provide a sensory representation of information detected on whiskers of the animal's snout (left).

fields are an aspect of the tangential (area) structure of cortex that emerges postnatally, and which are sensitive to whisker-related experience over the first days of life. For example, if a whisker is removed then the barrel field that normally corresponds to that whisker does not emerge, and neighboring ones may occupy some of the cortical space normally occupied by it (for review see Killackey, 1990; Killackey et al., 1994). In figure 5.6 we illustrate how the tangential divisions of the cortex arise as a result of similar divisions in structures closer to the sensory surface. In this case, it is almost as if the sensory surface *imposes* itself on to the brainstem, thence to the thalamus, and finally on to the cortex itself. The barrel field compartments emerge in sequence in these areas of the brain, with those closest to the sensory surface forming first, and the cortex patterns emerging last. While there is little evidence that barrel fields are prespecified in the cortex,[3] a map of sensory space comes to occupy the somatosensory cortex in a reliable way.

We should note that many of the experiments demonstrating that the tangential divisions of the cortex are not intrinsically prespecified have been conducted on rodents, and it is clear that there are at least some differences in the developmental timing between rodents and primates. In particular, while in the rodent thalamic input plays a major role in the areal specification of an undifferentiated protocortex, in the primate there is at least one case of areal specification in which thalamic input influences earlier stages of corticogenesis. Specifically, inputs from the visual thalamus may have an effect on the number of neurons produced in the proliferative zone that sends neurons to the area that will become primary visual cortex (see Kennedy & Dehay, 1993 for review). However, while it remains possible that some structural areal divisions in the primate cortex are regulated in this way, it is also clear that most aspects of areal differentiation occur much later in primates than in rodents, suggesting that there is even greater potential for influence by thalamic input.

This brings us to the issue of neural plasticity under pathological conditions. In this section we have reviewed the "default plan" in neural development. In the next section we discuss evidence regarding the extent and limits of plasticity that are possible within and between species. We will see that plasticity should not be viewed as a special adaptation to injury, but, argued in Chapter 3, rather as a phenomenon central to the understanding of normal development.

---

3. Some reports of prespecified boundaries between barrel fields (Cooper & Steindler, 1986) have turned out to be dependent on the pattern of thalamocortical afferents, and are not present before thalamocortical afferentation (see Schlaggar, Fox & O'Leary, 1993; Schlaggar & O'Leary, 1991 for review).

Plasticity

Early plasticity in vertebrates and infrahuman primates

As we pointed out in the previous section, virtually all areas of cortex are composed of six distinct layers of cells, each of which contains characteristic cell types. In addition, there are gross similarities in both interlaminar and extrinsic projections. In terms of gross structure, therefore, the cortical areas thought to subserve aspects of language (for example) start out looking very similar to those that subserve primary auditory representations. With this in mind it is hardly surprising that Sur and colleagues have reached the following conclusion:

> *(sensory) neocortex appears to consist of a basic structure held in common by all cortical areas, on which is superimposed a number of area-specific differences. A reasonable hypothesis is that similar aspects are intrinsically determined, perhaps before interaction with extrinsic influences (via afferent input) has occurred. Conversely, differences between areas may arise from extrinsic or afferent-induced factors, presumably at a later stage of development.* (Sur, Pallas, & Roe, 1990; p.228)

A number of experiments in recent years have shown that cortical regions (at least in the early stages) can take on a much larger range of representations than we would expect from the specialized representations that develop under normal (default) circumstances. In other words, there is a surprising amount of support for equipotentiality at the cortical level. These sources of evidence include the following:

1. Reduction of the size of thalamic input to a region of cortex early in life determines the subsequent size of that region (Dehay et al., 1989; O'Leary, 1993; Rakic, 1988).

2. When thalamic inputs are "rewired" such that they project to a different region of cortex from normal, the new recipient region develops some of the properties of the normal target tissue (e.g., auditory cortex takes on visual representations—Sur, Garraghty, & Roe, 1988; Sur et al., 1990).

3. When a piece of cortex is transplanted to a new location, it develops projections characteristic of its new location rather than its developmental origin (e.g., transplanted visual cortex takes on the representations that are appropriate for somatosensory input—O'Leary & Stanfield, 1989).

4. When the usual sites for higher cortical functions are bilaterally removed in infancy (i.e., the temporal region with primary responsibility for visual object recognition in monkeys), regions at a considerable distance from the original site can take over the displaced function (i.e., the parietal regions that are usually responsible for detection of motion and orientation in space—Webster, Bachevalier, & Ungerleider, in press).

Because these are very strong claims (with important implications for the connectionist framework developed in this book), let us consider each one in more detail.

Varying extent of sensory input

The effect on the cortex of manipulating the extent of sensory input (via the thalamus) to an area has been investigated in experiments where the thalamic input to an area of cortex is surgically reduced (Dehay et al., 1989).

Surgical intervention in newborn macaque monkeys can reduce the thalamic projections to the primary visual cortex (area 17) by 50%. This reduction results in a corresponding reduction in the extent of area 17 in relation to area 18. That is, the border between areas 17 and 18 shifts such that area 17 becomes much smaller. Despite this drastic reduction in the radial size of area 17, it is important to note that its laminar structure remains completely normal. Further, the area which is still area 17 looks identical to its normal structure, and the region which becomes area 18 has characteristics normally associated with that area, and none of those unique to area 17 (Rakic, 1988). Thus, there is (surprisingly) little effect of reducing the extent of sensory projections to area 17 on the subsequent layered structure of areas 17 and 18. The specific effect of this manipulation is to reduce the *area* of 17 relative to 18.

The output characteristics of areas 17 and 18 follow the shift in border between them. For example, while area 18 normally has many callosal projections to the other hemisphere, area 17 does not. The region which is normally area 17, but becomes area 18 in the surgically operated animals, has the callosal projection pattern characteristic of normal area 18. A reasonable conclusion reached on the basis of these observations is that the region of cortex that would normally mature into area 17 develops properties that are characteristic of the adjacent area 18 as a result of reducing its thalamic input. Thus, at least some radial area-specific characteristics of cortex appear to be regulated by experiential factors.

Redirecting input

Cross-modal plasticity of cortical areas has now been demonstrated at the neurophysiological level in several mammalian species (for review see Sur et al., 1990). For example, in the ferret, projections from the retina can be induced to project to auditory thalamic areas, and thence to auditory cortex. Following a technique initially developed by Frost (1990), one can arrange to keep exuberant connections that would ordinarily be eliminated (i.e., "extra" connections from the retina toward auditory cortex) by placing lesions in the normal visual cortex and in the lateral geniculate nucleus (the normal thalamic target of retinal projections). In addition, lesions are also placed such that auditory inputs do not innervate their normal thalamic target, the medial geniculate. Under these pathological conditions (i.e., elimination of retinal-to-visual connections that would ordinarily "win out"), retinal projections will re-route (i.e., exuberant connections will remain and continue their journey) to innervate the medial geniculate nucleus (MGN), which is of course the way station that normally handles auditory inputs. Projections from the MGN then project to the auditory cortex, in the normal fashion. The experimental question concerns whether the normally auditory cortex becomes visually responsive (i.e., in accord with its input), or whether it retains features characteristic of auditory cortex. The answer turns out to be that auditory cortex does become visually responsive. Furthermore, cells in what would have been auditory cortex also become orientation and direction selective, and some become binocular.

While these observations are provocative, they do not provide evidence that the auditory cortex as a whole becomes functionally similar to the visual cortex. It is possible, for example, that the visually driven cells in the auditory cortex would fire in isolation from the activity of their colleagues. That is, there may be no organization above the level of the individual neuron. In order to address this issue, we need evidence that there is a spatial map of the visual world formed across this area of cortex.

In order to study this issue, Sur and colleagues recorded from single neurons in a systematic way across the rewired cortex (Roe et al., 1990; Sur et al., 1988). These experiments revealed that the previously auditory cortex had developed a two-dimensional retinal map. In normal ferrets, the primary auditory cortex contains a one-dimensional representation of the cochlea. Along one axis of the cortical tissue, electrode penetrations revealed a gradual shift from responses to low frequencies to responses to high frequencies (see figure 5.7). Along the orthogonal dimension of cortex, frequency remains constant (the isofrequency axis). In contrast, the visual representation developed in the rewired animals occupies both dimensions (elevation and azimuth). The authors conclude:

> *Our results demonstrate that the form of the map is not an intrinsic property of the cortex and that a cortical area can come to support different types of maps.* (Roe et al., 1990)

Although these neuroanatomical and neurophysiological data support the idea that "auditory cortex can see," we don't know what the experience is like from the animal's point of view. Recently, Sur and colleagues have begun a series of behavioral studies to determine what use the ferret can make of visual input to

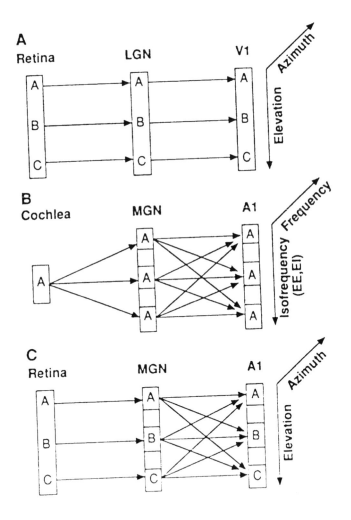

FIGURE 5.6 Schematic representation of cortical (A) Retinal information normally projects via the LGN to primary visual cortex, V1. (B) Auditory information from the cochlea normally projects via the MGN to primary auditory cortex, A1. In the "rewiring" experiments of Roe et al., 1990; Sur et al., 1988, retinal information is redirected to the MGN and A1, resulting in a reorganization of A1 so that visual information (elevation and azimuth) can be represented.

auditory cortex. They trained adult ferrets, rewired in one hemisphere at birth, to discriminate between visual and auditory stimuli presented to the normal hemisphere. After this they probed the functioning of the rewired hemisphere by presenting visual stimuli that activated only the rewired pathway. The ferrets reliably interpreted the visual stimulus as visual rather than auditory (see Pallas & Sur, 1993). If these results hold up, they provide support for Sur's conclusions about the functional (as well as structural) equipotentiality of cortical mapping.

Transplanting cortex

Additional evidence for cortical plasticity comes from studies in which pieces of cortex are transplanted from one region to another early in development. These experiments allow biologists to address the question of whether transplanted areas take on the structure/function of their *developmental origins*, or the structure/ function of the new location in which they find themselves.

Pieces of fetal cortex have been successfully transplanted into other regions of newborn rodent cortex. For example, visual cortex neurons can be transplanted into the sensorimotor region and vice versa. Experiments such as these, conducted by O'Leary and Stanfield (O'Leary & Stanfield, 1985; Stanfield & O'Leary, 1985) among others, have revealed that the projections and structure of such transplants develop according to their new spatial location rather than their developmental origins. For example, visual cortical neurons transplanted to the sensorimotor region develop projections to the spinal cord, a projection pattern characteristic of the sensorimotor cortex, but not the visual cortex. Similarly, sensorimotor cortical neurons transplanted to the visual cortical region develop projections to the superior colliculus, a subcortical target of the visual cortex, but not characteristic of the sensorimotor region. Thus, the inputs and outputs of a transplanted region take on the characteristics of their new location.

A further question concerns the internal structure of the transplanted region. Will it also take on the characteristics of its new home (i.e., "When in Rome, do as the Romans do...."), or will it retain the internal structure of its tissue source? The somatosensory cortex of the rat (and other rodents) possesses characteristic internal structures known as "barrel fields." Barrel fields are an aspect of the radial structure of the cortex, and are clearly visible under the microscope. Each of the barrels corresponds to one whisker on the rat's face. Barrels develop during postnatal growth, and can be prevented from appearing in the normal cortex by cutting the sensory inputs to the region from the face. Furthermore, barrel structure is sensitive to the effects of early experience such as repeated whisker stimulation, or whisker removal (Killackey et al., 1994; see also O'Leary, 1993). Hence we can ask whether transplanted slabs of visual cortex take on the barrel field structures that are typical of somatosensory cortex in the rat.

Schlaggar and O'Leary (1991) conducted a study of this kind, in which pieces of visual cortex were transplanted into the part of the somatosensory cortex that normally forms barrel fields in the rodent. When innervated by thalamic afferents, the transplanted cortex develops barrel fields very similar to those normally observed. Thus, not only can a transplanted piece of cortex develop inputs and outputs appropriate for its new location, but the inputs to that location can organize the internal structure of the cortical region.

Two caveats to the conclusion that the cortex is largely equipotential should be mentioned at this stage:

First, most of the transplant and rewiring studies have involved primary sensory cortices. Some authors have argued that primary sensory cortices may share certain common developmental origins that other types of cortex do not. Specifically, since an influential paper by Sanides (1972), some neuroanatomists have argued that the detailed cytoarchitecture of the cortex is consistent with their being "root," "core," and "belt" fields of cortex (Galaburda & Pandya, 1983; Pandya & Yeterian, 1985, 1990). These lineages of cortex differ slightly in the thickness of particular layers, the shapes of certain cell types, and also in their layered projection patterns to neighboring cortical regions. While the details of this theory are complex and remain somewhat controversial, it is worth noting that each of these cortical fields is claimed to facilitate a particular type of processing. For example, the "core" band of cortex is associated with primary sensory cortices, the "root" with secondary sensory areas, and the "belt" field with association cortices. Assuming that the structural differences between the putative lineages is due to phylogenetic, rather than ontogenetic, factors (and this is far from clear since these hypotheses are reached following study of adult brains), it is possible that levels of stimulus processing may be attracted to certain subtypes of cerebral cortex. Thus, it is possible that certain lineages of cortex which differ in detailed ways from other areas of cortex may be more suited for dealing with certain types of information processing. With regard to the transplant experiments discussed earlier, it may be that cortex is only equipotential within a lineage (e.g., primary-to-primary or secondary-to-secondary).

The second caveat to the conclusion that cortex is equipotential is that while transplanted or rewired cortex may look very similar to the original tissue in terms of function and structure, it is rarely absolutely indistinguishable from the original. For example, in the rewired ferret cortex studied by Sur and colleagues the mapping of the azimuth (angle right or left) is at a higher resolution (more detailed) than the mapping of the elevation (angle up or down) (Roe et al., 1990). In contrast, in the normal ferret, cortex azimuth and elevation are mapped in equal detail. We should also note that, with the exception of the study mentioned earlier, there is still very little physiological or behavioral evidence indicating that the transplanted tissue behaves the way the host area normally does. Hence, even though visual cortex can take on a barrel field structure, and auditory cortex can take on a retinotopic structure, these unusual preparations may exact a price. That is, they may never function in quite the same way as the cortical mappings that would have occurred under normal conditions. As Blakemore (personal communication, 1994) has pointed out, the cortical transplant studies carried out to date always involve tissue that has already received some thalamic inputs from its site of origin. That is, the transplants are not "virgins." This fact may place some constraints on their ability to adapt to a new site within the cortex.

Lesion studies of infant monkeys

We will end this section with one particularly dramatic example of large-scale cortical reorganization for higher cognitive processes in infant primates. Webster, Bachevalier, and Ungerleider (in press) have compared the effects of bilateral lesions in infant and adult monkeys trained in the visual nonmatch-to-sample task ("Pick the one that's different from the object you saw before, and you will get a reward"). Based on many years of work on higher visual centers in primates, most researchers in this field concur that a particular region of temporal cortex called TE plays a critical role in visual object recognition (i.e., the "what is it" portion of the visual system—Mishkin, Ungerleider, & Macko, 1983). By contrast, visual regions in the parietal regions of cortex appear to mediate the detection and prediction of movement (i.e., the "where is it" portion of the visual system). In adult monkeys, bilateral removal of area TE results in serious deficits in object recognition, as measured by the visual nonmatch-to-sample task (although it is worth noting that their performance is still above chance). By contrast, infant monkeys with homologous lesions perform very well on the same object recognition task (although their performance is reduced by approximately 5–10% compared with unoperated infant monkeys). Based on these results, Ungerleider and Bachevalier conclude that other areas of cortex seem to have taken over the function of visual object recognition in the lesioned infants. But what areas are responsible, and what is the mechanism that permitted this transfer to take place?

The most obvious candidate is an area adjacent to TE called TEO, which receives the necessary visual inputs and could take over the default functions of TE if exuberant connections were retained and/or additional sprouting took place to replace the missing circuitry. This form of "local" plasticity could involve new growth, retention of exuberant connections and/or reprogramming of spared tissue. But it would still constitute a relatively conservative form of general plasticity, within a region that is presumably well suited for support of the "what is it" function. In fact, some of the neuroanatomical evidence provided by Webster et al. does suggest that TEO receives atypical innervation in infant monkeys after bilateral lesions to TE. However, after a subsequent operation to produce bilateral lesions to TEO (so that the infant monkeys are now missing both area TE and area TEO), performance on the non-match-to-sample task was still very good (although it was reduced somewhat compared with performance after TE lesions only). In other words, area TEO may have made some contribution to the task, but apparently it has not taken over all responsibility for visual object recognition. The authors then investigated a further possibility: parietal areas that receive visual input, but which are ordinarily thought to play no role in visual object recognition, and are not adjacent to the damaged temporal zones (i.e., areas STP, PG and TF). When these parietal areas are bilaterally removed in adult monkeys, their loss normally has no significant effect on performance in the visual nonmatch-to-sample task. What would happen in monkeys who lost area TE in infancy? Webster et al. subjected their infant monkeys to a third operation, lesioning these dorsal-parietal regions bilaterally. This lesion finally managed to reduce performance on the nonmatch-to-sample task to the low levels observed in adult animals after a lesion restricted entirely to TE. They reached the tentative conclusion that distant (non-adjacent) areas that would ordinarily specialize in the "where" function have been recruited to assist in performance of the "what" function. An interesting question—which, to our knowledge, remains open—is whether the "where" function decreases in any way due to having to share resources.

We have gone into some detail with this last example, because it illustrates a crucial point. Higher cognitive functions like visual object recognition can be organized in a variety of ways. The mechanisms that underlie behavioral plasticity in the infant primate include "local" effects (i.e., sparing or growth of connections in adjacent areas), but they also include broad-scale forms of reorganization in which nonadjacent areas are recruited to serve a function for which they may be only minimally qualified (i.e., they do receive visual input of some kind). As we shall later, this provides a compelling animal model for research on recovery from focal brain injury in human children.

Plasticity in adult organisms

The evidence for plasticity that we have reviewed so far pertains entirely to variations that occur (or can occur under special circumstances) during brain development, before the adult endpoint is reached. It is widely believed (and undoubtedly true) that there is much less plasticity in the adult brain. However, this does not mean that plasticity has come to an end. At the very least, we know that some form of local structural change occurs whenever anything new is learned. But changes also occur on a larger scale. We now know that massive changes in afferent input (e.g., through amputation or immobilization of limbs) can result in a reorganization of the cortical maps associated with that input. For example, the cortical area that used to respond to input from a currently immobilized finger appears to be "taken over" by the adjacent fingers (Merzenich et al., 1983). In most of these experiments, the changes in question were rather small (spanning a few microns), which led some investigators to conclude that these were not true structural changes, but rather the "uncovering" of previously inhibited connections. However, a recent study of adult monkeys several years after limb amputation (the infamous Silver Spring monkeys—Pons et al., 1991) raises the possibility that cortical maps can reorganize over a much larger area, measured in centimeters. There is still some controversy about the mechanism responsible for this large-scale cortical reorganization. Does it reflect sprouting of new connections, reprogramming of connections that are already in place, reorganization at the thalamic level, or some combination of these? Whatever the basis for this change turns out to be, it now seems clear that the adult brain is capable of fairly large-scale structural and functional change—less plasticity that we find in the developing brain, but impressive nonetheless.

This conclusion probably comes as no surprise to Greenough and his colleagues, who have provided substantial evidence for dendritic sprouting in infant and adult rats as a function of experience (Greenough, Black, & Wallace, 1993). Infant rats who are raised in a rich environment (a circus of swings, trapdoors and other fun toys) develop thicker cortices with more dendritic branching, compared with rats who are raised in a dull and boring cage. More surprising, Greenough et al. have also shown significant dendritic growth in adult rats who receive their toys much later in life. Furthermore, adult rats who learn a specific motor skill (e.g., pulling pellets out of a hole with one forepaw) display significantly more dendritic sprouting in a subcortical region, the cerebellum. This result may be related to more recent studies by Merzenich and colleagues, showing that monkeys who are trained to detect subtle textural changes with a single finger display a clear expansion of the receptive fields in the somatosensory regions corresponding to that finger (Merzenich, in press). Similar results have been reported for adult humans by Karni et al. (1995), who used functional MRI to demonstrate specific changes in the size of the associated representations in primary motor cortex following extensive practice with a finger-tapping task.

Finally, Eva Irle (1990) has conducted a meta-analysis of 283 lesion studies involving adult primates, comparing the specific and general effects of lesion location and lesion volume on retention of old skills and learning of a new task. Figure 5.7 (redrawn from Irle, 1990) compares the effects of lesion volume on acquisition of a new skill, in primates and in rats (the rat data are taken from Lashley, 1950, in a classic study called "In search of the engram"). This figure illustrates two important points. First, monkeys are smarter than rats! There is a powerful linear relationship between lesion volume and loss of new learning in the rat population, regardless of lesion site. This was, in fact, the finding that led Karl Lashley to his controversial conclusions about equipotentiality and mass action. By contrast, lesioned primates maintain relatively high levels of new learning even after lesions that involve more than 40–50% of the cortex. Second, the relationship between lesion volume and new learning in primates is not linear. In fact, the primate data in Figure 5.7 reflect a significant U-shaped function, where performance on particular tasks actually improves (up to a point) with larger lesions. Irle has proposed a "fresh-start hypothesis" to account for this effect, suggesting that larger lesions may preclude any possibility of an old approach to a new problem, forcing the animal to come up with new solutions. For our purposes here, the

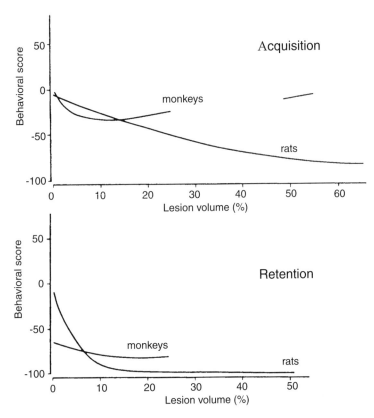

FIGURE 5.7 Effects of lesion volume on acquisition (top) and retention (bottom) of a new skill. There is a linear relationship between lesion volume and loss of new learning in rats, whereas monkeys maintain relatively high levels of learning even after 40–50% of cortex is lesioned.

point is that primate brains are rich and complex systems that can solve problems and learn new information in a variety of ways. Because of this complexity, adult primates can show more plasticity after cortical damage than we find in adult rats with comparable lesions.

Sensitive periods

If normal and abnormal development draw on the same neural mechanisms, as we have assumed here, then we are left with a new problem. Why does the organism's capacity to respond to new circumstances change so dramatically over time? This is the problem of sensitive periods. The term *sensitive period* refers to particular points during development in which the organism is especially sensitive to particular types of experience, and it has gradually come to replace the term *critical period* as it became evident that the end of these periods were neither as sudden nor rigid as previously thought. Reductions in plasticity (i.e., an end to the sensitive period) can be explained in at least two ways: maturational models (where reductions in plasticity are unrelated to learning itself) and entrenchment models (where reductions in plasticity are the product of learning). Classically, data for sensitive periods have been described and explained with discontinuous maturational models that require at least two causal factors: one that sets the first (sensitive) phase of learning into motion and determines its overall length and shape, and a separate factor (a stop signal) that brings this sensitive period to an end. It is usually assumed that both these factors (i.e., the sensitive phase, and the stop signal) are maturationally determined. That is, they are not a product of learning itself. However, as we have shown in Chapter 4, the same data are often better fit by a continuous decelerating function, like an upside-down learning curve. A function of this kind can often be explained by a single equation, which implies (in turn) a single, self-terminating process. Under this kind of a model, learning itself may be the terminating factor. The developing system starts to learn using whatever inputs are available, and as it learns it "spends" its plasticity. At some point, the old learning is entrenched to the point where certain kinds of new learning can no longer take place, i.e., a "point of no return."

And finally...the human brain

Human brain structure: Differences and similarities to other mammals

The facts and ideas that we have reviewed so far probably apply to all mammals, including humans. The principles that govern brain growth and regional specialization are the same, including the process of neurolation, principles of radial growth and migration, the role of thalamic afferents in the establishment of regional specialization, the role of competition and the importance of regressive events in the sculpting and parcellation of the brain. In addition, the following universals in brain structure hold across all mammals, including humans.

- The same basic cell types can be found across species (e.g., several different sizes and sizes of pyramidal cells, basket cells, chandelier cells, double bouquet cells etc.).

- There are approximately 60–100,000 cells per square millimeter in cortex (Braitenberg & Schüz, 1991). This generalization holds across most mammalian species, and across areas of cortex within species (though may be higher in striate cortex, see below).

- Connectivity is about 4 to 10,000 inputs per pyramidal cell, which make up the majority of cells.

- Cortex has 6 layers, with layers 1 and 4 being the main input layers.

- There are primary and secondary sensory areas in each of the three main sensory modalities (vision, audition, somatosensory).

- Sensory areas commonly project to entorhinal cortex, which projects to hippocampus, which assumes its characteristic form in mammals (detached dentate gyrus, CA fields).

All of these statistics leave us with a serious problem: What is it that makes us human? What are the brain structures that underlie language, art, mathematics, music, all the complex skills that make us special? Where are they? How and when do they arise in the course of human development?

The structural differences that have been found so far across mammals, primates and especially humans include the following:

- Larger animals (e.g., humans, cows, large rodents called capybaras) tend to have more folded cortex.

- Animals with large brain-body weight ratio (orthogonal to total body size) also tend to have a more folded cortex.

- In primates (compared with other mammals) there is an increase in the number of secondary areas in each of the main sensory modalities.

- Humans have more area devoted to secondary cortices (e.g., primary visual cortex is about 10% of cortex in most mammals, including monkeys, but is only 6% in apes and 2.5% in humans).

- Area V1 (striate cortex) is odd in humans and other primates, because it has roughly twice as many cells per unit area as any other region of cortex (an exception to the 60–100,000 cells per millimeter rule described above).

- Layer 2, and to some extent Layer 3, are slightly thicker in humans than in other primates (though apes may be intermediate).

- In humans, the secondary areas are bigger relative to their corresponding primary areas. For example, V2 is smaller than V1 in most mammals, about the same size as V1 in primates, and as much as 2–3 times the size of V1 in humans.

- Analogous to the increased ratio of secondary to primary cortex, humans also display a marked proportional increase in the size of the hemispheres of the cerebellum and related structures (e.g., the pontine nuclei, the principal nucleus of the inferior olive, a deep cerebellar nucleus called the dentate nucleus), compared with other cerebellar regions and their relatives. This fact directly parallels the high proportion of secondary sensory areas in human cortex, which receive input from corresponding areas of the cerebellum. In other words, there is a mirror-image relationship between the cortex and the cerebellum in the relative size of structures.

• In primates, there is a direct relationship between motor cortex and the motor neurons of the fingers, bypassing pattern generators and effectively lowering the hierarchical level of motor cortex for finger control. This parallels a similar development of forebrain connections in songbirds vs. ducks.

All of these differences (at least the ones found so far) are a matter of degree: variations in absolute and relative size, within and between areas, and variations in relative distance between systems. **There is no evidence that humans have evolved new neuronal types and/or new forms of neural circuitry, new layers, new neurochemicals, or new areas that have no homologue in the primate brain** (Deacon, 1990; Galaburda, 1994; Sereno, 1990). This conclusion might be less distressing to the notion of human uniqueness if it could be shown that the details of cortical wiring are inherited, in species-specific patterns. However, the evidence that we have reviewed throughout this chapter leaves little room for this kind of representational nativism. So we must seek our human uniqueness in the way that these basic building blocks are configured, their quantity, and/or the order and speed in which they unfold over time—which brings us to what is currently known about the timetable of brain development in humans.

Development of the human brain[4]

In this section, we will concentrate on how the human developmental timetable differs from the basic mammalian developmental schedule outlined earlier. But first we must admit to two problems that plague research on human brain development.

---

4. This section is adapted from passages in Bates, Thal and Janowsky, 1992.

First, the best available post-mortem neuroanatomical studies of human brains are based on relatively small samples within any given age, across a 0–90 year age range (e.g., Conel, 1939-1963; Huttenlocher, 1979; Huttenlocher & de Courten, 1987). Furthermore, those children who come to autopsy have often suffered from trauma or diseases that complicate generalizations to normal brain development. *In vivo* studies of adults using magnetic resonance imaging (Courchesne et al., 1987; Jernigan, Press & Hesselink, 1990) and positron emission tomography in infants and adults (Chugani, Phelps & Mazziotta, 1987; Petersen et al., 1988) hold considerable promise for the future (see also Simonds & Scheibel, 1989). But these techniques are expensive, and are usually restricted to children with clinical symptoms that justify neural imaging. Hence generalizations to normal brain development must be made with caution in these cases as well.

Second, because there are so few neuroanatomical data on humans, our current understanding of the principles that govern brain development are heavily based on studies of infrahuman species (e.g., Rakic et al., 1986). To generalize from animal models to human brain development, we have to make three kinds of assumptions:

1. We must assume that the sequence of major events in neural development is the same in humans and animals. This assumption appears to be well founded.

2. To compare specific phases of development in humans and animals, we must extrapolate from the shorter lifespans of another species into the longer human time frame. For example, the burst in synaptogenesis observed in 2-month-old monkeys presumably does not correspond to a 2-month burst in the human child; if we adjust for overall differences in rate of development and length of lifespan, a 2-month date in monkeys corresponds more closely to events that happen between 8–10 months in human children. These extrapolations are plausible. However, Gould (1977) and others have pointed out that evolutionary change often results from adjustments in the *relative* timing of *particular* events within

an animal's life span, as opposed to a general slowing of the developmental sequence. If this has been the case in hominid evolution, then extrapolations based on lifespan may be misleading.

3. In many cases, we need to assume that the specific brain regions under comparison in human and nonhuman species are structurally and functionally homologous. This assumption has been applied with considerable success in some domains (e.g., visual perception). However, the brain regions responsible for specifically human functions like language are still largely unknown, and may be more broadly distributed than previously believed. Hence it is difficult to know where to look for homologues in an animal model.

With these caveats in mind, our review is organized around the distinction introduced earlier, between progressive (additive) events and regressive (subtractive) events (see Table 5.2).

Additive events. The formation of neurons and their migration to appropriate brain regions takes place almost entirely within the period of prenatal development in the human. Except for a few brain regions that continue to add neurons throughout life (e.g., the olfactory bulbs), most of the neurons that we humans are ever going to have are present and in place by the 7th month of gestation (Rakic, 1975; Rodier, 1980).

Positron emission tomography studies of human infants (Chugani et al., 1987) show an adult-like *distribution* of resting activity within and across brain regions by 9–10 months of age. However, there are substantial differences between regions in the point at which this metabolic activity begins, starting with subcortical structures (active at birth), followed by occipital, temporal and parietal regions (which increase most rapidly between 3–6 months), with the most rapid increase in frontal lobe activity between 8–10 months. The functional significance of this 8–10 months shift in frontal metabolism is best considered together with evidence for changes in long-range connectivity that occur around the same time. Although cell formation and migration are complete at birth, neuroanatomical studies suggest that the long-range (axonal) connections among major brain regions are not complete until approximately 9 months of age (Conel, 1939-1963). Although these long-

TABLE 5.2

Age	Neural events	Linguistic and cognitive events
birth[a]	Completion of cell formation and cell migration	Establishment of a left-hemisphere bias for some speech and nonspeech stimuli
8–9 months	Establishment of long-range connections among major regions of cortex (including frontal and association cortex) Establishment of adult-like distributions of metabolic activity between regions	Word comprehension; suppression of nonnative speech contrasts; intentional communication by sounds & gesture; imitation of novel actions and sounds; changes in categorization and memory
16–24 months	Rapid acceleration in number of synapses within and across regions of cortex	Rapid acceleration in vocabulary & onset of word combinations, followed by a rapid acceleration in the development of grammar Concomitant increases in categorization, symbolic play & several other nonlinguistic domains
48 months	Peak in overall level of brain metabolism	Most grammatical structures have been acquired; period of stabilization and automatization begins
4 years-adolescence	Slow monotonic decline in synaptic density and overall levels of brain metabolism	Slow increase in the 'accessibility' of complex grammatical forms Slow decrease in capacity for second-language learning and recovery from aphasia

a. We arbitrarily start from birth, although of course there are both many neural events and considerable learning that take place during the latter part of fetal life.

range connections will not reach adult levels of myelination for a considerable period of time (see below), they are still probably capable of transmitting information.

Putting the metabolic and neuroanatomical evidence together, Bates, Thal and Janowsky (1992) suggest that the establishment of functional connections among active brain regions between 8 and 10 months of age in humans may provide a neuroanatomical basis for the large set of behavioral events that take place in the same time window. For example, establishment of the so-called "executive functions" of frontal cortex may play a role in the marked changes that take place at 9 months in tool use, intentional communication, imitation and hidden object retrieval (Diamond, 1988; Welsh & Pennington, 1988). These brain areas integrate information across modalities and may organize it within a temporal or spatial framework (Janowsky, Shimamura, & Squire, 1989; Milner, Petrides, & Smith, 1985; Shimamura, Janowsky, & Squire, 1990). Connections between association cortices and primary sensory areas may also contribute to cognitive changes that take place at this point, including the ability to learn categories through arbitrary correlations of features (e.g., Younger & Cohen, 1982), the suppression of sound contrasts that are not in the child's language input (Werker & Tees, 1984), the ability to recognize language-specific acoustic-articulatory patterns (at segmental, consonant-vowel level and at the suprasegmental or intonational level), and the ability to understand words (mapping of complex speech patterns onto the objects and events for which they stand). In short, a host of correlated changes in language, communication and cognition may depend upon the changes in regional connectivity and activity that are known to take place around 9 months of age in our species.

It is hard to find dramatic additive events in brain development that could explain the large-scale behavioral changes that take place after this 8–10 month watershed. One candidate that has been invoked often in the past is myelination. Myelination refers to an increase in the fatty sheath that surrounds neuronal pathways, a process that increases the efficiency of information transmission. In the central nervous system, sensory areas tend to myelinate earlier than motor areas. Intracortical association areas are known to myelinate last, and continue to myelinate into the second decade of life. Because myelination does continue for many years after birth, there has been a great deal of speculation about its role in behavioral development (Parmelee & Sigman, 1983; Volpe, 1987). However, interest in the causal role of myelination has begun to wane in the last few years, for two reasons. First, myelination brackets stages in behavioral development only in the most global sense, i.e., it takes place somewhere between 0–18 years. Because myelination is such a protracted process, is provides no clear-cut transitions that might provide a basis for major reorganizations in the behavioral domain. Second, we know that undermyelinated connections in the young human brain are still capable of transmitting information; additions to the myelin sheath may increase efficiency, but it is unlikely that they are the primary causal factor in brain organization of language or any other higher cognitive process.

This leads us to consider synaptogenesis, an event that occurs within the critical time window for early language development and associated cognitive functions. Between 9 and 24 months of age in human infants, the density of short-range synaptic connections within and across cerebral cortex reaches approximately 150% of the levels observed in the human adult. Although, as we discussed earlier, new synaptic branching can occur locally throughout life, total synaptic density begins to fall off after 24 months, going down to adult levels some time in the second decade. Because the most intense period of synaptogenesis corresponds closely to the milestones that define early language and cognition, it is reasonable to suppose that this neural event plays an important role in the mediation of behavioral change (Bates et al., 1992).

Unfortunately, it is difficult to determine exactly when this occurs. Although most experts agree that the period of synaptic exuberance is a major event in postnatal development, there is some controversy concerning its timing within and across brain regions. Based on evidence from human samples, Huttenlocher and colleagues report that the "synaptic peak" is reached at different points in different regions of cortex (Huttenlocher, 1990; Huttenlocher et al., 1982). For example, in striate cortex (a primary sensory area), synaptogenesis begins relatively early (around birth). Furthermore, this process has a slightly different timecourse within different layers of visual cortex. For instance, layers I–IVd reach maximum synaptic density by 4 months, layer V peaks at 11 months, and layer VI reaches its maximum density around 19–20 months (Huttenlocher, 1990; Huttenlocher & de Courten, 1987). Finally, Huttenlocher reports a relatively late peak in synaptogenesis in frontal cortex, around 24 months of age. If these estimates are correct, the timing of synaptogenesis in our species brackets the major milestones of language development from 9 months (the onset of word comprehension) to 24 months (the onset of grammaticization). And of course this means that synaptogenesis brackets all the other dramatic changes that take place in cognition and emotion within this age range (e.g., the ability to sort objects into categories, increases in the capacity to recall events, the emerging concept of self—Bates, 1990; Kagan, 1981).

A different view based on synaptogenesis in rhesus monkeys is offered by Rakic et al. (1986). Rakic et al. report that all areas of cortex appear to hit a peak in synaptic density around the same time— around 2–4 months in the monkey, corresponding (roughly) to 7–12 months in the human child. Contrary to Huttenlocher's findings, this suggests that there may be a common genetic signal to increase connectivity across all brain regions, simultaneously, regardless of their current maturational state. A sudden event of this kind stands in marked contrast to known region-by-region differences in the time course of cell formation, migration, myelination and metabolism (Conel, 1939-1963; Yakovlev & Lecours, 1967). Since Huttenlocher's work is based directly on human brains, and region-by-region differences in the timing of synaptogenesis are more compatible with the region-by-region timetables that have been observed for other neurodevelopmental events in the human, students of human development tend to prefer Huttenlocher's view to the Rakic proposal (e.g., Bates et al., 1992). But this is, of course, an empirical question.

As we noted earlier, Chugani's positron emission studies show a step-wise progression in the development of metabolic activity across the first year, with an adult-like distribution of regional activity (though not overall absolute activity) established by 10 months of age. However, recent evidence suggests that there are further developments in brain metabolism through the next few years of life. In particular, cortical regions show an extraordinary increase in metabolic activity (i.e., hypermetabolism) that peaks between 24–48 months of age. This increase parallels the burst in synaptogenesis described above, although it appears (in our view) to lag behind the peak in synaptic density by approximately two years. Chugani's *in vivo* metabolic work is illuminating for two reasons. First, his estimates of region-by-region metabolic activity are more compatible with Huttenlocher's timetable of synaptogenesis from primary sensory areas (visual cortex) to frontal association cortex. Second, there are striking similarities in the shape of the growth curves reported for brain structure (synaptogenesis) and brain function (metabolism), despite the two-year lag.[5]

Putting together the neuroanatomical and metabolic data at this level of development, we conclude that structural exuberance reaches a maximum between 9 and 24 months (most likely at different times in different regions), with a peak in total metabolic activity (functional exuberance) approximately two years later. Figure 5.8 illustrates these successive peaks, based on the Huttenlocher and Chugani data for human children (from Bates et al., 1992).

In addition to the caveats raised by the Huttenlocher/Rakic debate, there are behavioral reasons to interpret these attractive correlations with caution. Before we conclude that changes in neural capacity cause the well-known bursts in language and other cognitive domains, we must remember that almost all domains of skill development involve bursts and plateaux (e.g., learning to play ten-

---

5. Although we are impressed with the parallels between Chugani's findings and those proposed by Huttenlocher and colleagues, Chugani himself (personal communication, August, 1992) is convinced that his findings are more compatible with Rakic's proposals regarding a simultaneous burst of synaptogenesis across all brain regions. In fact, in view of the relatively small number of data points represented in Chugani's sample and in the sample studied by Huttenlocher et al., 1982, it is difficult to know whether the corresponding peaks in synaptogenesis and metabolism represent one event or a cascade of closely timed events.

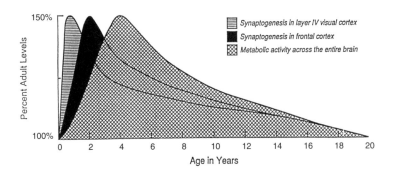

FIGURE 5.8 Relationship between age and various aspects of brain development (synaptogenesis in layer IV visual cortex and in frontal cortex, and metabolic activity across the entire brain). From Bates et al. (1992).

nis). Nonlinear learning curves can be observed at every stage of development, throughout the lifetime (see Chapter 4). At most, we might argue that the marked increases in neural connectivity and capacity that we see from 3–30 months *enable* language and cognitive development, permitting each successive skill to develop well enough to reach some critical mass. Under this interpretation, "bursts" in behavior may bear a relatively indirect relationship to "bursts" in brain capacity, a relationship that is mediated by experience. In fact, mastery or automatization of a skill system may actually result in a cutting back or sculpting of the neural substrate for that system—which brings us to a consideration of regressive events in human brain development.

Regressive events. With the possible exception of myelination, each of the additive events that we have just described is complemented by a subtractive process: Cell formation and migration are followed by cell death; axonal projections are followed by a phase of axon retraction; synaptogenesis is followed by a lengthy phase of synaptic degeneration; and bursts in metabolic activity are followed by a slow decline in overall levels of brain metabolism. There is substantial overlap in all these processes, and substantial interaction as well. Some forms of additivity continue through the lifespan (e.g., local synaptic branching). And some events (e.g., cell death) help to bring about others (e.g., axon retraction).[6] Overall, however, the picture that emerges from recent research on human brain development is one characterized by a particularly marked *overproduction* (a relatively sudden process), followed by *selective elimination* (a relatively slow process that spans a much longer period of time).

Although there is some disagreement about the timing of synaptogenesis in the human brain, Huttenlocher, Rakic and others do agree that the number of synapses declines markedly after this period of exuberant growth. In humans, synaptic degeneration continues for many years, and adult numbers are not reached until somewhere between 11 and 16 years of age in the cortex (Huttenlocher, 1990).

As noted earlier, Chugani and his colleagues also report a sharp rise in overall resting brain metabolism after the first year of life, with a functional peak approximately 150% above adult levels achieved somewhere around 4–5 years of age, i.e., roughly two years after the structural peak in synaptic density. After this point, overall levels of brain metabolism begin to decline, reaching the adult steady state somewhere in the second decade of life (see Figure 5.9). This metabolic decline may be a direct result of the structural reductions described above (i.e., elimination of neurons, axons and synaptic branches). However, it may also be the case that less metabolic activity is required after 4 years of age. Holding structure constant, the same activity may require less "mental energy" (and correspondingly less glucose metabolism) once a certain level of skill has been attained. A proposal of this sort could account for the

---

6. Axon retraction is a natural consequence of cell death; but cell death is not the only cause of axon retraction. Several studies have found that a single neuron may put forward more than one axon during development; the "extra" axons are retracted if and when an inappropriate or unsuccessful connection is made. For example, Innocenti and Clarke (1984) report that neurons which normally make a single axonal connection in adulthood put forward additional axonal branches during cortical development; these axons were retracted at a later point (see also Ivy, Akers, & Killackey, 1979).

two-year lag between structural and functional density illustrated in Figure 5.9.[7]

This period of hypermetabolism and subsequent decline may have important implications for recovery of cognitive functions after early brain injury. For example, Chugani et al. (1987) describe positron emission tomography results in a case of early hemispherectomy; they report that the remaining hemisphere was hypermetabolic after surgery, a hypermetabolism that remained at follow-up several years later. There are a number of explanations for this finding. It may be that the two hemispheres normally impose an inhibitory influence on each other, and that the hypermetabolism of the remaining hemisphere is due to a release from this inhibition. On the other hand, it may be that the hypermetabolism is due to an increase in neuronal firing as a compensation for having only one hemisphere. In other words, the remaining hemisphere may be literally forced to work harder, with a corresponding increase in the amount of energy consumed.

Putting these findings together, we suggest that the burst in brain metabolism observed between 2–4 years of age does indeed index a burst in learning activity, exploiting the huge increase in new connections that occurred shortly before. Conversely, the slow decrease in metabolism between 4–20 years of life may well index a reduction in activity corresponding to the reduction in synaptic connections observed in neuroanatomical studies. What are the implications of these subtractive events for human cognitive development? Changeux, Rakic and many others have underscored the importance of competition and functionally dependent survival as a general principle of brain development. Additive events set spatial and temporal boundaries on experience, i.e., they determine the range of possible outcomes, and the point at which certain forms of learning can begin. At the same time, the overabundance of cells, axons and synapses that has been observed in recent studies can be viewed as the raw material for development through competition— including competition from the cell next door, competition between

---

7. Of course, there are many other possible interpretations of these data, such as that the older children are less aroused by the scanning procedure.

regions, and competition among alternative solutions to the problems the child encounters in her physical and social world. Subtractive events (particularly synaptic degeneration) can be construed as the processes by which experience "sculpts" the raw material into its final configuration. Thus human brain development may represent a particularly strong example of the intricate codependance of maturation and experience.

The interaction between additive and subtractive events may help to explain results from recent studies of brain development in late childhood using magnetic resonance imaging (Jernigan & Tallal, 1990; Jernigan, Trauner et al., 1991). The authors did find small but significant increases in brain volume from 8–25 years of age, located primarily in the superior regions, especially superior frontal cortex. There were also significant increases in the cerebellar vermis, and in the ratio of neocerebellar to paleocerebellar structures.[8] Controlling for overall cranial volume, analysis of subcortical structures showed a significant *decrease* in the size of the basal ganglia and the posterior diencephalon, and a significant *increase* in the anterior diencephalon (which includes the hypothalamic nuclei and the septum). Based on animal studies looking at the effects of gonadal steroids on brain structure, Jernigan et al. speculate that this anterior diencephalic increase may reflect a specific effect of hormonal changes at puberty. Most interesting for our purposes here, the authors also report striking changes in the grey matter to white matter ratio in the cortex, with a substantial *thinning* of the cortical mantle between eight years of age and adulthood. They suggest that this thinning of the cortex is the product of the regres-

---

8. Interestingly, the ratio of neocerebellar to paleocerebellar structures appears to be abnormally high in children with Williams syndrome (Bellugi, Wang, & Jernigan, 1994; Jernigan & Bellugi, 1990; Jernigan, Bellugi, & Hesselink, 1989; Jernigan et al., 1993; Jernigan, Hesselink et al., 1991; Jernigan & Tallal, 1990) and abnormally low in children with autism (Courchesne, 1991; Courchesne et al., 1988). Given the alternative behavioral profiles displayed in Williams syndrome vs. autism, this raises the possibility that these cerebellar structures may play a much larger role in cognitive and social development than previously believed.

sive events that we have reviewed here, and represents a progressive sculpting of the brain by experience.

All of the additive and subtractive events just described for normal human brain development must be weighed against a growing literature on individual differences within the normal range. As increasingly sophisticated brain-imaging techniques are developed, it becomes increasingly evident that there is considerable variation in structure and function in normal adult subjects. For example, Gazzaniga and his colleagues (Oppenheim et al., 1989; Thomas et al., 1990; Tramo et al., 1994) reconstructed the cortical areas of two identical twins from MRI scans. Even in the case of genetically identical individuals, the variation in cortical areas was striking, with the occipital lobe occupying 13–17% of cortical area in one individual, and 20% in the other. Differences between individuals in brain structure also extend to brain functioning. For example, using a recently developed functional MRI technique, Schneider and colleagues studied the areas of activation following upper or lower visual field stimulation. While it had classically been assumed that the upper and lower visual field mapped on to the regions above and below the sulcus, there is in fact a lot of variation with some normal subjects showing an upper/lower visual field cut that straddles this structure (Schneider, Noll, & Cohen, 1993). And of course, these new forms of evidence for variability join an old literature on individual differences in handedness and hemispheric organization for language (e.g., Hellige, 1993; Kinsbourne & Hiscock, 1983).

In view of all this variability in normal adults, our efforts to construct a timetable for "normal" brain development in humans must be interpreted with caution. There is variability in the form of organization that is finally reached, and there is also ample variability in rate of development en route to these different end points (e.g., Bates, Dale, & Thal, 1995). With these caveats in mind regarding variability in the "default" plan, let us now turn to evidence on development and (re)organization under pathological conditions.

Plasticity and (re)organization in the human brain

Evidence for plasticity and alternative forms of organization comes from two major sources: experiments of nature that deprive individuals from normal sources of input (e.g., blindness, deafness, amputation), and focal injuries that destroy tissue that would normally play a critical role in language, spatial cognition and other higher cognitive functions. Let us consider each of these in turn, and then close this section with a few remarks on populations that fail to show the same degree of plasticity and recovery (e.g., mental retardation, specific language impairment).

Brain organization and sensory deprivation. Two lines of evidence for inter- and intrahemispheric plasticity in humans come from research on the congenitally deaf.

First, Neville and Lawson (1987) have shown that congenitally deaf adults display unusual patterns of brain activity over auditory cortex in response to stimuli in the visual periphery. This includes a bilateral expansion of event-related scalp potentials from occipital to auditory sites. To explain these findings, Neville (1990, 1991) points out that the young brain contains exuberant connections from the visual periphery to auditory cortex. In the normal course of brain development, these exuberant polysensory axons are eliminated. However, in the absence of input from the auditory system, exuberant connections may be retained (see also Frost, 1990). The bilateral occipital-to-auditory shift is present in deaf adults with and without experience in a visual-manual language (i.e., American Sign Language, or ASL). In those subjects who are fluent in a signed language (including hearing children of deaf parents), Neville and Lawson report that stimuli in the visual periphery also produce a larger response over the left hemisphere. To explain this particular aspect of their findings, they point out that fluent signers usually fixate on the speaker's face, using the visual periphery to monitor the rapid movements of the signer's hands. Hence the left-right asymmetry in response to peripheral visual stimuli may reflect a lifetime of signing experience.

A second line of evidence for plasticity in the face of sensory deprivation comes from studies of sign language aphasia in the congenitally deaf (Bellugi & Hickok, in press; Poizner, Klima, & Bellugi, 1987). It is now clear from this research that sign aphasias are correlated with injury to the left hemisphere, which implies that the left-hemisphere bias for language in our species is largely independent of modality. However, more detailed neuroanatomical studies of the left-hemisphere lesions that produce sign aphasia suggest that modality does have an affect on intrahemispheric organization for language (Klima, Kritchevsky, & Hickok, 1993). For example, these authors report cases of sign aphasia following injuries to dorsal-parietal areas of the left hemisphere—injuries that usually have little or no effect on spoken-language abilities in a hearing adult. In fact, this intrahemispheric pattern makes good anatomical sense, because these parietal areas include connections between somatosensory and visual cortex that ought to be involved in rapid and efficient use of a visual-manual language. It remains to be seen whether sign aphasias also require injury to the classic perisylvian areas of the left hemisphere, or whether injury to these sign-specific areas is sufficient. Indeed, given the surprising amount of variability in lesion-syndrome mapping that has been reported for hearing adults (Basso et al., 1980; Damasio & Damasio, 1992; Dronkers et al., 1992; Goodglass, 1993), it may be a long time before we understand the intrahemispheric circuitry responsible for signed vs. spoken language. But it does seem clear that early sensory deprivation and/or sign language experience can result in large-scale reorganization of the two hemispheres—in line with the animal model of Webster et al. that we described earlier.

All of the studies that we have reviewed so far pertain to individuals with auditory deprivation. Although there is (as far as we have been able to determine) much less information on the nature of brain development in congenitally blind humans, the little evidence that we have been able to find complements findings on the deaf. Wanet-Defalque et al. (1988) conducted a PET study of metabolic activity in blind human subjects. The scans were taken during a quiet resting state, with no specific task or form of stimulation other than the ambient sounds inside the room. However, their results showed a high degree of metabolic activity in the visual cortex of congenitally blind individuals, compatible with the idea that visual cortex has been recruited to serve other functions. A later PET study by Veraart et al. (1990) showed that this effect was diminished in individuals with late-onset blindness, suggesting that the metabolic activity observed in the congenitally blind is due to reorganization of cortex during a period of maximal plasticity. Related findings have been reported using event-related brain potentials to study sound location (Kujala et al., 1992).

There are also studies of cortical reorganization in human beings who have suffered amputation of a limb, suggesting that there is a marked degree of reorganization in somatosensory cortex—even for adults who underwent limb amputation relatively late in life (Ramachandran, 1993; Ramachandran, Rogers-Ramachandran, & Stewart, 1992). It is known, for example, that the face and the arm are mapped onto adjacent regions of somatosensory cortex. Adults who have lost all or part of an arm experience a "phantom limb" reaction to stimulation of specific areas on the surface of the face. Indeed, some individuals appear to have a detailed map of the hand laid out on the surface of the face. These findings suggest that the areas of sensorimotor cortex that receive input from the face now control the adjacent cortex that once subserved the arm (which now receives little or no afferent input).

We may conclude that evidence for plasticity following sensory deprivation is at least as great in humans as it is in laboratory animals with analogous (albeit more systematic) forms of sensory deprivation. This brings us to the study of humans with focal brain injury, experiments of nature that parallel the animal lesion studies reviewed earlier.

Brain organization after early focal brain injury. It has been known for some time that the human child can recover higher cognitive functions (e.g., language, spatial cognition) following acquired injuries that would lead to irreversible deficits in an adult (Hecaen, 1983; Satz, Strauss, & Whitaker, 1990; Woods & Teuber, 1978). These findings for children with focal brain injury contrast markedly with the slow development and persistent deficits that are associated with many congenital disorders (e.g., mental retardation, and/or certain congenital variants of specific language impairment). Hence there seems to be substantial plasticity for language and other higher cognitive functions in children with focal brain injury. This plasticity is apparently not available later on (at least not to the same degree), and it is not available in all forms of neurological impairment during childhood.

In fact, plasticity has been a controversial issue even within the literature on recovery from early focal brain injury. As reviewed by Satz et al. (1990), the theoretical pendulum has swung back and forth between two extreme views: *equipotentiality* (i.e., the human brain is completely plastic for higher cognitive functions, with no initial biases) and *irreversible determinism* (i.e., regional specification for higher cognitive functions is already established at birth, and any alternative form of organization following early brain injury will be grossly inadequate). The equipotentiality view received support from Lashley's famous experiments on maze learning in lesioned rats (Lashley, 1950), and it finds its high-water mark in Lenneberg's classic book *Biological foundations of language* (Lenneberg, 1967; see Basser, 1962 for a first-hand accounting of the data on which many of Lenneberg's conclusions were based). The equipotentiality view fell into disrepute in the 1970's, as evidence accumulated from studies of children with hemispherectomies suggesting that left-hemisphere removal always leads to selective deficits in language, especially for syntactic and phonological tasks (Dennis & Whitaker, 1976). Similar results were also reported for children with early focal brain injury, usually due to early strokes (Aram, 1988; Aram et al., 1985; Aram, Ekelman, & Whitaker, 1986, 1987; Riva & Cazzaniga, 1986; Riva et al., 1986; Vargha-Khadem et al., 1991, 1992; Vargha-Khadem, O'Gorman, & Watters, 1985; Vargha-Khadem & Polkey, 1992). These findings were compatible with studies of normal infants showing a left-hemisphere bias at birth in processing speech and other complex sounds (Molfese, 1989, 1990; Molfese & Segalowitz, 1988), and led many researchers to the conclusion that functional asymmetries for language in the human brain are established at birth, and cannot be reversed (Fletcher, 1994; Isaacson, 1975).

Most of the original sources claiming left/right differences following early unilateral damage (e.g., Dennis & Whitaker, 1976) are quite clear in stating that these differences are very subtle. That is, children with early left-hemisphere damage are not aphasic. Indeed, some of these effects are so subtle that their statistical reliability has been questioned (see Bishop, 1983, for a review). When correlations between language and side of lesion do appear, they usually are greater in children with left-hemisphere damage, supporting the idea that left-hemisphere tissue is particularly well suited to language (Aram, 1988; Aram et al., 1985, 1986, 1987; Riva & Cazzaniga, 1986; Riva et al., 1986; Vargha-Khadem et al., 1991; Vargha-Khadem, O'Gorman, & Watters, 1985). However, most of the children with left-hemisphere injury that have been studied to date are still well within the normal range (Aram & Eisele, 1992; Stiles & Thal, 1993; Vargha-Khadem & Polkey, 1992), and those that fall outside the normal range often have complications in addition to their focal brain injury (e.g., seizure conditions—Vargha-Khadem & Polkey, 1992).

So we may conclude that the brain is highly plastic for linguistic functions, although there are some clear constraints. Is this true for all higher cognitive functions? Are there differences between domains (i.e., language, spatial cognition, affect) in the degree and nature of this plasticity? And exactly how and when are these alternative forms of organization achieved? To answer these questions, Stiles and her colleagues are studying language and spatial cognition in children who have suffered a single unilateral injury to either the right or the left hemisphere, prenatally or before six months of life, confirmed by at least one radiological technique (Bates, Thal et al., 1994; Reilly, Bates, & Marchman, in press; Stiles & Thal, 1993; Thal, Marchman et al., 1991). The idea behind this project is to study the initial biases and subsequent (re)organization of brain and behavior *prospectively,* as they occur, in contrast with the retrospective studies that have characterized most research with this population. Although the findings produced by this group are quite complex, they contain some major surprises, suggesting a sharp difference in the principles that govern the development of spatial cognition and language. In particular, deficits in the development of spatial cognition look very much like the deficits observed in adults with left- vs. right-hemisphere injury, albeit more subtle and transient. By contrast, deficits in the development of language do not map onto the brain-behavior correlations observed in the adult aphasia literature, suggesting that early language learning and later language use draw on different brain systems.

Starting with results for spatial cognition, Stiles and colleagues have looked at the development of several different visual-spatial functions, including drawing, pattern analysis and block construction. The deficits revealed in these studies are often more subtle than we see in adults with homologous injury, and at each stage of development they eventually resolve into normal or near-normal levels of performance—only to reappear at the next level of difficulty on each task. Despite these quantitative differences, the visual-spatial deficits displayed by children with early left- vs. right-hemisphere injury are qualitatively similar to the results obtained in adults with homologous lesions.

An example of this last point can be found in Figure 5.9a and Figure 5.9b, which illustrates performance by left- and right-hemisphere-damaged adults and children who are asked to draw hierarchically organized stimuli (e.g., a letter D made up of small Ys) from a model and/or from memory. Figure 5.9a shows that adults with right-hemisphere damage display a deficit in spatial integration, i.e., they are able to reproduce the local, detailed elements in the figure but they fail to integrate those elements into a global pattern. Adults with left-hemisphere damage display a complementary deficit in spatial analysis, i.e., they can reproduce the global pattern but they tend to omit the local details (for reasons that will be apparent later on, we should underscore that this "local detail" problem usually involves individuals with left temporal damage). As Figure 5.9b shows, children with left-hemisphere injury go through a stage that strongly resembles the chronic deficit in spatial analysis displayed by left-hemisphere-damaged adults (as the children are followed over time, this deficit resolves). By contrast, children with right-hemisphere injury seem to perform rather well, compared with right-hemisphere-damaged adults. However, a detailed analysis of their performance yields evidence that the right-hemisphere children are not performing normally either. As Stiles et al. note, they seem to talk their way through the task (e.g., "A Y made of bs,

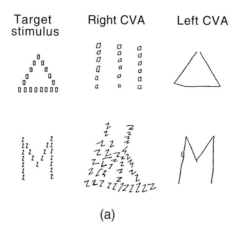

(a)

Memory Reproduction Task

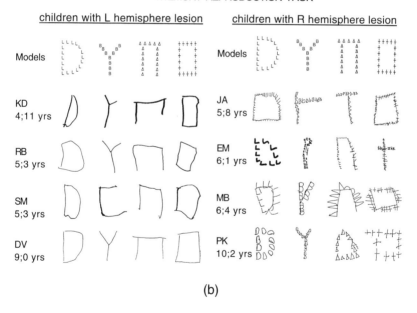

(b)

FIGURE 5.9 Examples of performance by adults (a) and children (b) with left- and right-hemisphere damage on a task which involves drawing hierarchically organized stimuli. The nature of the deficit depends both on site of damage as well as age.

a Y made of bs...."), and as a result, they produce a peculiar combination of upper and lower case characters that are not observed in normal children at any age. Stiles et al. conclude that children with focal brain injury show the same hemisphere-specific deficits in spatial cognition that are typically observed in adults. The patterns observed in children are more subtle, and development (i.e., recovery) does occur. But this recovery may involve strategies that are not observed under normal conditions.

In contrast with these results for spatial cognition, studies of language development in the same population have produced some major surprises. If results for language were qualitatively (if not quantitatively) similar to the patterns observed in adults with homologous injuries, exactly what would we expect? Based on the adult aphasia literature, Bates, Thal et al., 1994 derive three predictions for language delay in children with focal brain injury:

1. **Left-hemisphere specialization**: On virtually all measures of language comprehension and production (including symbolic and communicative gesture as well as language), children with left-hemisphere damage should show more marked delays than children with right-hemisphere damage.

2. **The Wernicke hypothesis:** Delays in language comprehension should be greater in children with injuries involving posterior regions of the left hemisphere (in particular, posterior temporal cortex).

3. **The Broca hypothesis:** Delays in language production should be greater in children with injuries involving left frontal cortex.

Bates, Thal et al. have put these three hypotheses to a test in a study involving 53 infants with early-onset focal brain injury, studied between 10 and 44 months of age (i.e., the period in which normal children make the transition from first words to grammar). All three hypotheses were roundly disconfirmed, as follows.

First, delays in word comprehension and in the use of symbolic and communicative gestures were more likely in children with right-hemisphere lesions. Nothing like this has ever been reported for adults with right-hemisphere injury, although two other studies have reported greater comprehension deficits in children with injuries on the right (Eisele & Aram, 1994; Wulfeck, Trauner, & Tallal, 1991). In fact, none of the children with damage to left temporal cortex (the presumed site of Wernicke's area) showed any sign of a delay in language comprehension!

Second, delays in expressive language (including measures of vocabulary and grammar) were more likely in children with injuries involving left temporal cortex. This result provides partial support for the left-hemisphere specialization hypothesis, but it is a direct reversal of the pattern we would expect if the Broca and Wernicke hypotheses applied to children. There was no evidence at any point in this age range supporting the idea that left frontal cortex (the presumed site of Broca's area) plays a special role in early language production. However, the authors do report a *bilateral* frontal effect in the period between 19 and 31 months. This is a particularly intense period in language development, in which most normal children display a sharp acceleration in vocabulary size, accompanied by the appearance of word combinations and the subsequent emergence of grammatical markers (see Chapters 3 and 4 for details). During this period, children with damage to *either* left frontal *or* right frontal cortex displayed significant delays in vocabulary and grammar (with no evidence for any dissociation between lexical and grammatical development, in contrast with claims that have been made for frontal damage in adults). Bates, Thal et al. offer several possible explanations for this bilateral frontal effect, including possible deficits in motor planning and/or working memory of the sort that are often reported for adults with frontal damage. However, the absence of a left-right asymmetry in frontal cortex is striking, particularly in view of the fact that left-right asymmetry is observed for injuries involving the temporal lobes. It seems that Broca's area may be less important than Wernicke's area (or, at least, less asymmetrical) during the initial stages of language learning, and that the differentiation in adults is the product of learning, not its starting point.

The specific effect of left temporal damage on expressive language continues up to approximately five years of age, in results by

Bates, Thal et al. for free speech and in a separate study of discourse development from 3–12 years of age by Reilly et al. (in press). However, in children tested after 5 years of age, Reilly et al. could not find any effect of lesion side or lesion site. In other words, the specific effects of left temporal damage observed between 1 and 5 years of age are no longer detectable after language is acquired. As a group, older children with a history of early focal brain injury performed significantly below the levels observed for normal controls, but their scores were still within the normal range (i.e., these children are not aphasic). Hence it appears that early brain damage does exact a price, but the price is very small compared to the deficits that are observed in adults with the same kind of focal brain injury. In this respect, results for language development are remarkably similar to those of Webster et al. for visual-spatial functions in infant monkeys with bilateral lesions to temporal area TE (i.e., the presumed site of the "what is it?" system).

Of course the Webster et al. results are based on bilateral lesions, while the human results that we have reviewed here are all based on unilateral damage. One likely explanation for recovery of language following left-hemisphere injury would be a switch to homologous areas in the right hemisphere, i.e., a kind of mirror-image effect. However, Rasmussen and Milner (1977) obtained surprising results from a study of 134 adults with a history of early unilateral brain damage. Because these individuals were candidates for neurosurgery to control seizures (a factor which, of course, complicates interpretation of their results), they were subjected to a sodium amytal procedure (i.e., the Wada test) to determine which hemisphere is lateralized for speech. (In this procedure, a single hemisphere is literally paralyzed for a period of approximately 15 minutes, following an injection to the carotid artery.) Results suggest that speech/language functions were lateralized to the right hemisphere in only 40% of the sample. Another 40% showed clear evidence for left-hemisphere dominance, while the remaining 20% displayed a form of organization in which different language tasks (spontaneous naming vs. repetition) were divided up between the two hemispheres. To be sure, there was a clear trend in their data indicating that the much-touted "shift to the right" was more likely if the patient's lesion included the classic left perisylvian zones (broadly defined to include much of the left frontal, temporal and parietal lobes surrounding the sylvian fissure). But there were clear exceptions to this trend, even with a generous definition of "language zones." In our view, the clearest take-home message from the Rasmussen and Milner study is that several different forms of inter- and intrahemispheric organization for language are possible in response to early focal brain lesion.

To summarize, children with early focal brain injury show a significant degree of plasticity for both language and spatial cognition. However, the deficits observed in spatial cognition are qualitatively similar in children and adults; the deficits observed in language differ markedly from one stage of development to another. To explain these differences, Stiles and Thal (1993) offer two interesting proposals. First, they suggest that plasticity may be greater for phylogenetically recent and species-specific functions like language, compared with visual-spatial functions that have evolved across the vertebrate line. Second, they suggest that the site-specific effects observed in language development from 1–5 years of age may be related to the site-specific effects observed in visual-spatial cognition. In particular, recall that children with left-hemisphere injury display deficits in the perception and production of fine-grained perceptual detail, while children with right-hemisphere injury find it difficult to integrate these details into a coherent whole. Stiles and Thal suggest that the specific disadvantage in expressive language displayed by children with left-hemisphere injury may reflect a deficit in the extraction of perceptual detail in the acoustic domain (see also Bates, Thal et al., 1994). Simply put, the child with left-hemisphere damage (in particular, damage to left temporal cortex) may find it difficult to perceive and store the "little sounds" and "little words" that are crucial for articulate speech.

But why should this *perceptual* effect have its greatest impact on *expressive* language? And why should right-hemisphere injury have a greater impact on word comprehension? According to Stiles and Thal, and to Bates, Thal et al. (1994), the explanation rests on the insight that language *learning* is not the same thing as language *use* in a mature adult. To understand the meaning of a new word, children have to integrate information from many different sources. Of course these sources include acoustic input, but they also include visual information (e.g., the shape and size of a named object), tactile information (what the object feels like in the child's hand, what it does and how it moves), memories of the immediately preceding context, memories of earlier encounters with the same objects and events, inferences about the current context, emotions, goals, feelings of success, failure and desire—in short, a range of experiences that define the initial meaning of a word and refine that meaning over time. To figure out what words mean for the first time, a detailed acoustic analysis may be less important than the ability to integrate information over many different sources. Because information integration draws heavily on processes supported by the right hemisphere, the delays in comprehension observed in some children with right-hemisphere damage may reflect a language-specific variant of the integrative deficit that Stiles and others observe in the visual domain. Conversely, the ability to extract perceptual detail may be more important when children have to turn their word comprehension into speech. For adults who have used their repertoire of words for many years, activation of existing motor templates is an automatic process. For infants who are struggling to create their first motor templates, perceptual analysis may be at least as important as motor control. In other words, the first phases of language production require a great deal of perceptual analysis, and perhaps for this reason, specific delays in expressive language are associated with early damage to left temporal cortex—the same system that supports the analysis, storage and recall of visual form.

Based on these findings, Bates, Thal et al. (1994) have proposed a developmental account of cortical organization for language in normal children, a complex variant of the afferent specification hypothesis proposed by Killackey (1990). Assuming the same taxonomy of innate constraints proposed in the introduction to this chapter (and in Chapter 1), Bates, Thal et al. suggest that the child does not begin with innate representations for language. However, the child does begin with strong computational constraints, and possibly some input/output constraints, that lead (under normal circumstances) to the familiar pattern of left-hemisphere specialization, with distinct roles for different areas within the left hemisphere. On their account, the child starts out with left-right asymmetries that are computational in nature, differences in speed and style of processing that show up across content areas. These asymmetries are initially restricted to the temporal lobes, including a predisposition for the extraction of perceptual detail on the left and a predisposition for integration across inputs on the right. Under default circumstances, the left temporal area "recruits" left frontal cortex into an integrated system that permits the fluent and efficient comprehension and production of speech. As a result of this centralization of language use, there may also be an asymmetrical distribution of the representations that support language. In the presence of early focal brain injury, the initial state is necessarily different, and the same cascade of default events cannot take place. Hence the early course of language development can be delayed in a number of ways, and the organization that finally emerges around 5 years of age is not always optimal. But it is still good enough to complete the task of language learning, and good enough to support language use within the normal range.

When plasticity fails. Why are some clinical groups unable to exploit the reorganizational mechanisms used by children with focal brain injury? An example comes from Thal, Wulfeck, & Reilly (1993), who compared children with focal brain injury to children with Specific Language Impairment of unknown origin (SLI). All children were tested between 4–9 years of age, in a widely used story-telling task called The Frog Story (Berman & Slobin, 1994). Measures included lexical diversity, morphological productivity, syntactic complexity and discourse coherence. On most of these measures, the focal lesion children did lag behind normal controls (cf. Reilly et al., in press), but they were still far ahead of SLI children in the same age range. In other words, language development proceeds less efficiently in children with SLI than it does in children with unilateral injuries involving the classical language zones— even though the SLI children do not display frank neurological symptoms, and show no evidence for cortical or subcortical lesions on magnetic resonance imaging (Jernigan, Trauner et al., 1991).

A similar puzzle emerges in studies of children with mental retardation. For example, older children and adolescents with a form of mental retardation called Williams Syndrome perform better on measures of language than they do on many aspects of nonverbal cognition (especially visual-spatial tasks—Bellugi, Wang, & Jernigan, 1994; Karmiloff-Smith, 1992b; Karmiloff-Smith & Grant, 1993; Wang & Bellugi, 1994). To be sure, most Williams children perform within the range that we would expect for their mental age, on most language measures (i.e., they are not linguistic savants). However, *some* individuals with Williams Syndrome perform *above* their mental age on at least a few language measures (e.g., mean length of utterance; number of words that can be generated that begin with a particular sound—Volterra et al., 1995). By contrast, children with Down Syndrome often perform far worse on many language measures than we would expect for their mental age (i.e., they appear to have a form of specific language impairment superimposed on their mental retardation—Chapman, 1993; Fabbretti et al., 1993; Fowler, 1993; Jernigan & Bellugi, 1994). These contrasting profiles are reminiscent of the dissociations between language and spatial cognition observed in adults with left- vs. right-hemisphere injury. And yet, magnetic resonance-imaging studies of Williams and Down Syndrome yield no evidence for focal brain injury, and no group differences in brain morphology along the left-right axis (Jernigan & Bellugi, 1994). Instead, the two groups differ in the relative size of frontal areas (proportionately larger in Williams), and in the absolute and relative size of cerebellar structures (e.g., disproportionately larger neocerebellar areas in Williams). In other words, profiles that resemble left- vs. right-hemisphere injury can be produced by brain abnormalities along a very different set of axes. In addition, electrophysiological studies of children with Williams Syndrome suggest that their relatively spared language may be achieved with unusual patterns of brain activity that are not observed in normal children (Neville, Mills, & Bellugi, 1994). In other words, Williams Syndrome may not represent a selective sparing of normal language in the face of mental retardation, but rather, a completely new and different mode of brain organization for language.

Taken together, these contrasting populations raise serious problems for the classical story of brain-behavior correlations. Children with injuries to the classical language zones go on to achieve normal (or near normal) language functions, and children with injuries to the left or right hemisphere move beyond their initial deficits to achieve an acceptable (though sometimes unusual) mode of functioning in visual-spatial tasks. Children with Down Syndrome and Williams Syndrome display contrasting dissociations in language and spatial cognition that look very much like the patterns that are often observed in adults with left- vs. right-hemisphere injury—and they do so in the absence of anything resembling a left-right difference in brain structure. And children with specific language impairment fail to develop normal language in the absence of frank brain lesions to either hemisphere. What can we conclude from this puzzling mosaic? Why is it that children with focal brain injury can overcome their deficits and exploit cortical plasticity, while children with Down Syndrome, Williams Syndrome and SLI cannot? And where do the contrasting patterns observed in Down Syndrome and Williams Syndrome come from, if they are not based on left-right contrasts in brain pathology?

We certainly do not have a ready answer to this puzzle, but there are a number of directions that can be explored. One approach has been to explain SLI and other persistent deficits in terms of diffuse cortical abnormalities that are not evident in structural MRI. For example, it has been suggested children with SLI and/or adults with congenital dyslexia have cytoarchitectonic abnormalities in several (possibly bilateral) regions of cortex, with special reference to the magnocellular stream in the visual system and its analogue in auditory cortex (e.g., Galaburda & Livingstone, 1993; Tallal, Sainburg, & Jernigan, 1991). On this argument, healthy cortical tissue cannot take over the functions normally subserved by impaired regions because there simply isn't enough healthy cortical tissue to go around.

Another approach attributes these congenital deficits to injuries outside of the cortex, in the cerebellum and/or various subcortical structures. On this account, healthy cortex cannot play its usual role in language and cognitive development because of blockage or improper gating of the inputs that are crucial for normal learning to take place. This emphasis on the developmental role of lower structures is important for the argument, since injuries to these regions do not cause the same specific deficits in language and/or spatial cognition when they occur in an adult.

Based on findings like these, we suggest that the familiar patterns of regional specialization for language and higher cognitive functions observed in human adults are not present at birth. Instead, they are progressively set up over time through a process of competition and recruitment that takes advantage of regional differences in computing style that are in place very early. Under default circumstances, these regional biases lead to those forms of neural organization that are usually revealed by lesion studies and by functional imaging of the normal brain. But other forms of organization are possible, and are amply attested in research on atypical populations (e.g., focal brain injury, SLI, Williams Syndrome, Down Syndrome).

Conclusion

Many of the facts that we have reviewed in this chapter are compatible with connectionism as it is currently practiced. But much of what we have learned about brain development provides a challenge to connectionist models of learning and change, a set of urgent priorities for future research. We will end this chapter with a brief summary of points where we believe that developmental neurobiology and connectionist theory converge.

A shared definition of knowledge and change

If we define knowledge in terms of fine-grained connectivity at the cortical level (see also Chapter 1), then the additive and subtractive events that we have reviewed here constitute a common basis for learning and change in human beings and in connectionist nets. What we know may be implemented in terms of weighted connections and potential states of activation across the brain. And every time we learn something new, our brains have a slightly different structure. There is no tape room in the brain, no storage bins that are separate from the mechanisms that compute information. The strict distinction between processing and knowledge is blurred. Connectionism offers a framework for explaining knowledge and learning that is compatible with these fundamental facts. Of course the connection machinery involved in the cortex is orders of magnitude more complex than any connectionist network currently under study. This quantitative difference undoubtedly carries important qualitative consequences, and it remains to be seen whether those principles of learning and change that work in a net with 10 to 400 units will scale up to a highly differentiated network with more than 10 billion units. But there are enough architectural and dynamic elements in common to encourage further, more detailed comparisons.

Representational plasticity

Connectionist modeling as currently practiced is consistent with the evidence that has accumulated in the last few years for cortical plasticity, in vertebrates in general and humans in particular. In principle, it should be possible to inherit specific and detailed patterns of cortical connectivity, a possibility that we referred to in Chapter 1 as *representational nativism*. The evidence that we have reviewed here suggests that the cerebral cortex is not built this way. There is some evidence for primitive innate representations in the midbrain (e.g., for very simple visual stimuli (Johnson & Morton, 1991)), but the cortex appears to be an organ of plasticity, a self-organizing and experience-sensitive network of representations that emerge progressively across the course of development. To understand what makes humans different from other species, we need to look at other forms of innateness, including *architectural constraints*, (including, for example, differences in unit, local, and global architectures), and *chronotopic constraints* (i.e., variations in the timing and topography of growth). In this respect, research on learning in artificial neural networks constitutes an important tool for exploring the development of structure and function in living brains.

The importance of noise

In human beings, the first years of learning and development take place in a nervous system that is noisy and unstable. The long-range axonal connections are not fully myelinated until the second decade of life, which insure a certain degree of cross-talk and static. There are also huge changes in the raw number of available units and connections, due to a combination of additive processes (in particular, the genetically timed burst in synaptogenesis that takes place between 0–4 years in human beings) and subtractive processes (cell death, axon retraction, and synaptic elimination). Although we know that some of these processes occur as a result of learning (e.g., strengthening of connections through Hebbian learning, experience-induced sprouting of dendritic branches at a local level, and elimination of connections and units through competitive learning), much of the change that takes place in the first few years of life is not contingent on experience. The noise and instability of early learning is good news for connectionism. In contrast with previous generations of research on machine learning (where noise is always a problem), noise can be a real advantage of learning in a non-linear neural network, because it protects the system from committing itself too quickly, i.e., from falling into local minima (partial, local solutions that can prevent further and more complete learning). The utility of noise and other developmental limitations on processing are explored in more detail in the next chapter.

These similarities between real and artificial neural development are seductive, but it has been argued that they are too superficial to hold up under further scrutiny. What will happen when we start to implement neural development in more realistic detail? This is an empirical question, and until that question has been addressed properly, rhetoric on either side of the issue is not going to take us very far. It is worth noting, however, that substantial progress has been made in the physical sciences by modeling complex phenomena in simpler systems (e.g., studies of ocean flow over topography in a small laboratory tank, or simulations of atmospheric turbulence under the stupefying assumption that the atmosphere over the earth is two-dimensional rather than three-dimensional). The crucial question is always: Does the simplified, idealized model system have properties that will generalize to the same problem at a larger scale? We think that a reasonable case has been made for a compelling similarity between human brain development and learning in a dynamic, developing neural network with the properties that we have outlined in this volume.

Interactions, all the way down

The developmental process is one of the most amazing mysteries in the universe. Each of us begins life as a single cell—a fertilized egg—and then proceeds to develop into an organism made up of more than 100 trillion cells, arranged in a very special manner. Where does the information that guides this process come from? Why do some cells develop into humans and others into turtles? Why do all humans so closely resemble each other?

The tempting answer is that this information must somehow all be contained in that initial cell. After all, if the information is not in the cell, where else could it be? But what does this mean, more precisely?

Early developmental theorists supposed that the initial cell contained a blueprint and this blueprint was what determined the final body plan (*bauplan*). No one has ever succeeded in locating such a blueprint. And on reflection, this hypothesis runs into problems on logical grounds as well. It does not seem possible that all the information needed to construct a body could fit in a single cell, at least not if it takes the form of a blueprint. (One estimate is that the human body contains about 5×10^{28} bits of information in its molecular arrangement, but the human genome only contains about 10^5 bits of information; Calow, 1976). Moreover, development is a process which is stochastic as much as it is deterministic, and this is hard to account for from the blueprint view. When we say that a form or a behavior is innate we really mean that, given normal developmental experiences, it is a highly probable outcome.

So we return to the initial question: If the information is not in that initial cell, where else could it be? The fact that it may not be possible to squeeze all the information needed to build a body into the genome does not mean that the information does not exist—

somewhere. In Chapter 3 we discussed the problem of why beehive cells have hexagonal shapes. We pointed out that the shape of the cell is the inevitable result of a packing problem: How to pack the most honey into the hive space, minimizing the use of wax (see Figure 3.1). The information specifying the shape of the cell does not lie in the bee's head, in any conventional sense; nor does it lie solely in the laws of physics (and geometry). Rather, the shape of the cell arises out of the interaction of these two factors. This example not only makes the point that the information is distributed over several sources; it also gives us a hint about how Nature has solved the problem of building complex bodies with minimal information: She cheats. The bee's genome needs only to specify part of a behavior (packing), relying on external geometric constraints to produce the final, more complex outcome.

We cannot emphasize enough that we do not question the powerful role that is played by our biological inheritance. What we question is the *content* and the *mechanism* of that inheritance. In the terminology we used in Chapter 1, this means (1) How domain- and representation-specific is innate knowledge, and (2) How direct is the link between gene and behavior? The strong tendency in cognitive science has been to assume that innate knowledge is highly domain- and representation-specific; and that the connection between gene(s) and behavior is direct. But as we pointed out in Chapter 1, although genes may sometimes produce morphological and behavioral outcomes with near certainty, those outcomes still remain probabilistic; and the behaviors are almost always opaque with respect to the enzymes and proteins which are the immediate products of genes. The outcomes also typically involve complex interactions, including molecular and cellular interactions at the lowest level, interactions between brain systems at the intermediate level, and interactions between the whole individual and the environment at the highest level.

Thus, our major thesis is that while biological forces play a crucial role in determining our behavior, those forces cannot be understood in isolation from the environment in which they are played out. This is true whether we are looking at genes or cells (in which case the environment usually consists of other genes or cells) or individuals (in which case the environment may include other individuals). The interaction between biology and environment is so profound that it is misleading to try to quantify the percentage of behavior attributable to both. Moreover, because evolution is opportunistic (or, as evolutionary biologists would prefer to say it, conservative), the nature of the interactions is rarely straightforward.

In this chapter we consider a range of cases which illustrate the sort of interactionist account we are trying to develop. We choose cases at various levels of organization. We begin with genes because they are, after all, the stuff of which innateness is made. We conclude with language, since this is a prime example of cognitive behavior. In each instance, we shall try to show that a complete understanding of the phenomenon can only be achieved by understanding the complex web of interactions which are necessary for producing the phenomenon. In our view, the role of the genetic component—call it the innate component, if you will—is to orchestrate these interactions in such a way as to achieve the desired outcome.

There is a second goal to this chapter. This book is about development, but so far we have considered only briefly (in Chapter 3) the most basic question of all that one might ask, which is why development occurs in the first place. It is not immediately obvious that development plays any positive adaptive role in an individual's life. After all, some species (precocial) are born more or less fully developed and hit the ground running. Given the vulnerability of being in a developmentally immature state and the enormous societal and parental resources which are devoted to rearing and acculturating the human young, a prolonged period of development would seem to be highly maladaptive. Why, then, do species such as ours allow development to go on for so long? In fact, we shall see that in many of the examples of interaction, the developmental process is critical. That is, development is not simply a phase "to be gotten through"; it is rather the stage on which these interactions are played out. To continue the metaphor, genes are the conductor who orchestrates these interactions.

The chapter is organized as follows. We have already discussed (in Chapter 1) the sorts of interactions which are typical of gene expression and cell differentiation. So in this chapter we concentrate on higher-level interactions. We begin with interactions between brain systems, using imprinting in chicks as an example. We discuss this process because it makes two points. First, it illustrates that behaviors may be innate and yet nonetheless involve considerable interactions. Thus, interaction and innateness are not incompatible. Rather, we argue that interaction is an important way in which behaviors may be innate. Second, this phenomenon demonstrates interactions at several levels of organization.

Then we turn to whole brain/environment interactions. We give two examples here. We describe a simulation which shows how the spatial localization of functions in a neural network can be achieved through timing (our "chronotopic constraints", from Chapter 1) rather than being explicitly stipulated in advance. This example also demonstrates that when Hebbian learning is subject to a spatio-temporal "maturational" process, things can be learned which can not be learned by Hebbian learning under normal circumstances.

Finally, we discuss the way maturation may interact with learning a complex domain. We show that there are cases where learning something hard may be easier when the learning device begins with limited resources (what we call "the importance of starting small"). Indeed, some complex domains may in fact not be learnable if one starts with a fully mature system.

To anticipate the punchline of this chapter: The main conclusion we come to is that part of the evolution of ontogenesis has involved taking advantage of interactions at increasingly higher levels. We shall suggest that organisms have evolved from ontogenetic development based on mosaic systems (molecular level interactions), to regulatory systems (cellular level interactions), to nervous systems (systems level interactions), to an increasing dependence on behavioral/cultural factors (environment-organism interactions). Each of these steps in the evolution of ontogenetic systems increases the time taken for the development of an individual of the species. In the case of our own species, this process has played a particularly crucial role. Or as Stephen Jay Gould (1977) has put it, the notable thing about the evolution of our species is that "human development has slowed down."

Brain systems level interaction

The development of most of the organs in the body can be accounted for in terms of some combination of mosaic and regulatory types of ontogeny (see Chapter 1). As we saw in the last chapter, however, brain development appears to rely more heavily on interactions which are characteristic of the regulatory style of development. Furthermore, a particular feature of nerves and nerve systems is that they permit cellular interactions of particular types at long distances. Thus, the brain is not confined to local cellular interactions the way most classic regulatory systems are. Nerve cells communicate with other cells at relatively long distances by means of axons, dendrites, and nerve bundle fibres, etc. This is not to deny that there are points in development, or even in the adult, where local cellular level interactions are important (see Edelman, 1988 for some examples of this). Indeed, recent research indicates that early in life cortical neurons may "communicate" with local neighbors in ways not observed in the adult.

However, we expect that as development proceeds there will be a general transition toward development involving higher levels of interaction. Thus, while local cellular interactions may be important, especially during the prenatal development of the brain, longer range brain system level interactions become increasingly important in postnatal development.

There are several sorts of such long-range brain system interactions? In the remainder of this chapter we shall discuss one particular kind of interaction involving *biasing*. Brain systems may be predisposed or biased toward storing or manipulating representations about certain classes of environmental stimuli. Furthermore, we can distinguish two broad classes of bias in such brain systems: selection biases which select or filter input to other systems during development, and architectural biases which restrict the type of information processing in the system. Let us consider examples which illustrate each sort of bias. The first example is chick imprinting; for this example we have both neurophysiological data and a connectionist model. The second example is a model which demonstrates how the timing of synaptic growth may help a network learn things which could not be learned easily otherwise.

Chick imprinting

Imprinting is the process by which newly hatched birds, such as chicks, become attached to the first conspicuous object that they see (for a recent review see Bolhuis, 1991). As we shall see, this learning process reflects interactions at a number of different levels: organism-environment, brain systems, cellular and molecular. After a brief overview of the interactions at all these levels, we will focus on interactions at the cellular level implemented as a connectionist model.

Considered at the behavioral level, imprinting obviously involves an interaction between the organism and its environment. Following prolonged exposure to an object, chicks will develop a strong attachment to it; they preferentially approach and stay close to the imprinted object. While in the laboratory situation chicks will imprint on a very wide range of moving objects, in a more natural context in which hens are present they nearly always imprint on to a particular mother hen. Some sort of organism-environment interaction taking place between the brain of the chick and the information in the external environment, leads to this specificity of outcome. To analyze this organism-environment level interaction further we can move to the brain systems level interactions.

The neural and cognitive basis of imprinting in the chick has been analyzed in some detail at the brain systems level (for reviews see Horn, 1985; 1991; Johnson & Morton, 1991). There is evidence for at least two independent brain systems that influence imprinting behavior. First, there is a predisposition for newly-hatched chicks to attend toward the head and neck region of conspecifics—that is, objects that possess configurations of features found in hens.

Systems of this type have been referred to as *Conspec* by Johnson and Morton (1991). Second, a learning system, subserved by a particular region of the chick forebrain known as the Intermediate and Medial Hyperstriatum Ventrale (IMHV), leads to the chick's imprinting on the attended object.

A variety of neurophysiological experiments have demonstrated that these two systems are dependent upon different parts of the chick forebrain. For example, damage to the IMHV does not affect the predisposition, only learning. Clearly, the two systems both influence the behavior of the chick, but the question remains how they interact. At least three forms of interaction between the two brain systems could exist. First, there might be no internal interaction between the brain systems, but an interaction mediated through the overall behavior of the chick in its environment; a second possibility is that *Conspec* directly activates learning in the IMHV when the appropriate objects are presented; or, a third possibility is that *Conspec* acts as filter through which information to be learned must pass.

Experimental data suggests that the first option is the most likely for these systems in the chick (Bolhuis et al., 1989, Johnson et al., 1992, see Johnson & Bolhuis, 1991 for review). That is, there appears to be no internal interaction between *Conspec* and the IMHV learning system. However, it is important to realize that just because there is no direct internal interaction between these systems does not mean that these systems do not interact! Indeed, the independent activities of *Conspec* and the IMHV very strongly and usefully interact via the behavior of the animal to lead chicks to, on the one hand, attend to hens and, on the other hand, imprint on the attended hens. This phenomenon is best examined at the next higher level of organism-environment interaction.

It is the chick's whole brain that interacts with its natural external environment, not a particular neural circuit or structure. In other words, systems level structures normally only interact with the external environment through the whole brain level. The use of special rearing and other experimental techniques, such as placing lesions in various brain systems, permits us to study the particular biases of individual systems. By use of such experimental methods, the properties of the two systems in the chick have been studied.

Earlier we suggested that the *Conspec* system is present from hatching. But this is a generalization which is not strictly speaking correct since the functioning of the *Conspec* system itself is dependent upon earlier interactions. Bolhuis et al. (1985) and Johnson et al. (1989) have shown that the early rearing environment of the chick can be manipulated such that the predisposition does not emerge. For example, if the chick is not allowed an opportunity to run about freely for at least a few minutes between 12 and 36 hours after hatching, the *Conspec* system does not develop. Note that this motor activity is essential, although non-specific and unrelated to species-specific imprinting. In this case, the learning system alone will determine the chick's preferences. While the exact mechanism of this effect remains unclear, one possibility is that running about in the environment elevates levels of the hormone testosterone, which then activates, or "validates" (Horn, 1985), the neural circuitry supporting *Conspec*.

Unlike the learning system, *Conspec* seems to have little or no ability to develop new representations as a result of interaction with the external environment. However, the action of *Conspec* affects what is learned by IMHV. *Conspec* exerts its influence indirectly, and via the organisms interaction with the external environment, by selecting the inputs to IMHV. *Conspec* is very selective with regard to the features of the external environment that engage it. Johnson and Horn (1988) established that the system is engaged by the correct configuration of features of the head and neck. The necessary combination of features is not species specific; the head and neck of any similar-sized bird or mammal will suffice. However, the absence of features such as the beak/mouth or eyes results in a significantly weaker behavioral response.

Turning to the learning system, Horn and collaborators have identified the expression of a particular gene, *c-fos*, within the IMHV which is closely related to the learning process (McCabe & Horn. 1994). This offers us a glimpse the molecular interactions underlying the learning process.

In addition to the selection of its input by *Conspec*, the learning system subserved by IMHV has a restriction on the elements of the environment that it can develop representations about even in the absence of *Conspec*. For example, objects have to be within a certain size range and be moving for the chick to be able to learn about them. A large static visual pattern is not an effective imprinting stimulus. In Chapter 1 we referred to such biases in neural systems as an computational constraints. That is, the architecture of the learning system is such that there are general restrictions on what it can learn about. Within these restrictions, however, the network is able to develop highly specific representations in response to environmental input. In order to investigate the nature of this biasing in more detail, O'Reilly and Johnson (1994) developed a connectionist model based on some features of the known neural architecture of IMHV.

A connectionist model of imprinting and object recognition

The O'Reilly and Johnson (1994) model provides a good example of how connectionism may help us to understand architectural biases in neural networks. The model is based on characteristics of the cytoarchitectonics of IMHV, shown schematically in simplified version in Figure 6.1. An important feature of this region of the forebrain is the existence of positive feedback loops between the excitatory principle neurons (PNs), and the extensive inhibitory circuitry mediated by the local circuit neurons (LCNs). O'Reilly and Johnson suggest that these properties lead to a hysteresis (the degree of influence of previous activations on current ones) of the activation state of PNs in IMHV, and that this hysteresis could provide the basis for the development of translation invariant object-based representations.

The general architecture of the model is designed around the anatomical connectivity of IMHV and its primary input area, the Hyperstriatum Accessorium (HA). The model, shown in Figure 6.2, is based on three sets of layers, each of which have lateral inhibitions within them. The input layer of the network, Layer 0, repre-

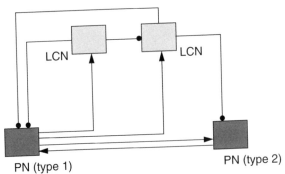

FIGURE 6.1 Simplified schematic of the circuitry of the IMHV in the chick. Excitatory contacts are represented by connections terminating in arrows; inhibitory connections are shown terminating in filled circles. Shown are the local circuit inhibitory neurons (LCNs) and their reciprocal connectivity with the excitatory principal neurons (PNs).

sents HA. HA then projects to one sub-population of IMHV PN cells, which are assumed to be Type 1 PNs. This is Layer 1 of the IMHV component of the model. Note that the laminar distinction in the model between these two component cells of IMHV is not intended to suggest that the cells are arranged as such in the IMHV itself, but rather serves to reflect the functional distinction between the two types of PN. The axons from the Type 1 neurons project to Type 2 projection neurons, as well as onto the local inhibitory neurons. This comprises the second layer of the model's IMHV component, Layer 2. The Type 2 PNs then send a bifurcating axon both back to the Type 1 PNs and the inhibitory cells (Layer 2), and to other areas, which are not modeled.

Within each layer of the model, strong lateral inhibition exists in the form of relatively large negative weights between all units in the layer. This reflects the presence of a large number of GABAergic inhibitory interneurons in IMHV (Tombol et al., 1988), and its relatively low levels of spontaneous activity. The strong inhibition in this layer of the model resulted in only one unit in each layer

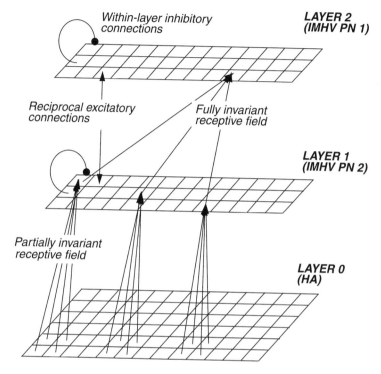

FIGURE 6.2 Network architecture in O'Reilly & Johnson model.

becoming active at any time (however, in other versions of the model several units are allowed to become active).

Training in the model consisted of presenting a set of feature bits (assumed to correspond to a given object) in different random spatial locations in Layer 0. Once the activation state of the network had reached equilibrium for each position, the weights between all units in the system were adjusted according to an associative Hebbian learning rule. The general properties of the Hebbian learning rule which are important for this model (and most other self-organizing models) are that it have both a positive and a negative associative character, which work together to shape the receptive fields of units both towards those inputs that excite them, and away from those that do not.

The degree of hysteresis (the influence of previous activation states on subsequent ones) was controlled by a decay parameter, which reset the activations to a fraction of their equilibrium values for the previous object position. The hysteresis caused units in the model that were active for a given position of a simulated object to remain active for subsequent positions. In combination with the Hebbian associative learning rule, this resulted in the development of units which would respond to a particular set of features in any of a certain range of different locations (or views of the object if it had been rotating), depending on how long that unit was able to remain active. Hysteresis in the activation states comes from the combined forces of lateral inhibition (which prevents other units from becoming active), and recurrent, excitatory activation loops, which cause whatever units are active to remain active through mutual excitation. These two forms of neural interaction are known to be present in IMHV, as was discussed above.

An important characteristic of the O'Reilly and Johnson model is that the learning system has an architectural bias such that it only develops representations of certain kind, namely, those in which objects are extracted and represented on the basis of the spatial-temporal continuity of component features. With this in mind, it is not surprising that chicks will not imprint on objects that do not move. Static objects are not actively filtered out, or ignored by attention. Instead, an architectural bias in the learning system simply means that spatially-invariant object representations are only developed of objects if they move.

When trained on a stimulus for around 100 epochs, some of the units in layer 2 of the network develop spatially invariant receptive fields for that object. The network then shows a "preference" for the object over a novel one in a similar way to chicks.

The parallels between chick imprinting and the behavior of the model go much further than this, however. Another set of simulations that O'Reilly and Johnson ran concerned the reversibility of acquired preferences and sensitive periods. It turns out that imprinted preferences for an object in chicks can sometimes be reversed by extended training on a second object. However, the chick still retains information about the first object since this object is still preferred over a completely new object. A critical factor determining whether this reversal of preference can take place is the extent of training on the first object.

Simulations in the model showed similar findings. First, the likelihood of being able to reverse the preference following training on the second object was shown to depend on (i) the length of training on the first object, and (ii) the length of training on the second object. In addition, the model showed a similar sharply timed "critical period" for reversibility dependent on the time of exposure to the first stimulus.

The second major finding in this set of simulations was that even when the preference for the first stimulus was reversed to the second object, the first stimulus was still preferred over a completely new object. This indicates that the network, like the chick, had retained a representation of the original training stimulus. This finding is of interest since many neural networks show catastrophic interference when re-trained on a new stimulus.

Another phenomenon associated with imprinting in chicks is generalization. For example, if a chick is trained on a blue ball it will show more preference for a blue box or a red ball, than for a red box. Generalization was also observed in the model since a stimulus that shared two out of three features in common with the original training stimulus was strongly preferred to a completely new object.

Interestingly, the hysteresis in the network model has some drawbacks for effective object recognition. It turns out that these problems can also be observed in chick object recognition. Fortunately, these failures to adequately develop object representations in chicks occur under laboratory circumstances unlikely to be encountered in the natural context. For example, if newly hatched chicks are exposed to two objects, A and B, in rapid succession (e.g., A shown for 15 seconds, then B for 15 seconds, then A, and so on) they develop a blended representation of the two objects that makes it very hard for them to learn other tasks involving discrimination between the two objects. In contrast, if they are exposed to the same two objects for the same total length of time, but in training blocks on ten minutes or more, then they develop two independent representations of the objects and are faster than normal to discriminate between them in other tasks. O'Reilly and Johnson conducted similar experiments with their network model. In the rapid alternating input condition the same units in layer 3 coded for the characteristics of both objects ("blended representation"). When the network was trained with blocks of trials of two stimuli different units coded for the features of each object.

To conclude, while the O'Reilly and Johnson model has no representational prespecification, the structure of the network in conjunction with a particular learning rule provides architectural constraints on the types of representations that are developed. Turning to the chick in its natural environment, we can see that multiple sources of constraints, both from within levels, and from other levels (molecular, organism-environment, etc.) ensure a particular outcome: a spatially invariant representation of the mother hen.

An architectural bias similar to that in the O'Reilly and Johnson model may account for some observations about the extent to which human infants develop object representations. Spelke and colleagues have conducted a series of experiments in which they have examined the extent to which human infants perceive objects when presented with visual arrays in which a rod appears to moves behind an occluder. Infants of four months and older perceive the two ends moving at either side of the occluder as being the two ends of a single rod. These and other results have been used to argue that infants possess an initial "object concept" composed of four principles- cohesion, boundaries, substance and spatio-temporal continuity (Spelke, 1988). The O'Reilly and Johnson model illustrates that at least one of these principles thought to underlie the object concept, spatio-temporal continuity, may be instantiated as an architectural bias in certain brain networks rather than as knowledge, or an "initial theory of the physical world." (Spelke, 1988, p.181). It should be evident that the model will develop a single "object" representation in the same ways as the infant is alleged to. Since both ends of the "bar" are moving at the same time, the units in Layer 2 will develop spatially invariant representations that "bind" the two ends together as one object. Note that, like the infant, the model would do this for a variety of motions of the two ends, such as rotation and vertical movement. Thus, a model similar to that developed by O'Reilly and Johnson may generate specific hypotheses about the implementation of object knowledge in infants.

Of course, there are still at least three other principles that remain to be accounted for. The principle of Cohesion for example, requires that surfaces of a body must be connected; in the case in question the two ends of the rod need to be "filled in" to be joined together as one object. This is not something that the current version of the O'Reilly and Johnson model achieves, since while some units learn to bind the two ends of the rod as one stimulus, there is no reason for a single continuous rod to be actually preferred over the two ends (as has been shown with infants).

The importance of time

In previous sections we have seen that one of the most obvious facts about development—that it unfolds over time—turns out to have not so obvious consequences. Altering the timing of interactions, be they between genes or cells or brain systems, can have dramatic consequences for developmental outcomes. Timing can make the difference between chicken teeth or beaks, or between octopi with shells or without (Chapter 1).

There are other effects which can be produced by the timing of developmental events, and which illustrate how maturational events (e.g., the presence or absence of growth factors, or the development of working memory) can interact with a system's ability to learn. We will examine two simulations. In both cases, the result is that the developmental process is crucial to successful learning, but the nature of the interaction appears to be paradoxical. Surprisingly, systems which attempt to learn starting in the mature state do not learn as well as those which begin learning while immature.

Spatial localization as an emergent property. In Chapter 5 we encountered a mystery. Experimental data suggest that the regional mapping of functions in the human cortex is not prespecified. Initially, the cortex appears to possess a high degree of pluripotentiality. Over time, however, a complex pattern of spatially localized regions develops, and the pattern of localization is relatively consistent across individuals. The mystery is how the specific functional organization of the cerebral cortex arises. We have seen network models in which self-organization occurs, so that is not the problem. For example, in the Miller et al. (1989) work, we saw that Hebbian learning could result in the development of ocular dominance columns (see Figure 2.15). The problem is that in those models, the specific patterning varies considerably (as it does in real life). What remains to be solved is how, at the next level of organization, systems made up of such components could be constructed, and constructed with a high degree of similarity across individuals. This similarity across individuals could be taken as highly suggestive of an innate component.

Shrager and Johnson, and subsequently Rebotier and Elman, building on earlier work by Kerszberg, Dehaene and Changeux (1992), have offered a preliminary account of at least one factor which might provide an answer to this question. Shrager and Johnson began with the assumption that the cortex is organized through a combination of endogenous and exogenous influences, including subcortical structuring, maturational timing, and the information structure of an organism's early environment. Their goal was to explore ways in which these various factors might interact in order to lead to differential cortical function and to the differential distribution of function over the cortex. They began with several simple observations.

First, Shrager and Johnson pointed out that although there are signals which pass through the cortex in many directions, subcortical signals (e.g., from the thalamus) largely feed into primary sensory areas, which then largely feed forward to various secondary sensory areas, leading eventually into the parietal and frontal association areas. Each succeeding large-scale region of cortex can be thought of as processing increasing higher orders of invariants from the stimulus stream. The image is that of a cascade of filters, processing and separating stimulus information in series up toward the integration areas.

Second, Shrager and Johnson note that (as we discussed in Chapter 5) a very striking aspect of development of the cerebral cortex is the initial overproduction and subsequent loss of neural connections, resulting in the relatively sparsely interconnected final functional architecture. This process of overproduction of synapses and subsequent (or simultaneous) thinning out of the arbor is thought to be key in cortical ontogeny. As Thatcher (1992) suggests, when initially heavily connected, the cortex is like a lump of stone which in the hands of the sculptor is shaped by removal of bits and pieces into its final form. Moreover, there appears to be a general developmental dynamic in which, grossly speaking, the locus of maximum neural plasticity begins in the primary sensory and motor areas and moves toward the secondary and parietal association areas, and finally to the frontal regions (Chugani, Phelps, & Mazziotta, 1987; Harwerth, Smith, Duncan, Crawford, & von Noorden, 1986; Pandya & Yeterian, 1990; Thatcher, 1992).

Thatcher (1992) analyzed the coherence of EEG waveforms in resting state children between the ages of 1;6 and 6;6. He computed the ages during which the change in EEG coherence was maximally positive. These were then defined as cortical growth spurts; that is, the ages during which different sections of the cortex are beginning to come together in terms of their electrical synchronization with one another. By using this metric, Thatcher was able to plot cortical growth spurts in terms of the areas that are beginning to act together at a particular age, and so compare short versus long distance synchronies, as well as watching the wave of cortical synchronization (that is, maximal coming into of synchrony of regions between EEG electrodes). In the 1.5–3 year spurt there is a lengthening along the rostral-caudal dimension and a rotation from lateral to medial, which is repeated again at 5.5–6.5 years. There is also a general rostral to caudal expansion of intracortical synchrony and a spurt of right frontal pole growth at age 5 years.

The iterative growth spurts and patterns of development during the postnatal period may reflect a convergence process which narrows the disparity between structure and function by slowly sculpting and shaping the brain's microanatomy to eventually meet the demands and requirements of an adult world. According to this notion, an individual's gross anatomical structure is established early in development and the postnatal sculpting process is used to fine tune anatomical structure to meet the needs of diverse and unpredictable environments. The sculpting process unlocks or tailors the functional potential of the stable gross anatomy according to individual needs and environmental demands.

Presumably, the cyclical patterning reflects a dialectical process which iteratively and sequentially reorganizes intracortical connection systems. (Thatcher, 1992; p. 47)

Given these observations, Shrager and Johnson posed the question, How might such a developmental wave of plasticity—in which different regions of cortex are more plastic at different points in time—affect the outcome of learning? To study this question, Shrager and Johnson developed a connectionist network which was designed to test the hypothesis that under certain regimes of wave propagation, we might expect a tendency toward the development of higher order functions in later parts of the cortical matrix. In this way, the model might account for spatial distribution of function in cortex without having to encode the localization directly.

The Shrager and Johnson model is shown in Figure 6.2. The model consists of an abstract "cortical matrix" composed of a 30 by 30 matrix of artificial neurons. Each neuronal unit has afferent and efferent connections to nearby units, and also receives afferent inputs from external signals designated A and B. For our purposes, we shall consider the case in which afferent and efferent connection weights are initially set at random, and in which the external signals A and B provide simultaneous inputs of 0 and 1, also at random.

The matrix weights are changed according to a Hebbian learning rule, so that connection strength grows between units whose activations are more highly correlated. Under the default conditions just described, the outcome after learning is that some units become active only when their A input is on; others become sensitive only to the presence of the B input; and still others become sen-

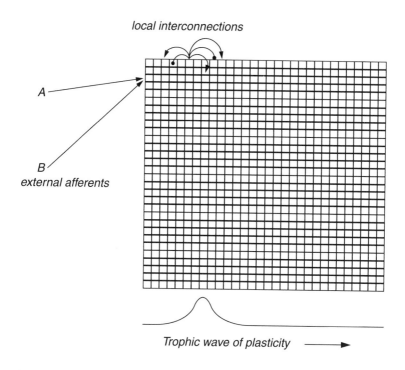

FIGURE 6.3 Shrager & Johnson model. Each unit has short local connections (excitatory and inhibitory) to close neighbors, and also receives afferents from the external afferents, A and B (shown here as excitatory, but initially set as excitatory or inhibitory at random). A tropic wave of plasticity spreads from left to right across the matrix.

sitive to A and B simultaneously (logical AND), or to either A or B (logical OR). A very large number of units are always off. No units develop which are sensitive to exclusive OR (XOR), which is not surprising since Hebbian learning does not typically lead to such higher order functions (see discussion in Chapter 2).

Shrager and Johnson then considered what might happen if the Hebbian learning is modulated by a trophic factor (TF) which passed through the matrix in a wave, from left to right. The effect of wave was that the columns of units underneath it, at any given point in time, were more plastic and therefore able to learn. During the first training cycle, for example, a modulation vector was produced for the 30-column matrix that might be [1.0, 0.86, 0.77, 0.66, 0.53,...,0.0, 0.0]. That is, TF transmission at location 1 in the matrix took place normally, whereas TF transmission at location 2 was reduced to 86% of what would have been moved, etc. On the next cycle, the wave moved to the right a small amount: [0.86, 1.0, 0.86, 0.77, 0.66,...,0.0, 0.0]. The progress of the wave thus modulated the transmission of trophic factor, leading to a dynamic plasticity in the cortical matrix. Leftward columns were plastic early and also lost their plasticity early on; whereas, rightward columns did not become plastic until later on, but were plastic toward the end of the simulation when most of the neurons were reaching asymptote on the stabilization and death curves.

Under certain regimes of wave propagation, Shrager and Johnson expected to observe a tendency toward the development of higher order functions in the cortical matrix. (Higher order functions are those which depend on both A and B inputs; lower order functions are those which depend solely on A or B.) The reason for this may be envisioned by considering two steps in the propagation of the wave from some leftward set of columns to the next set of columns to the right. We shall call these columns COL1 and COL2 (which is immediately to the right of COL1). COL1, initially more plastic than COL2, determines its function during receipt of input from A and B afferents, as has been the case all along. However, COL1 becomes fixated in its function relatively early, as the wave moves on to COL2. Now, however, COL2 is receiving input that is, in addition to the input coming from A and B afferents, includes the combined functions fixated by the earlier plasticity in COL1. Thus, COL2 has, in effect, three afferents: A, B, and COL1.

In fact, Shrager and Johnson found that the number of first order functions (A, ~A, B, and ~B) differed significantly from the number of second order functions (B-AND-~A, A-AND-~B, A-XOR-B, ~[A-AND-B], A-AND-B, A=B, A>=B, A<=B, and A-OR-B), when the wave was present, but not without the wave. Furthermore, as predicted, the density of higher order functions increased in regions of the matrix which were plastic later on, as determined by the propagation of the TF wave. Finally, when the propagation rate of the wave was tripled from the initial rate, a different picture emerged. Again, the first and second order functional densities were significantly different, but this time the mean values were inverted. In the slow wave case the second order functions were emphasized, whereas in the in fast wave case the first order functions were emphasized.

There is another result which is of great significance, and was the focus of a replication and extension by Rebotier and Elman (1995). In Chapter 2, we noted that Hebbian learning has a number of appealing characteristics. For one thing, Hebbian learning has a greater biological plausibility than back propagation learning. Also, Hebbian learning is a form of self-organizing behavior. The learning algorithm looks for correlations in neuronal activity patterns (or the stimuli which give rise to them) and does not require an external teacher. This addresses the problem which arises when the external teacher which is required by supervised learning algorithms such as back propagation of error is not available. On the other hand, Hebbian learning cannot be used to learn certain important problems. These include XOR and other functions in which classification cannot be done on the basis of correlations. This is unfortunate, because it means that the learning mechanism which is most natural on biological grounds seems to lack necessary computational properties.

Rebotier and Elman constructed a network of the form Shrager and Johnson devised and allowed Hebbian learning to take place through all parts of the network ("instant cortex"). Not surprisingly, Rebotier and Elman found no units which respond to the XOR of the inputs A and B. Rebotier and Elman then repeated the experiment, but this time allowed learning to be modulated by a spatial wave of trophic factor, which passed over the network from left to right. This time, a small percentage of units were found which computed XOR. These units tended to be on the right side of the network (i.e., the late maturing regions). The reason they could compute XOR is that they did not learn until later, after early units had developed with learn simpler functions such as AND and OR. These early learning units then became additional inputs to the later learning units. Since XOR can be decomposed into the AND and OR functions, this made it possible to learn a function which could not otherwise have been learned.

There are thus two important lessons to be learned from the Shrager and Johnson and the Rebotier and Elman studies. First, their model demonstrates how the differential functional architecture of the cortex might arise in early development as an emergent result of the combination of organized stimulus input, and a neurotrophic dynamic (whether produced by a natural wave of trophic factor or by some other endogenous or exogenous phenomenon). Second, development provides the key to another puzzle; the study shows how some complex functions which are not normally learned in a static mature system can be learned when learning is carried out over both time and space rather than occurring everywhere simultaneously. This is an important result which we develop further in the next study.

The importance of starting small. In most higher vertebrates, brain systems interact together as a whole brain with the external world. That is, several brain systems are likely to contribute to any behavior, and thus we have to go to the whole brain level to understand the interaction of the system in question in the external world. In the case of the chick, even a simple behavior such as imprinting involved at least two independent brain systems. Of course, in species with even simpler brains it is likely that there are cases were only a single brain system is responsible for a behavior of the whole organism (such as a reflex). In these cases, there is no need to discuss whole brain interactions; the brain systems level is sufficient. But for most of the more complex forms of behavior of interest to the cognitive scientist, multiple brain systems are likely to be involved. The challenge then, is to understand how interactions between brain systems produces the adaptive biases observed in behavioral development.

One way in which brain systems interact to produce adaptive whole brain interactions with the environment is by one brain system selecting an aspect of the environment for another. This what we have seen in the case of chick imprinting.

A similar sort of filtering can arise by other means as well. Turkewitz and Kenny (1982) have suggested that limitations on sensory channel capacities early in life can facilitate the development of other brain systems. Their hypothesis is that limitations in sensory or attentional capacities are thought to reduce the complexity of environmental information that impinges on a learning system. Thus the biasing is not accomplished by one brain system acting on another, but by timing of the maturational process itself. Let us consider a connectionist model in which "starting small" (by beginning learning with limited resources) turns out to play a critical role in the successful learning of a complex linguistic behavior.

One of the most important things human children must learn to do is communicate. Language learning is an enormously important behavior; it occupies a great deal of a child's time and takes place over many years. One of the fascinating features of this behavior is that its form is apparently so decoupled from its content. Manipulating words is not like manipulating a bicycle or using chopsticks or learning to walk. In these latter cases the form of the activity is directly related to its function. Language is different in this respect. It is a highly symbolic activity. The relationship between the symbols and the structures they form, on the one hand, and the things they refer to, on the other, is largely arbitrary.

This much is clear. But some theorists interested in language acquisition go further in their conclusions. They suggest that very little actual learning occurs during language development. Of course, most people do end up controlling at least one language, and languages of the world differ, so there is clearly some learning involved. But these theorists view this kind of learning as a minor sort of adjustment of parameters or a tuning of capabilities which are themselves genetically determined. The role of learning is minimal.

There are two basic arguments in favor of this position. First is the fact that human languages do not vary infinitely. The languages of the world differ, to be sure, but the differences by no means exhaust the range of possibilities. There are universally recurring patterns which suggest that there are strong biological constraints on language which are shared by all members of the species. As we have said previously, we have sympathy for this argument. The primary issue is what form those constraints take.

The second argument in favor of a "non-learning" approach to language is that language is simply too hard to learn given the data and the time available. One of the best known examples of this argument comes from a mathematical proof by Gold (1967). Gold was able to show that formal languages of the class which appear to include natural language can not be learned inductively on the basis of positive input only. A crucial part of Gold's proof relied on the fact that direct negative evidence (e.g., of the explicit form in which the parent tells the child, "The following sentence, 'Bunnies is cuddly', is not grammatical") seems virtually nonexistent in child-directed speech (but see MacWhinney, 1993). Since children eventually do master language, Gold suggested that this may be because they already know critical things about the possible form of natural language. That is, learning merely takes the form of fine-tuning.

Although we believe that there are many reasons why Gold's proof is not relevant to the case of natural language acquisition, we also agree that it would be a mistake to take the extreme opposite position and claim that language learning is entirely unconstrained. Children do not seem able to learn any arbitrary language, nor are non-human young able to learn human languages. The questions we are interested in are therefore, What are the constraints? and How are they implemented?

We believe that properties of the processing mechanisms which are engaged in language use (including our sensory-motor systems) provide one source of constraint; that another comes from the intrinsic logic which is imposed by the communicative function itself; and that characteristics of the sensory-motor systems by which language users experience and interact with the world and facts about the world itself provide additional constraints on the form and content of language. We also suspect that the developmental process itself helps to make language learning easier than it would be in the absence of development. That is the point made by the following simulation.

In Chapter 2 we described a simulation by Elman (1990) in which a simple recurrent network was trained to "read" sentences a word at a time. After every new word, the network was trained to predict what the next word would be. The form of the words themselves was arbitrary, and gave no clue as to their grammatical category or meaning. However, because both grammatical category and meaning are highly correlated with distributional properties, the network ended up learning internal representations which reflected the lexical category structure of the words.

The sentences in that simulation were all simple (monoclausal) and short. More realistic sentences would be more challenging; for example, they might be longer and might include embedded relative clauses. Interestingly, sentences with relative clauses possess exactly the sort of structural features which may make (according to Gold, 1967) a language unlearnable. Elman (1993) attempted to train a simple recurrent network using stimuli such as those shown in examples (1-5).

(1) Boy chases dog.
(2) Boys chase dog.
(3) Mary walks.
(4) Boy who dogs chase feeds cat.
(5) Girls who chases dogs hits cat.

(Words like "the", "a", etc., were not used in this artificial language.)

In order to successfully process these sentences the network had to learn to distinguish between various categories of words (e.g., nouns from verbs, singulars from plurals). It had also to learn that some verbs (e.g., in this language, "chase") were obligatorily transitive and required direct objects, whereas others (e.g., "walks") optionally permitted direct objects, and yet others ("exists") were always intransitive. And the network also had to learn how to deal with embedded material which was potentially disruptive. (Note, for instance, that in sentence (4), the first noun is singular but the first occurring verb is in the plural because it is part of an embedded structure. And although the verb "chase" is obligatorily transitive and usually takes a following direct object, in this sentence the next word is another verb, because the direct object has been mentioned before the verb. In (5), the plural "dogs" is followed immediately by a verb in the plural.)

In fact, Elman found that it was not possible to successfully train the network on such stimuli. Failure is rarely a conclusive result, of course, and there may be many reasons for the lack of success. But this result is just what one might have expected, given Gold's proof.

In order to discover just how much complexity the network could tolerate, Elman put aside the fully complex data set and attempted to train the original network with a subset of data containing only simple sentences. As expected (since this merely replicated the similar simulation reported in Chapter 2), the network was able to master the prediction task for these sentences relatively easily. Elman continued training the same network (i.e., keeping the weights that had be learned with the simple sentences) and gradually introduced more and more complex sentences. Surprisingly, the network assimilated these new data quickly and correctly. It turned out to be possible to ultimately finish with the original data set which had been unlearnable! Moreover, the network generalized its performance correctly to novel data of comparable (and even greater) complexity.

At first, this result seems to resemble the pattern of learning in human children, who also start with short and simple sentences. But there is an important disanalogy between the network's training and the actual conditions in which children learn. Children may initially *produce* only simple sentences, but they are certainly *exposed* to complex sentences from the start, unlike the network.

However, if the child's environment remains relatively constant during learning, the child herself changes in ways which might be highly relevant. For one thing, working memory and attention span in the young child may initially be restricted and increase over time. Could such a change facilitate learning?

In order to study a possible interaction between learning and changes in working memory, another new network was trained on the "adult" (i.e., fully complex) data which had initially been problematic. This time, at the outset of learning, the context units (which formed the memory for the network) were reset to random values after every two or three words. This meant that the temporal window within which the network could process valid information was restricted to short sequences. The network would of course see longer sequences, but in those cases the information necessary to make correct predictions would fall outside the limited temporal window; such sequences would effectively seem like noise. The only sequences which would contain usable information would in fact be short, simple sentences. After training the network in this manner for a period of time, the "working memory" of the network was extended by injecting noise into the context units at increasingly long intervals, and eventually eliminating the noise together.

Under these conditions, the performance at the conclusion of training was identical and as good as when the training environment had been manipulated. Why did this work? Why should a task which could not be solved when starting with "adult" resources be solvable by a system which began the task with restricted resources and then developed final capacities over time?

It helps to understand the answer by considering just what was involved when learning was successful. At the conclusion of learning, the network had learned several things: distinctions between grammatical categories; conditions under which number agreement obtained; differences between verb argument structure; and how to represent embedded information. As was the case in the simulation involving simple sentences, the network uses its internal state space to represent these distinctions. It learns to partition the state space such that certain spatial dimensions signal differences between nouns and verbs, other dimensions encode singular vs. plural, and other dimensions encode depth of embedding.

In fact, we can actually look at the way the network structures its internal representation space. Let us imagine that we do the equivalent of attaching electrodes to the network which successfully learned the complex grammar, by virtue of beginning with a reduced working memory. If we record activations from this network while it processes a large number of sentences, we can plot the activations in a three-dimensional space whose coordinates are the principal components of hidden unit activation space (we shall use the second, third, and eleventh principal components). The surface shown in Figure 6.2(b) shows the space which is used by the network.

As can be seen, the space is structured into distinct regions, and the patterning is used by the network to encode grammatical category and number. Once the network has developed such a representational scheme, it is possible for it to learn the actual grammatical rules of this language. The representations are necessary prerequisites to learning the grammar, just because these internal representations are also play a role in encoding memory (remember that the hidden unit activation patterns are fed back via the context units). Without a way to meaningfully represent the (arbitrarily encoded) inputs, the network does not have the notational vocabulary to capture the grammatical relationships. Subjectively, it's the same problem we would have if we try to remember and repeat back words in an unfamiliar language-it all sounds like gibberish. Note that this creates a bit of a problem, however. If the network needs the right internal representations to work with, where are these to come from? The truth is that these representations are learned in the course of learning the regularities in the environment. It learns to represent the noun/verb distinction because it is grammatically relevant. But we just said it couldn't learn the grammar without having the representations. Indeed, this chicken and egg problem is exactly the downfall of the network which starts off fully developed (but lacking the right representations). If we look at the internal space of this network after (unsuccessful) training, shown in Figure 6.2(a) we see that the space is poorly organized and not partitioned into well-defined areas. The network which starts off with limited resources, on the other hand, actually is at an advantage. Although much of what it sees is now "noise," what remains—the short, simple sentences—are easier to process. More to the point, they provide a tractable (because they are short, and impose fewer demands on a well-developed representation/memorial system) entry point into the problem of discovering the grammatical patterns and categories latent in the environment. Once these catego-

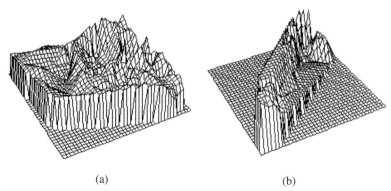

(a)   (b)

FIGURE 6.4 View of hidden unit space (in three of 70 dimensions) of a network which fails to learn the grammar (a), and which succeeds (b). The surfaces are plotted by passing a large number of test sentences through each network and recording the hidden unit activation vector following each word. In the case of successful learning, the hidden unit state space is structured and can be interpreted in terms of various dimensions of relevance for the task (e.g., noun vs. verb, singular vs. plural, etc.). In the case of the unsuccessfully trained network, the state space is poorly organized and no clearly interpretable dimensions are found.

ries have been induced, they provide the bootstrap by which the network can go on, as its working memory improves, to deal with increasingly complex inputs and refine its knowledge.

Seen in this light, maturational limitations take on a very positive character. If a domain to be mastered is complex, it helps to have some clues about where to start. Certainly the solution space for inducing a grammar from the data is extremely large, and finding the right grammar might be an intractable problem. It makes sense therefore that children (or networks) might need cues to help guarantee they discover the right grammar. The question is, what do these cues look like?

One possibility is that children (or networks) might be prewired in such a way that they know about concepts such as "noun" and "verb" at birth. We might endow them as well with special knowledge about permissible classes of structures, or grammatical operations on those structures. The role of experience would be to help the learner figure out which particular structures or operations are true of the language being learned. This is the hypothesis of Parameter Theory (Chomsky, 1981).

The simulation here, like the Hebbian network we discussed earlier, shows another solution to the problem of finding the needle in the grammatical haystack. Timing the development of memory has the effect of limiting the search space in exactly the right sort of way as to allow the network to solve a problem which could not be solved in the absence of limitations.

So if we understand that limitations can impose constraints, and take constraints as a form of innate predisposition, then we are entitled to say that the network described here is "innately constrained" to discovering the proper grammar. But of course, this is a very different sort of innateness than envisioned by the pre-wired linguistic knowledge hypothesis.

Is there any evidence that this positive interaction between maturational limitations and language learning plays a role in children, as it seems to in networks? Elissa Newport has suggested that indeed, early resource limitations might explain the apparent critical period during which languages can be learned with native-like proficiency. Newport calls this the "less is more" hypothesis (Newport, 1988, 1990).

It is well-known that late learners of a language (either first or second) exhibit poorer performance, relative to early or native learner. What is particularly revealing is to compare the performance of early (or native) learners when it is at a comparable level to that of the late learners (i.e., early on, while they are still learning). Although gross error scores may be similar, the nature of the errors made by the two groups differs. Late learners tend to have incomplete control of morphology, and rely more heavily on fixed forms in which internal morphological elements are frozen in place and therefore often used inappropriately. Young native learners, in contrast, commit errors of omission more frequently. Newport suggests that these differences are based in a differential ability to analyze the compositional structure of utterances, with younger language learners at an advantage. This occurs for two reasons. Newport points out that the combinatorics of learning the form-meaning mappings which underlie morphology are considerable, and grow exponentially with the number of forms and meanings. If one supposes that the younger learner is handicapped with a reduced short-term memory, then this reduces the search space (because the child will be able to perceive and store a limited number of forms). The adult's greater storage and computational skills work to the adult's disadvantage. Secondly, Newport hypothesizes that there is a close correspondence between perceptually salient units and morphologically relevant segmentation. With limited processing ability, one might expect children to more attentive to this relationship than adults, who might be less attentive to perceptual cues and more inclined to rely on computational analysis. Newport's conclusions are thus very similar to what is suggested by the network performance: there are situations in which maturational constraints play a positive role in learning. Counterintuitively, some problems can only be solved if you start small. Precocity is not always to be desired.

Tieing it all together

We have sketched out a number of different phenomena in this chapter, and the data range from molecular genetic facts to models of high-level cognitive function. We believe however that a coherent picture emerges from these examples. What conclusions are we led to?

Interactions occur at all levels

First, interactions are pervasive and occur at all levels of development, even down to the genetic level. Although there are cases of single-action genes, we find that even such (presumably) simple phenotypic characteristics as eye color may require the concerted action of many genes. Even more startling is the discovery that what and where a gene is, at any given point in time, depends on the joint activity of several factors. What counts as an exon or an intron, and which sequences of base pairs work together to synthesize proteins, is often the outcome of a time-varying interaction between a large number of molecules inside the cell nucleus which are constantly in flux, snipping, rearranging, deleting, and pasting together bits and pieces of genetic material as required in order to produce a functioning gene.

This dependence on interactions is repeated at successively higher levels of organization. Indeed, as we look at more complex behaviors, we suspect that there will rarely if ever be cases in which developmental interactions do not play a crucial role in the formation of the behavior. From an evolutionary perspective, of course, this is hardly surprising. Evolution selects for the fitness of whole individuals. What evolution sees is the integrated behavior of those individuals and not a collection of discrete phenotypic characteristics.

At the same time, it is true that animals, in which the developmental process resembles the construction of a number of independent pieces (similar to a mosaic), do exist; C. Elegans is one such well-studied example. This style appears to impose built in limitations on the complexity which can be achieved. This brings us to our second point.

Interactions increase the complexity possible in development

The blueprint view of the genome, in which the genetic material somehow contains a literal image of the target animal, is easy to reject. Nothing remotely resembling such a blueprint has ever been discovered. Nor is such a blueprint even logically possible, since there is simply not enough space in the genome to contain a full and complete description of the adult. Those animals in which there is, if not a blueprint, a straightforward and relatively direct relationship between genome and phenotype (as in mosaic species such as the nematode, C. Elegans) arguably represent the upper bound of complexity which is possible given this sort of a tight genetic control on development.

An alternative view is that of the genome as computer program. This is an appealing hypothesis because we know that programs have enormous expressive power. Simple programs can generate huge outputs. Moreover, there are sequences of the genome which seem to work like BEGIN and END statements in a program. The very phrase "genetic code" invites us to think of computer code. However, we think this view is off the mark as well.

To be sure, the genome resembles a program in the limited sense of controlling the execution of some process. The shortcoming of the metaphor is that computer programs are essentially self-contained. One can examine a program and—looking only at the code—make a reasonable guess about what it will do. This is not possible with genetic material. The relationship between DNA base triples and amino acids may be direct; but the assembly of amino acids into proteins, the timing of when specific genes are expressed, and the effect of a gene's products are highly context-sensitive.

However, there is much more going on here than the opacity of the relationship between gene and behavior. What is important is not that the relationship between gene and behavior is merely complicated or indirect or difficult to analyze. What is at stake is the expressive power of the gene.

It is more useful to view genes as catalysts rather than codes or programs. What do we mean by this distinction? Programs are (more or less) informationally self-contained. Catalysts, on the other hand, are embedded in an environment of natural laws and processes. In isolation, a catalyst does nothing. Its information content is zero. But in its functioning context, a catalyst can make things happen which would not happen on their own. So a catalyst's information content is environmentally dependent which means that it is potentially enormous, embracing whatever "information" there is in the environment.

In the case of gene products, we are typically talking about catalysts which accelerate biochemical reactions. The acceleration can be by orders of magnitude. The *catalase* gene produces an enzyme which allows reactions to occur at rates greater than 1,000,000 per second. The gene does not define the conditions for reaction. These conditions, which are specified by the laws of biochemistry, are given and are external to the biological organism. What genes do is to harness those laws by ensuring that critical components are present at the right time, and then nudging the reaction forward. One cannot therefore look at a gene in isolation and know what it does. One can learn this only by watching it in action and seeing how it functions *in situ*.

There are two important consequences to such a view. First, as we have just suggested, it may often be difficult to know what a gene does unless one understands that its activity is part of an interactive process. Second, the information content of the gene is now increased enormously. The gene is no longer a self-contained entity but able to exploit processes and reactions whose existence are governed by the laws of biochemistry. The gene is like the conductor leading an orchestra; it makes no music on its own but with the proper participants can produce things of enormous beauty and complexity.

Why development takes time

At first blush, immaturity is not particularly adaptive. The immature animal finds itself in a state of great vulnerability. It is susceptible to predation and highly depending on others for care and protection. The care-takers pay a price as well, since their time and energy are now committed to protecting and rearing the young for as long as the immature state lasts.

So given the apparent disadvantage of the immature state, one wonders why development should occur at all? Why should species such as our own which have run a long evolutionary gauntlet have such a long development period in their life cycle? Why not be like the nematode, up and running in a few days? Or at least we might be like our close cousin, the chimpanzee, who reaches maturity within a few years? Two answers come to mind.

First, we have argued that the developmental process is essentially dependent on interactions, and that the outcome of many developmental interactions is highly time-dependent. Different outcomes can be produced from the same interaction, depending on when the interaction occurs. We emphasize that the importance of timing in fact goes to the heart of development. Indeed, changes in developmental timing play a major role in evolutionary change (see Gould, 1977; and McNamara & McKinney, 1991, for dramatic examples of this). Time is thus an essential requirement for interactions to take place. Cutting short development also cuts short the opportunity for higher level and more complex interactions to occur.

But the importance of time goes deeper than the fact that interactions cannot all take place simultaneously. We suggest that there are some problems whose very solution depends critically on timing. We have seen two examples of this.

In the Hebbian network simulation, we found that a network which did not normally develop units that detected high order functions in stimuli would do so when learning was modulated over space and time. In the language learning simulation, a grammar which was not learnable by an adult network was mastered by a network whose working memory was initially limited and which increased over time. In these examples time has the natural effect of decomposing a difficult problem. The simpler parts are learned first and form the basis for bootstrapping the learning of the more complex functions.

These examples also address a puzzle we raised at the end of Chapter 5. There, we suggested that the patterns of regional specialization for language and higher cognitive functions found in human adults are not present at birth. We argued that the spatial distribution of function is set up over time through a process of competition and recruitment that takes advantage of regional differences in computing style that are in place very early. We are now in a position to elaborate and extend the account of how cognitive functions might come to be localized in different brain regions without being specified either at birth or explicitly in the genome. We identify three important factors which contribute to the indirect specification of cortical function.

(1) First, certain problems have intrinsically good solutions. The problem space itself constrains what are likely outcomes. For example, in Figure 6.5 we show three different architectures which could be used to learn XOR. In the most common architecture, shown in Figure 6.5a, one of the hidden units learns the AND function and

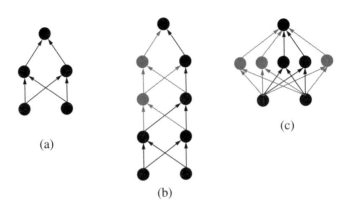

FIGURE 6.5 Three architectures for learning XOR. In the first network, one of the hidden units typically learns the AND function and the other learns the OR function. When additional units or layers are added, networks often end up ignoring the extra resources in favor of the AND/OR solution.

the other learns OR. XOR is detected in cases where the OR units gives a positive response and the AND unit gives a negative response.

What happens if we provide extra hidden units, either with additional layers (Figure 6.5b) or by extending the width of a single layer (Figure 6.5c)? When we analyze the solution found by one of these other networks, it turns out that the extra units and connections are essentially ignored, and the networks stick with the AND/OR solution. (Sometimes this occurs through weights decaying to 0 and other times the AND and OR functions are distributed across multiple hidden units.) There's no keeping a good solution down, it seems.

Finding that different networks end up learning the same solution thus does not necessarily imply that the solution has been pre-wired. None of these networks have innate knowledge that leads to solving the XOR function with AND and OR units. The fact that this solution recurs, even when we begin with different initial architectures, arises because the problem itself constrains the kinds of solutions which are likely to be found. In this case, we may say that the external environment (that is, the problem space) provides the constraints.

(2) It is also possible for solutions to be weakly specified by virtue of computational characteristics of the processing mechanism, be it brain or network. We emphasize that what we mean by "computational characteristics" are *domain independent* properties. We mean by these the kinds of differences we described in Chapters 1 and 5, including packing density (i.e., number of neurons within a particular zone), size and speed of neurons (e.g., large and fast-acting magno cells vs. small, numerous, but slow-acting parvo cells), gross differences in amount and type of connectivity (i.e., regions with high vs. low "fan in" and "fan out"), degree of excitation, inhibition and self-modulation, and neurochemical properties that are still poorly understood.

If we assume that the brain is an enormous and highly differentiated neural network, with many parts of the system activated in parallel, then it is reasonable to assume that development is in part based on competition. Through this competition, regions of the brain attract those inputs they handle particularly well, and they are recruited for those tasks which require a particular form of computation.

This point is well-illustrated in the modular networks developed by Jacobs and his colleagues, and described in Chapter 2 (see Figure 2.11). Recently, Jacobs and Kosslyn (in press) have used this approach to show through connectionist simulations how regional differences in the kinds of information processed by an area (e.g., categorical information vs. metric coordinate information) might arise from differences in the size of the receptive field attended to by different areas of the brain. In the Jacobs and Kosslyn simulations, what is important in determining which part of a network learns which task is the size of the visual field of the processing units. Units with large receptive fields take over the job of making a categorical judgment (e.g., whether a dot is above or below a bar), and units with small receptive fields preferentially learn to make metric coordinate judgments (e.g., whether a dot is within four metric units the bar). What is innate is not that one part of the network is prespecified to extract categorical vs. coordinate information per se; rather, different types of tasks are better done by units with different receptive fields. Kosslyn and Jacobs also found the complementary effect: Not only do certain types of units attract certain tasks, but given a specific task and the ability of adapt the receptive field size, a network would modify the receptive field size in a way which best suited the particular task.

(3) Finally, our example of the Hebbian network illustrates yet another way in which cortical function may come to be assigned to different brain regions. Differential rates of maturation (in this case, a spatio-temporal wave of growth factor which alters the plasticity of connections) result in localization of low-order functions in early-learning "cortex" and high-order functions in late-learning areas. We believe that evolution has selected for a scheduling of cortical maturation which in fact results in a fairly complex organization in which early developing regions act as filters which extract low-order invariants from their afferents, and then pass their outputs to later learning areas. The result is a cascade of filters from primary cortical centers to areas which process successively higher-order and more abstract information. The chronology, rather than content of development, may be closely regulated, yielding an outcome which is tightly constrained and relatively invariant across individuals. This view is what we called in Chapter 1 a form of "chronotopic nativism."

CHAPTER 7

Rethinking innateness

Where does knowledge come from?

This was the question we began with. As we suggested at the outset, the problem is not so much that we do not know what the sources of knowledge are. The problem is rather in knowing how these sources combine and interact. The answer is not Nature *or* Nurture; it's Nature *and* Nurture. But to say that is to trade one platitude for another; what is necessary is to understand the nature of that interaction.

We stated at the beginning that including the word "innate" in our title gave us misgivings. There can be no question about the major role played by our biological inheritance in determining our physical form and our behaviors. We are not empiricists. What troubles us about the term innate is that, as it is often used in the cognitive and developmental sciences, it suggests an overly simplistic view of how development unfolds. To say that a behavior is innate is often taken to mean—in the extreme case—that there is a single genetic locus or set of genes which have the specific function of producing the behavior in question, and only that behavior. To say that a behavior is innate is thus seen as tantamount to explaining the ontogeny of that behavior. In fact, nothing has been explained. And when innate is used more reasonably to refer in general to our genetic endowment, the term ends up being almost vacuous, given the enormous complexity of interactions which are involved in the expression of that endowment.

So why did we use the word at all? In part, it is precisely because the concept of innateness is in such wide use that we wanted to highlight it in the title. There is no avoiding innateness. It has both widespread use in the scientific community, and, unfortunately, figures prominently in the recurring debates about the supposed genetic bases for intelligence. We chose therefore to take the bull by the horns.

But our purpose is more ambitious. We are in fact interested in understanding those phenomena which are often said to be innate. In an important sense we agree that an enormous amount of our behavior *is* innate—only what we mean by the word is quite different than what is often meant. Our goal in this book has been to provide a framework and some tools for understanding the details of the mechanisms by which nature maximizes the likelihood of achieving good solutions to difficult problems. This task is complicated by the fact that all nature cares about is outcomes. Optimal and efficient solutions count for little; tidiness and elegance count for even less.

Despite this, we feel that neuroscientific and modeling techniques have advanced to the point where we can begin to make sense of the developmental process in some concrete detail. We regard this book as a tiny step in this direction. What we propose is not yet a theory, but may hopefully be the beginning of one. It might best be thought of as a framework, a set of conceptual tools for thinking about issues such as development and innateness.

In this final chapter we wish to tie together the various strands laid out in earlier chapters. We begin by summarizing what we see as the primary lessons that can be drawn from the work we have presented. Some of these lessons are negative—warnings about the possible overinterpretation and misinterpretation of phenomena which have been taken as strong indicators of innate bases for behavior. Other lessons are positive—new ways of understanding these same phenomena. Finally, we turn to the future. We recognize that we have but touched the tip of a very large iceberg and much remains to be done. In the final section we identify what we see as some of the more pressing issues and the more promising avenues for future work.

So, what have we learned? And what have we tried to do? The following are points we would emphasize:

• It is important to distinguish the *mechanisms* of innateness from the *content* of innateness. There need not be a one-to-one mapping between the two. We suggest that for higher-level cognitive behaviors, most domain-specific outcomes are probably achieved by domain-independent means.

•The relationship between mechanisms and behaviors is frequently nonlinear. Dramatic effects can be produced by small changes.

•What appear to be single events or behaviors may have a multiplicity of underlying causes, some of which may be distant in time.

•Knowledge ultimately refers to a specific pattern of synaptic connections in the brain. In this very precise sense, we argue that there is probably no higher-level knowledge which is innate.

•The developmental process itself lies at the heart of knowledge acquisition. Some complex behaviors may not be acquirable without passing through a developmental pathway.

•Connectionism provides a useful conceptual framework for understanding emergent form and the interaction of constraints at multiple levels; connectionism should definitely not be thought of as radical empiricism. Connectionism is still in its infancy, however, and development provides a rich set of phenomena which challenge existing technology.

•Development is a process of emergence. Connectionism provides a conceptual vehicle for studying the conditions under which emergent form arises and the ways in which emergence can be constrained.

Let us consider each of these points in turn.

A crucial point: Mechanism and content are not the same thing

In Chapter 1, we discussed innateness in terms of mechanism and content. If there is one single point we would insist on over others, it is that this distinction is crucial to understanding innateness. In our opinion, enormous confusion has been generated by failure to draw this distinction, with the result that the domain specificity of cause is conflated with the domain specificity of outcome.

In talking about **mechanisms**, we have suggested that constraints may operate at three levels: representations, architectures, and timing.

The strongest and most specific form of constraint—and the one which is most direct with regard to outcome (see discussion of "knowledge," below)—is representational, expressed at the neural level in terms of direct constraints on fine-grained patterns of cortical connectivity. This is, we think, the only kind of neural mechanism capable of implementing the claim that detailed knowledge of (for example) grammar, physics or theory of mind are innately specified.

Architectural constraints come next. In our analysis, architectural constraints can be further divided into three sublevels: unit-level architectural constraints (corresponding to the physical structure and computing properties of individual elements within a region, including neuronal types, their relative density, kinds of neurotransmitters, inhibitory vs. facilitative capacity); local architectural constraints (corresponding to the patterns of connectivity that define types of regions, and the general characteristics of layering and connectivity within a region); and global architectural constraints (corresponding to the ways in which local brain regions are connected to one another, and to the input/output pathways connecting the brain to the rest of the body).

Finally, we come to constraints on the timing of developmental events within and across regions. Time is a powerful mechanism for developmental change, and dramatic changes in species or individuals can result from small changes in the timing of developmental sequences.

As we pointed out in some detail in Chapter 5, representational constraints (the strongest form of nativism) are certainly plausible on theoretical grounds, but the last two decades of research on vertebrate brain development force us to conclude that innate specification of synaptic connectivity at the cortical level is highly unlikely. We therefore argue that representational nativism is rarely, if ever, a tenable position. This may be the single most controversial position we take in this book. To explain why mice do not become men, and vice-versa, we are probably going to have to work out a scenario involving constraints on architecture and timing. Neural networks can, we think, play a very useful role in exploring these hypotheses.

This taxonomy of mechanisms is logically independent of **content**. In particular, as we pointed out in Chapter 1, claims about innate mechanisms must be separated from the much-debated issue of domain specificity. Once we have separated the two, however, they can be used together as we try to understand how change comes about. Do we have innate mechanisms (representations, architectures, timing) that evolved in the service of specific content, including species-specific domains like language, music, faces or theory of mind? It is certainly possible, indeed likely, that uniquely human activities have played some role in the evolution of a uniquely human brain. This does not mean, however, that we are entitled to leap directly from "special" content to "special" mechanisms. When we say that a given ability is "special," we are talking about unique or unusual properties that could hold at many different levels: (a) in the structure of the task or problem to be solved, (b) in the unique solutions that we develop to deal with that task, (c) in the representations that underlie our ability to solve the problem (and hence our knowledge of the solution), (d) in the learning device and/or processing mechanism that made it possible for us to acquire those representations, and perhaps (e) in the genetic code that leads to the development of these mechanisms. The mere demonstration that a given behavior is "special," i.e., unlike anything else that we do, does not constitute evidence for (or against) the specificity of the mechanism that learns and produces this behavior. We believe that connectionist models are particularly useful as we try to figure out just how much specificity of content we have to build into our machines, and at what level, in order to simulate something that looks like human learning.

Dramatic effects can be produced by small changes

When we see a dramatic change in behavior, we are tempted to make several inferences. First, we tend to suppose that the dramatic effect has an equally dramatic cause. We might believe that a new mechanism has kicked in (e.g., the learning of a rule), or perhaps that an existing mechanism has atrophied (e.g., the loss of a language acquisition device). Sometimes we also conclude that the cause is proximal in time, i.e., behavior changes because the child is now ready for that change.

In fact, we have offered many examples in this book of cases in which a single mechanism gives rise to very different overt behaviors at different points in time. The best known example is the well-studied case of the English past tense, where a single learning mechanism operating on a constant or incrementally expanding data base goes through dramatic changes at the behavioral level— the kind of change that was once ascribed (without further ado) to a switch from rote memory to the application of a rule. A single mechanism, operating incrementally, need not undergo abrupt *internal* changes in order to produce abrupt *external* changes in behavior. Furthermore, the sources of change may be quite distant from the changes themselves. This includes temporal distance (in cases where the causal factors long precede the changes that we ultimately observe) and logical distance (defined in terms of levels of organization). In short, things do not always happen right at the point where we notice a change. A basic lesson here is that we need to distinguish carefully between overt external behavior, and the presumed internal changes which affects behavior. The mapping may not be straightforward.

Multiplicity underlying unity: A single event has many causes, and the same event can come about in many ways

We have just argued for cases in which incremental changes in a single mechanism can lead to qualitatively different outcomes over time. This goes against an implicit assumption in much developmental research, i.e., that qualitative change at the behavioral level must reflect a qualitative shift in underlying mechanisms. One also finds a tendency in some developmental work in just the opposite direction, where unitary outcomes are attributed invariably to a unitary cause.

(Notice that if one does this, then it is very natural to infer that domain-specific outcomes have domain-specific causes. Thus, for example, if there is selective impairment in the production of regular morphology, it is tempting to suppose that the "regular morphology box" has been broken. But as we saw in Marchman's work, diffuse damage to an entire network can lead to greater deficits in processing regular forms, compared with irregulars.)

The reality is that what might be thought of as single events or behaviors can often be produced by multiple interacting mechanisms. The chick-imprinting model, described in Chapter 6, is a good example of this. The outcome in the network—imprinting on an object—requires the interaction between a nonspecific "subcortical" biasing system and a "cortical" learning system.

We have also seen that the very same event can be brought about in a variety of different ways. Noise and lesioning, for instance, can both lead to the same or similar degraded performance in networks and in humans.

On knowledge

As soon as one asks "*Where* does knowledge come from?" one has also to confront the difficult question, "*What* is knowledge, anyway?" What do we really mean when we say that a child knows something? One of the benefits of thinking in mechanistic terms (such as the connectionist framework we have described) is that it not only requires us to be explicit about such questions, but also provides conceptual tools for answering them.

As the term is currently used in the developmental literature, "knowledge" potentially conflates different kinds of mechanisms (representations, architectures, constraints on timing) and different kinds of content (from task to gene). When a developmentalist argues that knowledge is innate (e.g., Spelke, 1994), we think it would be useful to specify exactly what this means.

We have proposed that the term "knowledge" should be used to refer to the representations that support behavior, and we have proposed operational definitions of the term "representation" that could be implemented in real brains (i.e., the fine-grained patterns of cortical activity, which in turn depend on specific patterns of synaptic connectivity) and in neural networks (i.e., patterns of activations over units in an n-layered system, which depend on the patterns of weighted connectionist between units). Under this definition, a child might still have strong constraints operating at other levels (e.g., architectural or timing) which might highly favor the emergence of specific knowledge. The knowledge itself, however, would not be innate and would require appropriate interactions to develop.

We recognize that some might reject this definition in favor of a different view, and we certainly respect that decision. But we would like to invite our colleagues to engage in a similar exercise, laying out their definition of "knowledge" in explicit, systematic terms that could be implemented in some kind of biologically plausible machine.

Why development?

We asked in Chapter 1 why development occurs at all, especially given its (apparently) maladaptive consequences (immaturity, vulnerability, dependence, consumption of parental and social resources, etc.). In fact, there are species (e.g., ungulates) which are born in a relatively mature state—they are "up and running" (literally) from the start. So it is all the more curious that so-called "higher species" should tend to have protracted periods of immaturity. And of all primates, humans take the longest to mature.

According to the perspective we have developed here, there are a number of good reasons why development should occur.

First, a long period of development allows greater time for the environment (both sociocultural and physical) to play a role in structuring the developing organism. Second, the view we have proposed is that development is the key to the problem of how to get complex behaviors (in the mature animal) from a minimal specification (in the genes). It's Nature's solution to the AI "scaling problem."

What we are not

Connectionism is sometimes viewed (usually by its critics, but sometimes by its proponents) as a return to the *tabula rasa,* i.e., to a behaviorist conception of the infant as a blank page waiting for Nurture's pen. That is, we insist, simply the wrong view. We are neither behaviorists nor radical empiricists. We have tried to point out throughout this volume not only that the *tabula rasa* approach is doomed to failure, but that in reality, all connectionist models have prior constraints of one sort or another. What we reject is *representational* nativism. Connectionism can provide an invaluable albeit incomplete set of tools for reasoning about the nature of interaction, as we set out to unravel the complex interplay of multiple internal and external constraints that underlie the diverging developmental pathways of mice and men.

Emergent form

We are also not the only nor the first to acknowledge the importance of emergent form, or to argue against a strict dichotomy between innate and learned. The new dynamic framework for studying development that we have outlined in this book represents a new contribution to an old tradition in developmental psychology and developmental biology that includes Baldwin, Bateson, D'Arcy Thompson, Oyama, Piaget, Vygotsky, Waddington, Wimsatt, and many, many others.

Our proposals also have a great deal in common with recent writings by other developmentalist colleagues who underscore the value of dynamical systems theory as a formal implementation of the old notion of emergent form (e.g., Smith & Thelen, 1993; Thelen & Smith, 1994; van Geert, 1994; see Port & van Gelder, 1995, for a recent collection of papers on dynamical systems and cognition). Indeed, we are all engaged in a similar enterprise. What we have tried to do in this volume is to bolster these arguments with additional details from developmental neurobiology, and to provide a more concrete and precise set of proposals for the implementation of emergent form in neural networks. We also recognize that our emphasis on representations is not shared by all those who work with dynamical systems.

Models: brain or behavior

We have tried to develop an intermediate position with regard to the role of modeling, an approach which stands somewhere in between pure brain modeling and purely cognitive modeling. To be sure, for many of the higher cognitive processes of interest to us (and to most developmental psychologists), we are forced to simplify away from the details of wet brains. Our hope is that such models will embody abstract but realistic principles of learning and change that make contact with cognitive issues, while preserving some degree of neural plausibility. This is, we acknowledge, a hard choice, and we hope that future research in this area will take place at many levels from brain to cognition, with efforts to translate across levels whenever possible.

Does anyone disagree?

Does anyone really disagree with this interactionist view? Are we railing against a Straw Nativism embraced by no one but pilloried by all?

We think not. Some radical nativist proposals have been offered in the last few years, growing in strength and number in concert with rising public interest in genes for complex outcomes from cancer to divorce. To reassure worried readers that our concerns are justified, we present a sampling of statements from developmental psychologists and linguists that embody, we think, a strong version of what we have called "representational nativism." Many of these quotes come from the field of linguistics, where the Nature-Nurture debate has generated great heat and limited light. But examples from other areas are readily available.

Starting with nativist theories of cognition and conceptual change, we reiterate what we pointed out above, that the word "knowledge" is frequently used but loosely defined. If we interpret their use of the word "knowledge" in the representational sense outlined above, then the following quote from Carey and Spelke (1994) can be viewed as a manifesto in favor of representational nativism:

> *We argue that human reasoning is guided by a collection of innate domain-specific systems of knowledge. Each system is characterized by a set of core principles that define the entities covered by the domain and support reasoning about those entities. Learning, on this view, consists of an enrichment of the core principles, plus their entrenchment, along with the entrenchment of the ontology they determine. In these domains, then, we would expect cross-cultural universality: cognitive universals akin to language universals.* (p. 169)

This approach is taken several steps further in the following quote from Spelke (1994):

> *Although debates continue, studies of cognition in infancy suggest that knowledge begins to emerge early in life and constitutes part of humans' innate endowment....* (p. 431) *Unlike later-developing knowledge, initial knowledge appears to capture what is most true about the*

entities that a child perceives, not what is most obvious about those entities.... (p. 438) If the same initial principles underlie perception and reasoning, however, then the principles could not be learned, because the child would have no other way to parse the stream of experience into the relevant entities. Initial knowledge may emerge through maturation or be triggered by experience, but learning and processing do not appear to shape it. (p. 439)

Leslie (1992) has provided an application of this approach to the infant's "theory of mind" (i.e., beliefs about the contents of other people's minds relative to one's own—see also Leslie, 1994a,b), suggesting (a) that there is an innate knowledge system to deal with mental states, and (b) that the absence of such a system is responsible for childhood autism:

I have argued that the normal and rapid development of theory-of-mind knowledge depends on a specialized mechanism that allows the brain to attend to invisible mental states. Very early biological damage may prevent the normal expression of this theory-of-mind module in the developing brain, resulting in the core symptoms of autism (p. 2)

This mechanism is essentially innate and in some sense, a specific part of the brain. (p 20)

Leslie (1994b) places this theory-of-mind module within a more general framework that assumes innate and modular knowledge in many different domains:

To the extent that there are mechanisms of domain-specific development, then a deeper notion of domain is possible—one that is less software dependent, less profligate, and more revealing of the design of human cognition. This kind of domain specificity reflects the specialization of mechanisms in core cognitive architecture...the core contains heterogeneous, task-specialized subsystems. Vision is an obvious example of a specialized subsystem with a specialized internal structure. The language faculty is another (p. 120)*...a mechanics module...the infant's processing of the physical world appears to organize rapidly around a core structure representing the arrangement of cohesive, solid, three-dimensional objects embedded in a system of mechanical relations, such as pushing, blocking, and support.* (p. 125)

Leslie's claims about innate domain-specific mechanisms seem to coincide with what we have called representational nativism.

Indeed, it would be difficult to implement such detailed claims about innate domain-specific knowledge without assuming fine-grained patterns of cortical connectivity. However, these particular authors are never explicit about the kind of neural mechanism that they envision as implementations of their nativist claims. Indeed, there may be room here for a reinterpretation in terms more compatible with the less direct but more neurologically plausible architectural and chronotopic constraints that we have put forward in this book.

In contrast with the potentially ambiguous claims offered by nativists working on cognitive development, claims within linguistics and child language are often quite explicit regarding the degree and kind of representational detail that must be innate.

In an article with the provocative title "Language acquisition in the absence of experience," Stephen Crain (1991) offers the following remarks:

> *In linguistics, one finds a near consensus on the need for constraints in explaining the development of linguistic knowledge. In addition, linguists generally find it reasonable to suppose that constraints are innate, domain-specific properties. A distinguishing feature of recent linguistic theory, at least in the tradition of generative/transformational grammar, is that it postulates universal (hence, putatively innate) principles of grammar formation, rather than characterizing the acquisition of language as the product of general cognitive growth (Chomsky, 1971; 1975). This theoretical framework is often referred to as the theory of Universal Grammar, a theory of the internal organization of the mind/brain of the language learner.*

> *We began by observing that recent developments in linguistic theory (the postulation of universal constraints of language acquisition), together with the observation that children's linguistic experience is quite limited (the absence of carefully sequenced input or negative evidence), reinforce the view that syntactic knowledge is in large part innately specified. What is innately given is knowledge of certain restrictions on the meanings that can be mapped onto sentences as well as restrictions on the sentences that can be used to express meanings. This knowledge is encoded in constraints. The problem for the learner is that there are no data available in the environment corresponding to the kinds of negative facts that constraints account for.*

Crain then goes into detail on the following examples, arguing that children must have innate constraints to help them infer that "He" cannot refer to the Ninja Turtle in sentence 1b.

1 a) The Ninja Turtle danced while he ate pizza.
  b) He danced while the Ninja Turtle ate pizza.
  c) While he danced the Ninja Turtle ate pizza.

It seems quite clear to us that innate constraints on the interpretation of examples of this kind must be very detailed indeed, requiring genetic specification of cortical wiring at the synaptic level (corresponding roughly to whatever it is we know after we have learned a language).

Lightfoot (1989) is even more explicit about the role of genetics in specifying innate linguistic knowledge, revealed in the following passages:

> *In the last thirty years, generative grammarians have been developing a selective theory of language acquisition. We have sought to ascertain what information must be available to children independently of any experience with language, in order for the eventual mature linguistic capacities to emerge on exposure to some typical "triggering experience." Cutting some corners, we have assumed that this unlearned information is genetically encoded in some fashion and we have adopted (1) as our explanatory model:*
>
> *(1) a. trigger (genotype→phenotype)*
> *   b. primary linguistic data (Universal Grammar → grammar).*
>
> *The goal is to specify relevant aspects of a child's genotype so that a particular mature state will emerge when a child is exposed to a certain triggering experience, depending on whether the child is raised in, say, a Japanese or Navajo linguistic environment. (1.b) reflects the usual terminology, where "Universal Grammar" (UG) contains those aspects of the genotype directly relevant for language growth, and a "grammar" is taken to be that part of a person's mental make-up which characterizes mature linguistic capacity....Under current formulations of linguistic theory (e.g., Chomsky, 1981), the linguistic genotype, UG, consists of principles and parameters that are "set" by some linguistic environment, just as certain receptors are "set" on exposure to a horizontal line.*

Based on the facts about brain development reviewed in Chapter 5, we think it quite unlikely that Universal Grammar could be encoded so directly in the genotype. However, many lines of evidence have been cited to support such a claim. Because language is the domain in which nativist arguments have been worked out in greatest detail, it is probably worth our while here to follow some of those arguments through, examining them from the perspective that we have outlined in this book.

Twelve arguments about innate representations, with special reference to language

Throughout the volume we have stressed that our position is not anti-nativist, but that it is essential in developmental cognitive science to (1) specify the level of innateness that is invoked for a given function (i.e., representations, architectures, timing), (2) distinguish between innateness and domain specificity, and (3) distinguish between innateness and localization. Yet one is repeatedly struck by scientists' focus on a single level (what we identify as representational nativism) in an attempt to identify genes that will explain particular behaviors. This has been particularly evident with respect to arguments in favor of a gene or a specific set of genes for language.

Let us examine the different lines of evidence that are frequently offered in favor of genetic control over the cortical representations that constitute linguistic knowledge, that is, arguments in favor of domain-specific innate representations for language. These arguments include: Species specificity, genetically based language disorders, studies of lesioned brains, activation studies of grammar in the normal brain, structural eccentricity of language, poverty of the stimulus, linguistic universals, modularity of processing, dissociations, critical periods of language learning, and robustness under different learning conditions. We invite the reader to reevaluate these claims in the light of our arguments throughout the book.

1. Species specificity

Clearly no one denies that human beings are the only creatures that are able to learn and use grammar in its full-blown form. While studies of symbol use and language-like behavior in chimpanzees and other nonhuman primates are interesting, all of the animals studied to date stop far short of the language abilities displayed by a normal human child at 3 years of age. However, it does not automatically follow that grammar itself is under strict genetic control because, as we have shown repeatedly in this volume, behavioral outcomes can be the product of multiple levels of interaction in which there is no representational prespecification. To argue for innateness, a specific link between genetic variation and some grammatical outcome must be demonstrated. Can such a link be shown to obtain in the case of language disorders?

2. Genetically based language disorders

In recent years, a number of popular reports in the press, radio and television have suggested that a link between genetic variation and grammatical outcome has indeed been found. These reports were based on preliminary work regarding a specific language impairment in one large family in London (i.e., the KE family), first documented by Hurst et al. (1990). As Hurst et al. noted, this impairment runs through the KE family in patterns that are typical of Mendelian genetic transmission (i.e., autosomal dominance). The following quote from a review by a distinguished British psychologist illustrates the attraction for many scientists of the notion that humans are special and that the human genome contains a "gene for grammar."

> *[There is] one peculiar family whose members, as a result of a defective gene, cannot form plurals, even though in all other respects they speak normally.* (Sutherland, March 7, 1993)

The startling form of highly domain-specific grammatical impairment implied by this quote would indeed (if it were true) constitute strong evidence for a genetic effect restricted to grammar.

Furthermore, given the Mendelian nature of the pattern of heritability described for the KE family, it would not be unreasonable to conclude that a single gene is responsible. The implications that such a finding would have for the nature and evolution of grammar were recognized by two eminent evolutionary biologists in a recent review entitled "The major evolutionary transitions."

> *Perhaps the most convincing evidence both for the belief that grammatical competence is to some degree independent of general learning ability, and for the possibility of functional intermediates between no grammar and perfect grammar, comes from studies of hereditary variation in linguistic competence. One remarkable case involves a family in which a so-called feature-blind dysphasia seems to be inherited in a mendelian fashion, a single dominant gene being responsible (Gopnik, Nature 344, 715, 1990). Members cannot automatically generate plurals and past tense. Although they understand the meaning of plural and past perfectly well, they have to learn each new case anew: "Paint" and "painted," "book" and "books" must be learned separately (in the case of exceptions such as "go" and "went," normal individuals must do the same). To be sure, this is not a genetical violation of one of Chomsky's rules, but it demonstrates that there can be something useful between perfect grammar and protolanguage: it also holds out the hope that we will in the future be able to dissect language genetically, as we are today dissecting development.* (Szathmary & Smith, March 16 1995; p. 231)

How sound is the evidence on which Sutherland and Szathmary/Maynard Smith base their powerful conclusions? To evaluate the actual data in some detail, it is necessary to trace the "gene for grammar" back to its original sources, and follow the relevant literature up to the present day. In doing so, several things become immediately apparent.

First, this international cause célèbre is actually based on a short and very premature report—not wholly false, but incomplete. The story begins with a brief letter to *Nature* in 1990 (Gopnik, 1990), in which a series of preliminary tests of the KE family's language abilities were reported. Gopnik concluded at the time that members of the family suffered from a deficit that is restricted primarily to the grammatical rules that underlie regular inflectional morphology (e.g., regular past-tense markings like kiss ↦ kissed, and regular pluralizations like dog ↦ dogs), sparing the ability to memorize words and their meanings, and the ability to memorize irregular morphemes (e.g., irregular past-tense markings like go→went, and irregular pluralizations like mouse→mice). An extended version of this study was published the following year in the journal *Cognition* (Gopnik & Crago, 1991). Based on their data for the KE family and other dysphasic subjects in Montreal, Gopnik and Crago conclude with the following:

> *It is not unreasonable to entertain an interim hypothesis that a single dominant gene controls for those mechanisms that result in a child's ability to construct the paradigms that constitute morphology.* (p. 47)

Although these conclusions were couched in terms of tentative hypotheses and have since been weakened as more comprehensive data on both the language and nonlinguistic cognition of this family became available (Gopnik et al., in press), the attractions of the Gopnik & Crago data were rapidly taken up by others in the field. For example, in a review article in *Science*, Pinker (1991) argued for a neurally and perhaps genetically based double dissociation between regular and irregular grammatical morphemes. Pinker contrasts the dissociation in the KE family (where regulars are supposedly more impaired than irregulars) with a complementary deficit reported in Williams Syndrome. Integrating this apparent double dissociation with other arguments for a distinction between regular and irregular forms (see Chapter 3 for more details), Pinker marshalled the following argument:

> *Focusing on a single rule of grammar, we find evidence for a system that is modular, independent of real-world meaning, non associative (unaffected by frequency and similarity), sensitive to abstract formal distinctions (for example, root versus derived, noun versus verb), more sophisticated than the kinds of rules that are explicitly taught, developing on a schedule not timed by environmental input, organized by principles that could not have been learned, possibly with a distinct neural substrate and genetic basis.* (p. 253)

Of course, if the data were as clear-cut as originally supposed, then Pinker's claims would indeed constitute a strong case for the domain-specific innateness of the human language faculty (see, also, Pinker, 1994b). Although Pinker makes it clear that he does not believe in a single gene for grammar, he does argue that the specific profile reported for the KE family reflects a specific genetic program for language that cannot be derived by simply reconfiguring less specific cognitive systems. Despite Pinker's own disclaimer of a single gene for grammar, reviews of his influential book in popular newspapers and magazines brought the grammar gene concept back to center stage in its original form.

A sharply contrasting view of the KE family comes from a group of British researchers at The Great Ormond Street Hospital for Children in London, where the family has been under study since 1988. No one disputes the conclusion that this disorder has a genetic base, a finding that was first reported by Hurst et al. (1990), who conclude that the pattern of inheritance in the KE family may be due to a single gene (i.e., autosomal dominance). The contrast between Gopnik and the Great Ormond Street data revolves not around the innateness of this disorder, but its domain specificity. In a letter of rebuttal to Gopnik and her colleagues published in *Nature,* Vargha-Khadem and Passingham (1990) note that the grammatical symptoms displayed by members of the KE family are just one part of a much broader symptom complex:

> *Gopnik has focused on only one aspect of the disorder, and it is inaccurate to conclude that the speech and language problem stems simply from an impairment in manipulating feature markers.* (Vargha-Khadem & Passingham, p. 226; see also Fletcher, 1990)

In 1995, a comprehensive report of the London group's results was published, including extensive tests of language and nonlanguage abilities (Vargha-Khadem et al., 1995). Controlling for age, they showed that affected members of the KE family perform significantly worse than unaffected members on all but one of 13 language tests (including tests of phonology, grammar and lexical abilities). In addition, the affected members are significantly lower on *both* verbal and nonverbal IQ (about 20 points in both cases). Moreover, they perform very poorly on a nonlanguage test of oral-facial praxis (i.e., production and imitation of simple and complex movements of the tongue and mouth). In fact, their oral/facial apraxia is so severe that some members of the family are completely unintelligible to strangers. In contrast to the claims that have been made in some of the scientific literature and in the popular press, the affected members of the KE family are significantly worse than the unaffected members on *both* regular and irregular morphology, with no evidence for a dissociation between the two. The disparity between the Vargha-Khadem et al. and the Gopnik and Crago findings is explained by the fact that the former are based on a much larger set of experimental morphological contrasts than the original four past-tense items used by Gopnik and her colleagues: Two regulars (kissed and walked) and two irregulars (went and was). Yet these subsequent clarifications about the broad array of deficits actually present have gone largely unnoticed in public discussions of the KE family.

The search for a genetic base to grammar has been prominent among some developmental investigators too (e.g., Matthews, 1994; Rice, in press; van der Lely, 1994). Consider the syndrome known as Specific Language Impairment, or SLI. SLI is usually defined in terms of expressive (and perhaps receptive) language abilities being one or more standard deviations below the mean, and below the same child's nonverbal Performance IQ, in the absence of evidence for mental retardation, frank neurological impairment, abnormal hearing, social-emotional disorders and/or social-demographic conditions that could explain the disparity between language and other cognitive functions (Bishop, 1992; Bishop & Rosenbloom, 1987). Hence, by definition, SLI is thought to represent an impairment that affects language and nothing else. Furthermore, it has been repeatedly demonstrated that grammatical morphology is particularly vulnerable in children with SLI (Johnston & Kamhi, 1984; Leonard, 1992), albeit with little evidence for the claim that regular morphemes are more vulnerable than irregulars (see Marchman, Wulfeck, & Weismer, 1995, for a discussion of this point). Most important for our purposes here, a number of studies have shown that SLI and associated disorders (especially dyslexia) tend to run in families (Bishop, 1992; Bishop & Rosenbloom, 1987; Pennington, 1991; Tallal, Ross, & Curtiss, 1989; Tallal et al., 1991).

At face value, this looks like evidence for genetic specification of grammar. However, as with the case of the KE family, detailed studies of children with SLI have shown that the syndrome is not restricted to language, nor to grammar within language.

For instance, Tallal and her colleagues have amassed a large and compelling body of evidence suggesting that children with SLI suffer from a deficit in the processing of rapid temporal sequences of auditory and (perhaps) visual stimuli (Tallal, 1988; Tallal, Stark, & Mellitts, 1985). Other specific nonlinguistic deficits implicated in SLI include symbolic play in younger children (Thal et al., 1991), aspects of spatial imagery in older children (Johnston, 1994), specific aspects of nonverbal attention (Townsend et al., 1995), and a wide range of neurological soft signs (e.g., Trauner et al., 1995). More recently Tallal and colleagues (Tallal et al., 1995) have shown that by short-term training of SLI children on oral language in which phonemic transitions are lengthened to facilitate processing, dramatic improvements are found at all levels of the language system, suggesting once again that it is not morphosyntax that is specifically impaired, but some more general deficit in the processing of rapid sequential material.

In fact, problems with receptive processing of grammatical morphemes are observed in an even wider range of populations, including anomic aphasics who display no grammatical deficits in their spoken language (Bates, Devescovi et al., 1994; Devescovi et al., in press), and a subset of elderly patients who are hospitalized for nonneurological disorders (Bates, Friederici, & Wulfeck, 1987). Receptive agrammatism has even been induced in college students who are forced to process sentences under stress (e.g., perceptual degradation and/or cognitive overload: Bates, Devescovi et al., 1994; Kilborn, 1991; Miyake, Carpenter, & Just, 1994).

It appears, then, that grammatical morphology is selectively vulnerable under a wide range of conditions, genetic and environmental. A parsimonious account of all these findings would be that grammatical morphology is a weak link in the processing chain of auditory input, one that is highly likely to be impaired when things go awry. None of these examples points necessarily to specific genes for grammar.

Why did the 1990 Gopnik letter to *Nature* find such a wide audience while the 1990 rebuttals by Vargha-Khadem and Passingham and Fletcher, and their subsequent full-scale report, fall on deaf ears? Why is the notion of specific "genes for grammar" so attractive? There seems to be a deep-rooted desire to believe that humans are not only unique (every species is, after all) but that our uniqueness arises from a quantum leap in evolution. The grammar gene(s) satisfies this desire. In reality, the fact that we share almost 100% of our genes with nonlinguistic species suggests that our language capacity is more likely to be the not-so-simple result of multiple constraints operating at different levels in ways argued throughout this book.

Localization

As stressed in Chapter 5, localization and innateness are not the same thing. For example, studies using positron emission tomography (PET) have revealed specific areas of the brain that are very active in response to real words and to nonsense words that follow the rules of English spelling, but not to words that violate English spelling rules (Petersen et al., 1988). Although this clearly suggests localization, surely all would agree that a prespecified brain circuit for English spelling is a highly unlikely candidate for genetic determination. In the same vein, PET studies have revealed specific areas of the brains of chess masters that glow at specific points in the game (i.e., Nichelli et al., 1994), and yet no one (least of all Nichelli et al.) would argue that chess is innate, at least not in any nontrivial sense. These localized areas for spelling or chess are the *result* of progressive specialization following massive experience, not of innate specifications. Yet evidence of localization is often used to buttress claims about innateness, on the assumption that innate systems have inherited their own dedicated neural architecture (Fodor, 1983; Pinker, 1994b). Kandel, Schwartz, and Jessell (1995) express these assumptions in a chapter on brain and language:

> *Chomsky postulated that the brain must have an organ of language, unique to humans, that can combine a finite set of words into an infinite number of sentences. This capability, he argued, must be innate and not learned, since children speak and understand novel combinations*

*of words they have not previously heard. Children must therefore have
built into their brain a universal grammar, a plan shared by the grammars of all natural languages.* (p. 639)

Does the human brain contain at birth specific neural regions
that are dedicated exclusively to the representation and processing
of grammar? Two complementary lines of evidence are relevant to
this question: studies of language ability in adults and children
with focal brain injury, and studies of language-related brain activity in normal individuals.

3. Localization I: Lesion studies

It has been known for some time that aphasic disorders in adults are
strongly correlated with lesions to the left hemisphere in right-
handed individuals. This is not in dispute. But during the 1970's
and early 1980's, a stronger claim was made: Deficits in grammar
are associated with lesions to specific areas of left frontal cortex,
resulting in agrammatic Broca's aphasia (Caramazza & Berndt,
1985; Heilman & Scholes, 1976; Zurif & Caramazza, 1976; for an
extension of this assumption to early child language, see Greenfield,
1991).

A key to this argument lies in the generality of the agrammatic
deficit, in both comprehension and production. This hypothesized
syndrome would include the well-documented impairments of
expressive language that characterize Broca's aphasia (i.e., tele-
graphic speech with omission of inflections and function words), as
well as the receptive processing of the same grammatical elements—a receptive deficit that can only be detected when the
patient is prevented from using semantic information to interpret
sentences.

For example, Broca's aphasics typically perform quite well in
the interpretation of passive sentences like "The apple was eaten by
the boy," but they perform poorly on semantically reversible sentences like "The boy was chased by the girl," sentences that force
these patients to rely exclusively on grammatical information.
Based on findings like this, it was proposed that agrammatic
Broca's aphasics have lost all access to grammatical knowledge,

with the sparing of semantics. A complementary analysis was offered to explain the comprehension deficits, empty speech and severe word-finding deficits associated with fluent Wernicke's aphasia: All these symptoms could be interpreted in terms of a by-product of a central deficit in semantics, with the sparing of grammar.

The proposed double dissociation between grammar and semantics seemed to unite a broad array of symptoms under a single description, while providing neurological evidence for a modular distinction between grammar and semantics that had been proposed on independent grounds within generative linguistics. This appealing theory has, however, been challenged by subsequent data, including (1) case studies showing that some agrammatic Broca's aphasics perform normally on receptive grammar tasks (Miceli et al., 1983; Tyler, 1992), (2) a large number of studies showing that receptive grammatical deficit is in fact observed in many different clinical groups and, as mentioned above, in normal adults who are forced to process sentences under adverse conditions (Bates, Devescovi et al., 1994; Blackwell & Bates, 1995; Kilborn, 1991; Miyake, Carpenter, & Just, 1994); (3) the finding that "classic" agrammatic patients are able to make fine-grained judgments of grammaticality (Linebarger, Schwartz & Saffran, 1983; Shankweiler et al., 1989; Wulfeck, 1988); and (4) cross-linguistic studies showing that so-called agrammatic aphasics retain detailed features of their native grammar, evident in many different aspects of their expressive and receptive language (Bates (Ed.), 1991; Menn & Obler, 1990).

At this point in the history of aphasia research, few investigators still cling to the idea that grammatical knowledge is localized in or around Broca's area. Instead, the grammatical deficits of nonfluent patients are usually explained with reference to processing deficits that are only indirectly related to grammar itself. This transition in thinking about the nature of agrammatism is illustrated in the following quote from Edgar Zurif, one of the original proponents of central agrammatism in the 1970's (Zurif et al., 1993):

> *The brain region implicated in Broca's aphasia is **not** the locus of syntactic representations per se. Rather, we suggest that this region provides processing resources that sustain one or more of the fixed*

> *operating characteristics of the lexical processing system characteristics that are, in turn, necessary for building syntactic representations in real time.* (p. 462)

The story is rendered still more complex by a growing body of evidence showing that children with left-hemisphere injury are able to acquire language abilities within the normal range (Aram, 1992; Eisele & Aram, 1995; Reilly, Bates, & Marchman, in press; Vargha-Khadem et al., 1991, 1992). As we noted in Chapter 5, this does not mean that the brain is equipotential for language at birth, because specific correlations between language deficits and lesion site are observed in the first stages of language development (Bates, Thal, et al., 1994; Reilly et al., in press; Thal et al., 1991). However, these correlations do not map onto the lesion-symptom relations that are typically observed in adults (e.g., comprehension deficits in children with right-hemisphere injury; expressive deficits that are specifically associated with the putative receptive areas of left temporal cortex). Furthermore, there is no evidence for a dissociation between grammar and semantics in these children. The same left temporal lesions that lead to a delay in expressive vocabulary are also implicated in the delays that are observed during the emergence of grammar.

Although it is possible that a specific grammar region may emerge from lesion studies at some future time, the evidence to date suggests that grammatical knowledge is broadly distributed in the adult brain. Indeed, recent evidence from split-brain patients suggests that detailed judgments of grammaticality can be made in the right hemisphere (Baynes, 1990; Baynes & Gazzaniga, 1987; Zaidel, 1990), suggesting that grammatical knowledge is distributed across both hemispheres.

4. Localization II: Activation studies of grammar in the normal brain

Tools for the study of language-associated brain activity include positron emission tomography (PET), functional magnetic resonance imaging (fMRI), and event-related scalp potentials (ERP). All of these methods have been applied to the study of processing at the single-word level (Garnsey, 1993; Petersen et al., 1988), and PET and ERP have also been used in studies that attempt to disentangle semantic and grammatical processing (Hagoort et al., 1993; King & Kutas, 1992; Mazoyer et al., 1993; Mecklinger et al., 1995; Münte, Heinze, & Mangun, 1993; Neville et al., 1991; Osterhout & Holcomb, 1993). Although most studies show greater activation in the left hemisphere during language tasks, there is little or no consensus across studies regarding the regions within the left hemisphere that have greatest responsibility for phonological, semantic and/or grammatical processing (Poeppel, in press).

An initial round of ERP studies comparing the brain potentials associated with semantic vs. grammatical violations did lead some investigators to propose that grammatical violations are associated with specific patterns of brain activity (Hagoort et al., 1993; Mecklinger et al., 1995; Neville et al., 1991; Osterhout & Holcomb, 1993). Candidate syntactic potentials have included an early negative wave that is largest over left anterior scalp (i.e., the so-called N280—Neville et al., 1991), and a slow positive wave that extends to around 500-800 msec following a grammatical error (i.e., the syntactic positive shift, SPS or P600—Hagoort et al., 1993; Osterhout & Holcomb, 1993). However, more recent studies have shown that the N280 is neither specific to function words nor specific to grammatical/syntactic errors. Instead, this negative component is present for all words, with a latency that reflects the word's length and frequency of usage (King & Kutas, 1995). In fact, both the N280 and the N400 appear to lie along a continuum that cuts across word class.

In the same vein, studies varying the probability of violations (i.e., the percentage of items that contain such a violation) have shown that the P600 behaves in many respects like a well-known positive component (the P300) that is observed with violations in many different verbal and nonverbal tasks (Coulson, King, & Kutas, 1995). Certain types of grammatical (morphosyntactic) violations elicit both an N400 and a P600 simultaneously (Kluender & Kutas, 1993, in press; Münte et al., 1993). In addition, one other component (the left anterior negativity or LAN, presumably different from the N280 although it overlaps spatially and temporally) that has been linked to syntactic violations has been observed even in the absence of violations, specifically in the context of perfectly grammatical sentences (wh- questions) that make heavy demands on working memory (Kluender & Kutas, 1993, in press). Both the multiplicity of "grammar" and "semantic waves" and the substantial overlap between them indicate that grammar and semantics are not each uniquely associated with "signature" components of the ERP. Thus, while it is fair to say that different violations are associated with different patterns of brain activity, these differences do not constitute evidence for a special, autonomous grammatical processor.

In fact, the number of so-called language-specific areas are multiplying almost on a daily basis. Every new functional imaging study seems to bring another language area to our attention. This is true not only for PET, fMRI, and ERP, but also for studies of brain activity in epileptic patients with electrode grids placed directly on the cortical surface. For example, new language areas have been discovered in the fusiform gyrus, in left and right basal temporal cortex (Lüders et al., 1986, 1991), an area that has never previously been linked to language impairments in lesion studies. But it is important to stress that this proliferation of areas is not unique to language; it has been occurring in every area of perception and cognition as fineness of technical measures improves (e.g., Orban et al., 1995).

This all leads to the conclusion that domains like language do not live within well-defined borders, at birth or at any other point in development. Existing evidence for the localization of language (such as it is) provides little support for the idea that children are born with neural mechanisms that are prespecified for and dedicated solely to language processing.

5. Structural eccentricity

Beyond the biological/neurological arguments, investigators of a strong Nativist persuasion have also invoked the structural eccentricity of language to argue for its domain-specific innateness. Language, it is argued, is so different from other behavioral systems that it cannot be explained by general mechanisms of learning and cognition. It is true that language involves a host of opaque and arbitrary structural facts, a point that Chomsky has made in numerous publications. There is, for example, no obvious communicative account of the fact that:

*We expected Bill to like each other.

is ungrammatical whereas

Each of us expected Bill to like the other(s).

is grammatical, since both sentences are perfectly comprehensible (Chomsky, 1975, p. 101). Why is the fact that language is often expressed in structures which have no obvious communicative advantage relevant to the innateness debate? It is relevant because eccentricity of form is used as evidence for eccentricity in the underlying processors responsible for the learning and processing of those forms. And yet we have offered numerous examples throughout this volume in which peculiar input-output mappings were learned by a connectionist network with no prespecified representations. Furthermore, such eccentricities may arise for reasons which have nothing to do specifically with domain-specific (linguistic) representational constraints, but rather with peculiarities at the level of architecture or timing which may be domain neutral in their content (but still having specific consequences for different domains). We conclude that the mere existence of structural eccentricities tells us little about either the level of the constraint which gives rise to them, or their domain-specific or domain-general character.

6. Poverty of the stimulus

Perhaps the most compelling behavioral evidence for innateness comes when a newborn gazelle leaps to its feet and runs at birth, or when a spider makes a perfect web on the first try with no prior opportunity to observe web-weaving by another spider. Such examples constitute evidence for performance in the absence of experience and learning. Human language does not fall directly within that class of behaviors, because children take at least three years to get a full-blown grammar up and running. However, a cornerstone of strong nativist claims about language rests on the idea that grammars cannot be learned, no matter how long it takes, because children do not receive enough evidence from the input to support the kinds of generalizations that they ultimately draw. This is, of course, the well-known poverty-of-the-stimulus argument.

Gold's theorem (Gold, 1967) has been used as the foundation for arguments of this kind, a formal proof that grammars of a certain class are unlearnable in the absence of negative evidence (i.e., in the absence of evidence that certain sentences are illegal in the language). But of course, as we have noted elsewhere (e.g., Chapters 2 and 3), Gold's theorem rests on some very strong assumptions about the nature of grammar and the nature of the learning device, e.g., (1) that the grammar to be learned consists of strings of discrete symbols governed by discrete rules, (2) that the learning device itself is an hypothesis-testing device that makes yes/no decisions on each trial, and (3) that whole grammars are tested one at a time against each incoming sentence (for detailed discussions, see Pinker, 1979; Wexler & Culicover, 1980). If these assumptions do not hold (and they are implausible for human children), then Gold's theorem is simply not germane. Connectionist simulations of language learning can be viewed as empirical tests of learnability claims, based on very different assumptions about the nature of grammatical knowledge and the nature of the learning device. Some of these simulations have already shown that the impossible is in principle possible, although a lot of important work obviously still lies before us.

7. Universals

Another common argument for innateness revolves around the existence of language universals. Despite their differences, the grammars of natural languages have many properties in common. Because these universals could not have arisen by chance, they are offered as prima facie evidence for innate grammatical knowledge (Roeper, 1988).

However, in this volume we have seen a number of examples of connectionist networks that discover similar and stable solutions to well-defined problems; an example we have used often is XOR. Another such example comes from models of the visual system. It is very common for units that have center-surround and line orientation receptive fields to emerge in simulations in which 3-D images are mapped onto a 2-D array, and the task for the network is to reconstruct the original 3-D object from the lower dimensional representation (e.g., Lehky & Sejnowski, 1988). These solutions do not "look like" the problem they solve, but they are ubiquitous, if not universal. The are solutions which are contained in the structure of the problem space—as long as the processing mechanism has the appropriate constraints for solving the problem.

A similar story may underlie the universals that have been reported for grammar. The grammars of natural languages may be thought of as solutions to an even more daunting dimension reduction problem, in which multi-dimensional meanings must be mapped onto a linear (one-dimensional) output channel (the mouth). The fact that these grammars may not always obviously resemble or reflect the underlying content of the message may be irrelevant to the question of where these solutions come from.

8. Modularity of processing

In the mature speaker/listener, certain aspects of language processing take place rapidly and efficiently, and they often appear to be impervious to contextual factors and/or conscious strategies. These processing characteristics are the hallmarks of modularity as it is defined by Fodor (1983), and the existence of modularity has been used by many theorists as another form of evidence for the innateness of "special purpose" processors (e.g., Gardner, 1983).

However, as Fodor himself has noted (e.g., Fodor, 1985), the same processing characteristics have been demonstrated in laboratory studies of perceptual-motor learning. Any arbitrary skill can achieve "automaticity" if it is practised often enough, under highly predictable conditions (Posner & Snyder, 1975; Shiffrin & Schneider, 1977) —which means that the skill becomes very fast, efficient, hard

to think about, and impervious to interference once it gets underway. Hence modularity may be the result of learning rather than its
cause, i.e., "Modules are made, not born" (Bates, Bretherton, & Snyder, 1988, p. 284), a process of progressive modularization, not initial modularity (Karmiloff-Smith, 1986, 1992a).

9. Dissociations

We have already pointed out that localization and innateness are
logically and empirically independent. Therefore, arguments based
on the dissociations that result from focal brain injury are not necessarily germane to the innateness debate. However, it is sometimes
argued that dissociations give us insights into the organization of
"virtual architecture" or "mental organization," independent of the
neural substrate (Marshall, 1984; Shallice, 1988). In the same vein,
developmental dissociations have been used to argue for the "natural" or "innate" boundaries of the mind (Cromer, 1974; Gardner,
1983; Smith & Tsimpli, 1995). To the extent that language can be dis-
sociated from other cognitive systems, it can be argued that language is a "natural kind."

Although this is a plausible argument (see Karmiloff-Smith,
1992a, Chapter 2, for an extended discussion), dissociations do not
necessarily reflect innate or domain-specific boundaries, because
they could result from deficits that are indirectly related to the final
outcome. This is clear from a number of connectionist simulations
in which a general-purpose learning device "acquires" some aspect
of language and then suffers a "lesion" to the network (Hinton &
Shallice, 1991; Marchman, 1993; Martin et al., 1994; Plaut, 1995).
These simulations often display dissociations that are strikingly
similar to those reported in the literature, and this in the absence of
prespecified (innate) structure and/or localization of domain-specific content. In other words, dissociations can be explained in a
number of ways. Although they provide potentially useful information about the seams and joints of the resulting cognition, they do
not alone constitute compelling evidence for innateness.

10. Maturational course

Fodor (1983) has argued that biologically relevant and domain-specific modules like language follow their own peculiar maturational course. It is true that language has its own characteristic onset time, sequencing, and error types, and the course of language learning includes many examples of U-shaped functions and discontinuous change. Does this constitute evidence for genetically timed maturation? Only if one assumes (as some theorists do) that learning is linear, continuous and subject to immense variability in onset time and sequencing—a claim that is increasingly difficult to defend. We have furnished many examples throughout this volume of nonlinear learning in neural networks, including discontinuous shifts, U-shaped functions, novel overgeneralizations and recovery from overgeneralization. These phenomena have been shown to occur in the absence of prespecified representations, although their emergence always depends on the nature of the initial architecture, the nature of the input, and the time course of learning (cf. Chapters 3, 4 and 6). The line between learning and maturation is not obvious in any behavioral domain (see especially Chapter 5). Language is no exception in this regard.

11. Critical periods

This is a special case of the maturational argument above, referring to constraints on the period in development in which a particular kind of learning can and must take place. The quintessential examples come from the phenomenon called "imprinting," which is argued to differ from domain-general learning in four ways: stimulus specificity, absence of external reinforcers, irreversibility, and temporal boundedness. Although the original imprinting phenomena have been challenged, imprinting persists as a metaphor for constraints on language learning. Three kinds of evidence have been invoked to support the idea that there is a critical period for language learning in humans: "wild child" cases (e.g., deprived children like Genie, or Izard's Wild Boy (Curtiss, 1977; Lane, 1976), age-related limits on recovery from brain injury (Lenneberg, 1967), and age-related limits on second-language learning (Johnson & Newport, 1989).

Although we agree that there are age-related changes in the ability to learn and process a first language, in our opinion these changes are not specific solely to language (e.g., they occur for many other complex skills introduced at different points in life). Furthermore, we have shown that changes of this kind may be the result of learning rather than its cause. Within a nonlinear neural network, learning results in concrete changes in the structure of the network itself. Across the course of learning (with or without the appropriate inputs for a given task), the weights within a network become committed to a particular configuration (a process that may include the elimination of many individual elements). After this "point of no return," the network can no longer revert to its original state, and plasticity for new tasks is lost. Marchman (1993) has demonstrated this kind of critical-period effect for networks lesioned at various points in grammatical learning, and the same explanation may be available for other critical-period phenomena as well. What is important is that the seeming critical-period does not require (although it does not preclude) an extrinsic, genetically driven change in learning potential. It has also recently been shown that the so-called critical period even for second-language learning may turn out to involve differences in critical types and amounts of experience rather than actual age of acquisition (Bialystok & Hakuta, 1995).

12. Robustness

Normal children rush headlong into the task of language learning, with a passion that suggests (to some investigators) an innate preparation for that task. Several lines of evidence attest to the robustness of language under all but the most extreme forms of environmental deprivation (Curtiss, 1977; Sachs et al., 1981). Examples include the emergence of "home sign" (rudimentary linguistic systems that emerge in the gestures produced by deaf children of hearing parents—Goldin-Meadow & Mylander, 1984) and creolization (the transformation of a pidgin code into a fully grammaticized language through intermarriage and constant use of the pidgin in natural situations, including first-language acquisition by children—Bickerton, 1981; Sankoff, 1980). A particularly striking example of creolization comes from a recent study of the emergence of a new sign language among deaf adults and children who were brought together in a single community in Nicaragua less than a dozen years ago (Kegl, Senghas, & Coppola, 1995). It has been argued that the rapid and robust emergence of grammar with limited input constitutes solid evidence for the operation of an innate and well-structured bioprogram for grammar (Bickerton, 1981). We would agree that these phenomena are extremely interesting, and that they attest to a robust drive among human beings to communicate their thoughts as rapidly and efficiently as possible. However, these phenomena do not require a preformationist scenario (i.e., a situation in which the grammar emerges because it was innately specified). We argued in Chapter 1 and above that grammars may constitute the class of possible solutions to the problem of mapping nonlinear thoughts onto a highly constrained linear channel. If children develop a robust drive to solve this problem, and are born with processing tools to solve it, then the rest may simply follow because it is the natural solution to that particular mapping process.

*   *   *

We end this section with a brief comment on the issue of the use of the term "innate" and social responsibility. At this writing, interest in innate ideas and innate constraints on cognition has reached another high-water mark. This is evident in popular books on "human instincts" (e.g., Pinker, 1994b), but it is also evident in books that argue for racial differences in intelligence (Herrnstein & Murray, 1994).

Of course, these two approaches to innateness are not the same. One can obviously argue for the innate basis of characteristics shared by all human beings while rejecting the notion that individual or subgroup differences are immutable. But this neat division runs into difficulty as we move from behavioral description to the elucidation of an underlying mechanism. The problem is that genetic differences and genetic commonalities come from the same source. If we ascribe a complex and highly specific ability to some direct genetic base, then we have opened the door for genetic variation and the disturbing sociopolitical implications that ensue.

As scientists, we cannot and should not hide from facts. If our data force us to an unpopular or unhappy conclusion, then we must live with the consequences, playing out our moral beliefs on the plane where such beliefs belong (i.e., in civic responsibility and political action). At the moment, cognitive scientists and neuroscientists have a long way to go before our findings achieve the status of immutable and irrefutable truth.

However, the words we use to explain our interim findings still have important consequences. The word "innate" means different things to different people, some refusing to use it at all. To some people, "innateness" means that the outcome in question cannot and should not be changed.

We disagree with such a position and have taken pains throughout this volume to be as explicit as possible about the multiple levels of complex interactions at which something might be innate. If scientists use words like "instinct" and "innateness" in reference to human abilities, then we have a moral responsibility to be very clear and explicit about what we mean, to avoid our conclusions being interpreted in rigid nativist ways by political institutions. Throughout we have stressed that we are not anti-nativist, but that we deem it essential to specify at precisely what level we are talking when we use terms like "innate." If our careless, underspecified choice of words inadvertently does damage to future generations of children, we cannot turn with innocent outrage to the judge and say "But your Honor, I didn't realize the word was loaded."

Where do we go from here?

Connectionism is not a theory of development as such. Rather, it is a tool for modeling and testing developmental hypotheses. We hope to have convinced readers during the course of this volume that connectionism provides a rich, precise, powerful and biologically plausible framework for exploring development, and in particular for rethinking the complex issue of innateness. We are excited about the possibilities for new ways of understanding development which are opened up by the framework we have outlined here. But we are also painfully aware of how little we understand and how much remains to be studied. It is with neither cynicism nor discouragement that we acknowledge what is usually true in science: At any given point in time it will be the case that we only know 5% of what we want to know, and 95% of that will eventually turn out to be wrong. What matters is that we might be on the right track, that we are willing to discard those ideas which are proven wrong, and that we continue searching for better ideas. In this final section we would like to identify what we see as particularly important challenges. Thus, this section represents for us a program for the future.

Multi-tasking in complex environments

Some of the most influential early connectionist models of psychological phenomena were motivated by a desire to understand how interactions between various knowledge sources and how context might affect processing (e.g., the word-reading model of McClelland & Rumelhart, 1981). And connectionism is often seen as being quite opposed to modularist theories (although as we pointed out in Chapter 2, there is no necessary reason why connectionist models should not be modular; the more interesting question is what the content and the ontogeny of the modules is).

It is therefore ironic that most current models are in fact highly task-specific and single-purpose (see discussion in Karmiloff-Smith, 1992a, Chapter 8). Most of the past-tense verb-learning models, for instance, do not attempt to integrate the knowledge of past-tense morphology with other parts of the morphological system. Yet clearly the child does not learn language according to some scheme whereby Monday is past tense, Tuesday is relative clauses, etc. Not only is language learned in an integrated fashion, but language learning is highly dependent on the acquisition of behaviors in non-verbal domains.

We believe that in order to study the development of complex behaviors, it will be crucial to have models which have greater developmental and ecological plausibility. These must be models which are capable of carrying out multiple behaviors (in computer jargon, multi-tasking) in complex environments which require behaviors which are coordinated and integrated.

Active and goal-oriented models

Most models are passive. They exist in an environment over which they have no control and are spoon-fed a preprogrammed diet of experiences.

Babies, on the other hand are active. (They may be spoon-fed, but consider how often the spoon ends up in their hair or on the floor.) To a large extent, they select their environment by choosing what they will attend to and even where they will be. Thus, there is an important distinction between what is input (presented to a child) and what is uptake (processed). Sometimes this difference is not under control (as when a child's physical or cognitive level of maturity precludes processing an input), but at other times it is.

Furthermore, children have agendas. Their behaviors are typically goal-oriented; they do something (or learn something) because it moves them closer to some goal. Thus if we want to understand why some behaviors develop sooner or later, it is often important to understand the goals towards which the behaviors are directed. Contrast this with the typical network, which has no internal drives or focus.

Where do these goals come from? Who writes the child's agenda? It seems to us that these are not solely developmental questions and that to answer them we must also attend to phylogenetic (i.e., evolutionary) considerations. We are encouraged by the recent interest (not only among connectionists) in artificial life and evolutionary models. We do not believe that the many basic goal-oriented behaviors are learned or taught. Rather, they are evolved. So these new evolutionary models may eventually be profitable avenues for studying ontogenetic development.

Social models

From an early age, the infant's behaviors are highly social. Many of the earliest behaviors are related to the infant's desire to interact with its caretakers. Later in life, social interactions play a critical role in shaping and stimulating the child's development.

In a similar vein, we would like to see models which have a more realistic social ecology. There are many behaviors which have strong social components. Language is an obvious case: It is conventional (i.e., relies on socially agreed-upon forms), and it is one of the most useful currencies for social interaction. There are also phenomena such as the development of a child's awareness that others have private mental states which cannot be directly observed (Theory of Mind), which cannot be modeled by networks in isolation.

Recent work in cognitive science may provide a good theoretical basis for developing social connectionist networks. Hutchins' "distributed cognition" asserts that many cognitive phenomena occur at the level of group interactions. Thus, in just the same way that one cannot understand what makes a termite nest function without looking at behaviors in the aggregate (individual termites are fairly dumb when considered in isolation), there are many human behaviors which are manifest at the level of group activity. Hutchins himself has developed models of communities of networks (Hutchins & Hazlehurst, 1991), and we see this as a fruitful direction for future developmental work.

Higher-level cognition

The models we have discussed throughout the book do not touch on planning, reasoning, and theory building. Indeed, Karmiloff-Smith has argued that connectionist models account particularly well for some aspects of development, but stop short of addressing important developmental facts regarding higher cognitive functions (Karmiloff-Smith, 1992a, 1992c). It is well documented that children go beyond successful behavioral outcomes to form theories about how, for instance, language and the physical world function. Future models will have to focus on ways in which connectionist simulations that embody implicit knowledge about tasks could eventually create explicit knowledge that could be transported from one domain to another in the service of reasoning and theory building. We do not believe there is any reason in principle not to believe that such models are possible—but their present lack presents an important challenge for the future.

More realistic brain models

We stated at the outset that we take biological plausibility as an important source of constraint on our models. This does not require that we limit ourselves to what is currently known about the biology, simply because too much remains unknown. And models can play a useful role in motivating the search for new empirical data. But we also take seriously the charge that models not stray too far from what is at least plausible (if not known), and our models must stay abreast of new developments.

The search for greater realism must necessarily be carried out with a greater understanding into what *matters*. Brains are clearly very complicated things. But it is often not clear what part of that complexity contributes directly to function, and what part is simply there as an outcome of evolution's necessarily Rube Goldberg approach—the result of random variations which have proven to be adaptive, but which do not necessarily make for an optimal or even tidy system. One challenge is the need to identify the relevant features of brains, and then find a level of modeling which continues to make contact with the anatomy and physiology, but which can also make contact with behavior.

A second challenge is the development of models of the brain which take into account the *entire* (well, at least several parts!) of the brain. The O'Reilly and Johnson model of imprinting which we discussed in Chapter 6 is a good example of the type of model we have in mind.

A final note

This brings us to our end, but obviously we also regard this as a beginning. For the six of us, the journey which we have followed to get to this point has been exciting, stimulating, and often personally challenging.

This is not a book which any one of us could have written alone. And we are certainly different people as a consequence of having written it together. We have not always agreed at the outset of our discussions, but we see this as a strength. Each of us comes with a different background and frequently a different perspective. These differences, we feel, have played an important role in making this book what it is. Rather than trying to submerge disagreements, we have respected our differences of opinion, and have attempted to forge a synthesis which goes beyond the initial differences to a new perspective which reconciles them.

We hope that you, the reader, may feel some part of the excitement which we feel at the new prospects for understanding just what it is that makes us human, and how we get to be that way.

References

Annett, M. (1985). *Left, Right, Hand and Brain: The Right Shift Theory.* Hillsdale, NJ: Erlbaum.

Antell, E., & Keating, D.P. (1983). Perception of numerical invariance in neonates. *Child Development, 54,* 695-701.

Aram, D.M. (1988). Language sequelae of unilateral brain lesions in children. In F. Plum (Ed.), *Language, Communication, and the Brain* (pp. 171-197). New York: Raven Press.

Aram, D.M. (1992). Brain injury and language impairment in childhood. In P. Fletcher & D. Hall (Eds.), *Specific Speech and Language Disorders in Children.* London: Whurr Publishers.

Aram, D.M., & Eisele, J.A. (1992). Plasticity and function recovery after brain lesions in children. In A. Benton, H. Levin, G. Moretti, & D. Riva, (Eds.), *Neuropsicologia dell'età evolutiva [Developmental Neuropsychology]* (pp. 171-184). Milan: Franco Angeli.

Aram, D., Ekelman, B., Rose, D., & Whitaker, H. (1985). Verbal and cognitive sequelae following unilateral lesions acquired in early childhood. *Journal of Clinical & Experimental Neuropsychology, 7,* 55-78.

Aram, D.M., Ekelman, B., & Whitaker, H. (1986). Spoken syntax in children with acquired unilateral hemisphere lesions. *Brain and Language, 27,* 75-100.

Aram, D., Ekelman, B., & Whitaker, H. (1987). Lexical retrieval in left- and right-brain-lesioned children. *Brain and Language, 28,* 61-87.

Aram, D.M., Morris, R., & Hall, N.E. (1992). The validity of discrepancy criteria for identifying children with developmental language disorders. Journal of Learning Disabilities, 25(9), 549-554.

Baillargeon, R. (1987a). Object permanence in 3.5- and 4.5-month-old infants. *Developmental Psychology, 23,* 655-664.

Baillargeon, R. (1987b). Young infants' reasoning about the physical and spatial properties of a hidden object. *Cognitive Development, 2,* 170-200.

Baillargeon, R. (1993). The object concept revisited: New directions in the investigation of infant's physical knowledge. In C. E. Granrud (Ed.), *Visual Perception and Cognition in Infancy* (pp. 265-315). London, UK: Erlbaum.

Baillargeon, R. (1994). How do infants learn about the physical world? *Current Directions in Psychological Science, 3,* 133-140.

Baillargeon, R., Graber, M., Devos, J., & Black, J. (1990). Why do young infants fail to search for hidden objects. *Cognition, 36,* 255-284.

Baillargeon, R., & Hanko-Summers, S. (1990). Is the top object adequately supported by the bottom object? Young infants' understanding of support relations. *Cognitive Development, 5,* 29-53.

Baldwin, D.A. (1989). Establishing word-object relations: A first step. *Child Development, 60,* 381-398.

Barrett, M.D. (1986). Early semantic representations and early word usage. In S.A. Kuczaj & M.D. Barrett (Eds.), *The Development of Word Meaning.* New York: Springer-Verlag.

Barto, A.G., & Anandan, P. (1985). Pattern recognizing stochastic learning automata. *IEEE Transactions on Systems, Man, and Cybernetics, 15,* 360-375.

Barto, A.G., Sutton, R.S., & Anderson, C.W. (1983). Neuronlike adaptive elements that can solve difficult learning control problems. *IEEE Transactions on Systems, Man, and Cybernetics, 15,* 835-846.

Basser, L. (1962). Hemiplegia of early onset and the faculty of speech with special reference to the effects of hemispherectomy. *Brain, 85,* 427-460.

Basso, A., Capitani, E., Laiacona, M., & Luzzatti, C. (1980). Factors influencing type and severity of aphasia. *Cortex, 16,* 631-636.

Bates, E. (1990). Language about me and you: Pronominal reference and the emerging concept of self. In D. Cicchetti & M. Beeghly (Eds.), *The Self in Transition: Infancy to Childhood.* Chicago and

London: University of Chicago Press.

Bates, E. (Ed.). (1991). Special issue: Cross-linguistic studies of aphasia. *Brain and Language, 41(2).*

Bates, E., Appelbaum, M., & Allard, L. (1991). Statistical constraints on the use of single cases in neuropsychological research. *Brain and Language, 40,* 295-329.

Bates, E., Benigni, L., Bretherton, I., Camaioni, L., & Volterra, V. (1979). *The Emergence of Symbols: Cognition and Communication in Infancy.* New York: Academic Press.

Bates, E., Bretherton, I., & Snyder, L. (1988). *From First Words to Grammar: Individual Differences and Dissociable Mechanisms.* New York: Cambridge University Press.

Bates, E., & Carnevale, G.F. (1993). New directions in research on language development. *Developmental Review, 13,* 436-470.

Bates, E., Dale, P.S., & Thal, D. (1995). Individual differences and their implications for theories of language development. In Paul Fletcher & Brian MacWhinney (Eds.), *Handbook of Child Language* (pp. 96-151). Oxford: Basil Blackwell.

Bates, E., Devescovi, A., Dronkers, N., Pizzamiglio, L., Wulfeck, B., Hernandez, A., Juarez, L., & Marangolo, P. (1994). Grammatical deficits in patients without agrammatism: Sentence interpretation under stress in English and Italian. *Brain and Language, 47(3),* 400-402.

Bates, E., Friederici, A., & Wulfeck, B. (1987). Comprehension in aphasia: A crosslinguistic study. *Brain and Language, 32,* 19-67.

Bates, E., McDonald, J., MacWhinney, B., & Appelbaum, M. (1991). A maximum likelihood procedure for the analysis of group and individual data in aphasia research. *Brain and Language, 40,* 231-265.

Bates, E., & Thal, D. (1991). Associations and dissociations in child language development. In J. Miller (Ed.), *Research on Child Language Disorders: A Decade of Progress.* Austin, TX: Pro-Ed.

Bates, E., Thal, D., Aram, D., Eisele, J., Nass, R., & Trauner, D. (1994). From first words to grammar in children with focal brain injury. To appear in D. Thal & J. Reilly, (Eds.), *Special issue on Origins of Communication Disorders, Developmental Neuropsychology.*

Bates, E., Thal, D., & Janowsky, J. (1992). Early language development and its neural correlates. In I. Rapin and S. Segalowitz (Eds.), *Handbook of Neuropsychology, Vol. 7: Child Neuropsychology* (pp. 69-110). Amsterdam: Elsevier.

Bates, E., Thal, D., & Marchman, V. (1991). Symbols and syntax: A Darwinian approach to language development. In N. Krasnegor, D. Rumbaugh, R. Schiefelbusch, & M. Studdert-Kennedy (Eds.), *Biological and Behavioral Determinants of Language Development*. Hillsdale, NJ: Erlbaum.

Baynes, K. (1990). Language and reading in the right hemisphere: Highways and byways of the brain? *Journal of Cognitive Neuroscience, 2(3),* 159-179.

Baynes, K., & Gazzaniga, M.S. (1987). Right hemisphere damage: Insights into normal language mechanisms? In F. Plum (Ed.), *Language, Communication and the Brain.* New York: Raven Press.

Bechtel, W., & Abrahamsen, A. (1991). *Connectionism and the Mind: An Introduction to Parallel Processing in Networks.* Oxford: Basil Blackwell.

Becker, S., & Hinton, G.E. (1992). Self-organizing neural network that discovers surfaces in random-dot stereograms. *Nature, 355(6356),* 161-163.

Bellugi, U., & Hickok, G. (in press). Clues to the neurobiology of language. In R.Broadwell (Ed.), *Neuroscience, Memory and Language. Decade of the Brain Series, Vol. 1.* Washington, DC: Library of Congress.

Bellugi, U., Wang, P.P., & Jernigan, T.L. (1994). Williams Syndrome: An unusual neuropsychological profile. In S. Broman & J. Grafman (Eds.), *Atypical Cognitive Deficits in Developmental Disorders: Implications for Brain Function.* Hillsdale, NJ: Erlbaum.

Berko, J. (1958). The child's learning of English morphology. *Word, 14,* 150-177.

Berman, R.A., & Slobin, D.I. (1994). *Relating Events in Narrative: A Cross-linguistic Developmental Study* [in collaboration with Ayhan Aksu-Koc et al.]. Hillsdale, NJ: Erlbaum.

Best, C.T. (1988). The emergence of cerebral asymmetries in early human development: A literature review and a neuroembryological model. In D.L. Molfese & S.J. Segalowitz (Eds.), *Brain*

Lateralization in Children: Developmental Implications. New York: Guilford Press.

Bhide, P.G., & Frost, D.O. (1992). Axon substitution in the reorganization of developing neural connections. *Proceedings of the National Academy of Science USA, 89(24),* 11847-11851.

Bialystok, E., & Hakuta, K. (1994). *In Other Words: The Science and Psychology of Second-Language Acquisition.* New York: Basic Books.

Bickerton, D. (1981). *The Roots of Language.* Ann Arbor, MI: Karoma.

Bishop, D.V.M. (1983). Linguistic impairment after left hemidecortication for infantile hemiplegia? A reappraisal. *Quarterly Journal of Experimental Psychology, 35A,* 199-207.

Bishop, D.V.M. (1992). The underlying nature of specific language impairment. *Journal of Child Psychology and Psychiatry, 33,* 3-66.

Bishop, D.V.M., & Rosenbloom, L. (1987). Childhood language disorders: Classification and overview. In W. Yule & M. Rutter (Eds.), *Language Development and Disorders* (Clinics in Developmental Medicine, No. 101/102, 16-41). Oxford: Blackwell.

Blackwell, A., & Bates, E. (1995). Inducing agrammatic profiles in normals: Evidence for the selective vulnerability of morphology under cognitive resource limitation. *Journal of Cognitive Neuroscience, 7(2),* 228-257.

Bliss, T.V., & Lømo, T. (1973). Long-lasting potentiation of synaptic transmission in the dentate area of the anaesthetized rabbit following stimulation of the perforant path. *Journal of Physiology (London), 232,* 331-356.

Bolhuis, J.J. (1991). Mechanisms of avian imprinting: A review. *Biological Reviews, 66,* 303-345.

Bolhuis, J.J., Johnson, M.H., Horn, G., & Bateson, P. (1989). Long-lasting effects of IMHV lesions on the social preferences of domestic fowl. *Behavioral Neuroscience, 103,* 438-441.

Borer, H., & Wexler, K. (1987). The maturation of syntax. In T. Roeper & E. Williams (Eds.), *Parameter setting* (pp. 123-172). Dordrecht, Holland: Reidel.

Brainerd, C.-J. (1978). The stage question in cognitive-developmental theory. *Behavioral and Brain Sciences, 1(2),* 173-213.

Braitenberg, V., & Schüz, A. (1991). *Anatomy of the Cortex: Statistics and Geometry.* Berlin: Springer-Verlag.

Brothers, L., & Ring, B. (1992) A neuroethological framework for the representation of minds. *Journal of Cognitive Neuroscience, 4(2),* 107-118.

Brown, R. (1973). *A First Language: The Early Stages.* Cambridge, MA: Harvard University Press.

Burton, A.M., Bruce, V., & Johnston, R.A. (1990). Understanding face recognition with an interactive activation model. *British Journal of Psychology, 81,* 361-380.

Bybee, J.L., & Slobin, D.I. (1982). Rules and schemas in the development and use of the English past. *Language, 58,* 265-289.

Calow, P. (1976). *Biological machines: A cybernetic approach to life.* London: Arnold.

Caramazza, A., & Berndt, R. (1985). A multicomponent view of agrammatic Broca's aphasia. In M.-L. Kean (Ed.), *Agrammatism* (pp. 27-63). Orlando: Academic Press.

Carey, S. (1982). Semantic development: The state of the art. In E. Wanner & L.Gleitman (Eds.), *Language Acquisition: The State of the Art.* Cambridge University Press.

Carey, S., & Spelke, E. (1994). Domain-specific knowledge and conceptual change. In L.A. Hirschfeld & S.A. Gelman (Eds.), *Mapping the Mind: Domain Specificity in Cognition and Culture* (pp. 169-200). Cambridge, UK: Cambridge University Press.

Carpenter, G.A., & Grossberg, S. (1987). ART2: Self-organization of stable category recognition codes for analog input patterns. *Applied Optics, 26,* 4919-4930.

Carpenter, G.A., & Grossberg, S. (1988). The ART of adaptive pattern recognition by a self-organizing neural network. *Computer,* March 1988, 77-88.

Changeux, J.P., Courrège, P., & Danchin, A. (1973). A theory of the epigenesis of neural networks by selective stabilization of synapses. *Proceedings of the National Academy of Sciences USA, 70,* 2974-2978.

Changeux, J., & Danchin, A. (1976). Selective stabilization of developing synapses as a mechanism for the specification of neuronal networks. *Nature, 264,* 705-712.

Chapman, R.S. (1993, July). Longitudinal change in language production of children and adolescents with Down syndrome.

Paper presented at the Sixth International Congress for the Study of Child Language, Trieste, Italy.

Chauvin, Y. (1988) *Symbol acquisition in humans and neural (PDP) networks.* Doctoral dissertation, University of California, San Diego.

Chomsky, N. (1971). *Syntactic Structures.* The Hague: Mouton.

Chomsky, N. (1975). *Reflections on language.* New York: Parthenon Press.

Chomsky, N. (1980). *Rules and Representations.* New York: Columbia University Press.

Chomsky, N. (1981). *Lectures on Government and Binding.* New York: Foris.

Chomsky, N. (1986). *Knowledge of Language: Its Nature, Origin, and Use.* New York: Praeger.

Chugani, H.T., Phelps, M.E., & Mazziotta, J.C. (1987). Positron emission tomography study of human brain functional development. *Annals of Neurology, 22,* 487-497.

Churchland, P. M.(1995). The engine of reason, the seat of the soul: A philosophical journey into the brain. Cambridge, MA: MIT Press.

Churchland, P.S., & Sejnowski, T.J. (1992). *The Computational Brain.* Cambridge, MA/London: MIT Press.

Clark, A. (1989). *Microcognition: Philosophy, Cognitive Science, and Parallel Distributed Processing.* Cambridge, MA: MIT Press.

Clark, E.V. (1973). What's in a word? On the child's acquisition of semantics in his first language. In T.E. Moore, (Ed.), *Cognitive Development and the Acquisition of Language.* New York: Academic Press.

Clifton, R., Rochat, P., Litovsky, R., & Perris, E. (1991) Object representation guides infants' reaching in the dark. *Journal of Experimental Psychology: Human Perception and Performance, 17,* 323-329.

Cohen-Tannoudji, M., Babinet, C., & Wassef, M. (1994). Early determination of a mouse somatosensory cortex marker. *Nature, 368(6470),* 460-463.

Cole, B.J., & Robbins, T.W. (1992). Forebrain norepinephrine: Role in controlled information processing in the rat. *Neuropsychophar-*

macology, 7(2), 129-142.

Conel, J.L. (1939-1963). *The Postnatal Development of the Human Cerebral Cortex* (Vols. 1-6). Cambridge, MA: Harvard University Press.

Cooper, N.G.F., & Steindler, D.A. (1986). Lectins demarcate the barrel subfield in the somatosensory cortex of the early postnatal mouse. *Journal of Comparative Neurology, 249(2),* 157-69.

Corballis, M.C., & Morgan, M.J. (1978). On the biological basis of human laterality: I. Evidence for a maturational left-right gradient. *Behavioral and Brain Sciences, 1,* 261-269.

Cottrell, G.W., & Fleming, M.K. (1990). Face recognition using unsupervised feature extraction. *Proceedings of the International Neural Network Conference* (pp 322-325), Paris, France. Dordrecht: Kluwer.

Coulson, S., King, J.W., & Kutas, M. (1995). In search of: Is there a syntax-specific ERP component? *Cognitive Neuroscience Society Second Annual Meeting Poster Abstracts, 9.* Davis, CA: Cognitive Neuroscience Society.

Courchesne, E. (1991). Neuroanatomic imaging in autism. *Pediatrics, 87(5),* 781-790.

Courchesne, E., Hesselink, J.R., Jernigan, T.L., & Yeung-Courchesne, R. (1987). Abnormal neuroanatomy in a nonretarded person with autism: Unusual findings with magnetic resonance imaging. *Archives of Neurology, 44,* 335-341.

Courchesne, E., Townsend, J. & Chase, C. (1995). Neuro-developmental principles guide research on developmental psychopathologies. In D. Cicchetti & D. Cohen (Eds.), *A Manual of Developmental Psychopathology, Vol. 2* (pp. 195-226). New York: John Wiley & Sons.

Courchesne, E., Yeung-Courchesne, R., Press, G., Hesselink, J.R., & Jernigan, T.L. (1988). Hypoplasia of cerebellar vermal lobules V and VI in infantile autism. *New England Journal of Medicine, 318(21),* 1349-1354.

Crain, S. (1991). Language acquisition in the absence of experience. *Behavioral and Brain Sciences, 14,* 597-611.

Cromer, R.F. (1974). The development of language and cognition: The cognitive hypothesis. In B. Foss (Ed.), *New Perspectives in*

Child Development. Harmondsworth: Penguin.

Curtiss, S. (1977). *Genie: A Psycholinguistic Study of a Modern-Day Wild Child.* New York: Academic Press.

Damasio, A. (1989). Time-locked multiregional retroactivation: A systems-level proposal for the neural substrates of recall and recognition. *Cognition, 33,* 25-62.

Damasio, A., & Damasio, H. (1992). Brain and language. *Scientific American, 267,* 88-95.

Daugherty, K., & Seidenberg, M. (1992). Rules or connections? The past tense revisited. *Proceedings of the 14th Annual Meeting of the Cognitive Science Society.* Hillsdale, NJ: Erlbaum.

Deacon, T.W. (1990). Brain-language coevolution. In J.A. Hawkins & M. Gell-Mann (Eds.), *The Evolution of Human Languages, SFI Studies in the Sciences of Complexity* (Proc. Vol. X). Addison-Wesley.

Dehay, C., Horsburgh, G., Berland, M., Killackey, H., & Kennedy, H. (1989). Maturation and connectivity of the visual cortex in monkey is altered by prenatal removal of retinal input. *Nature, 337,* 265-267.

Dennis, M., & Whitaker, H. (1976). Language acquisition following hemidecortication: Linguistic superiority of the left over the right hemisphere. *Brain and Language, 3,* 404-433.

Devescovi, A., Bates, E., D'Amico, S., Hernandez, A., Marangolo, P., Pizzamiglio, L., & Razzano, C. (in press). *An on-line study of grammaticality judgment in normal and aphasic speakers of Italian.* To appear in L. Menn, (Ed.) *Special Issue on Cross-linguistic Aphasia, Aphasiology.*

Devescovi, A., Pizzamiglio, L., Bates, E., Hernandez, A., & Marangolo, P. (1994). Grammatical deficits in patients without agrammatism: Detection of agreement errors by Italian aphasics and controls. *Brain and Language, 47(3),* 449-452. Fletcher, P. (1990). Speech and language defects. *Nature, 346,* 226.

Diamond, A. (1988). Abilities and neural mechanisms underlying $A\overline{B}$ performance. *Child Development, 59(2),* 523-527.

Diamond, A. (1991). Neuropsychological insights into the meaning of object permanence. In S.Carey & R. Gelman (Eds.), *The Epigenesis of Mind: Essays on Biology and Cognition* (pp. 67-110).

Hillsdale NJ: Erbaum.

Dore, J. (1974). A pragmatic description of early language development. *Journal of Psycholinguistic Research, 4,* 423-430.

Dromi, E. (1987). *Early Lexical Development.* Cambridge and New York: Cambridge University Press.

Dronkers, N.F., Shapiro, J.K., Redfern, B., & Knight, R.T. (1992, February). *The third left frontal convolution and aphasia: On beyond Broca.* Paper presented at the Twentieth Annual Meeting of the International Neuropsychological Society, San Diego, CA.

Dronkers, N.F., Wilkins, D.P., Van Valin, Jr., R.D., Redfern, B.B., & Jaeger, J.J. (1994). A reconsideration of the brain areas involved in the disruption of morphosyntactic comprehension. *Brain and Language, 47(3),* 461-463.

Durbin, R., & Rumelhart, D.E. (1989). Product units: A computationally powerful and biologically plausible extension to backpropagation networks. *Neural Computation, 1,* 133.

Edelman, G.M. (1987). *Neural Darwinism: The Theory of Neuronal Group Selection.* New York: Basic Books.

Edelman, G.M. (1988). *Topobiology: An Introduction to Molecular Embryology.* New York: Basic Books.

Eisele, J., & Aram, D. (1994). Comprehension and imitation of syntax following early hemisphere damage. *Brain and Language, 46,* 212-231.

Eisele, J., & Aram, D. (1995). Lexical and grammatical development in children with early hemisphere damage: A cross-sectional view from birth to adolescence. In Paul Fletcher & Brian MacWhinney (Eds.), *The Handbook of Child Language* (pp. 664-689). Oxford: Basil Blackwell.

Elman, J.L. (1990). Finding structure in time. *Cognitive Science, 14,* 179-211.

Elman, J.L. (1993). Learning and development in neural networks: The importance of starting small. *Cognition, 48,* 71-99.

Elman, J.L., & Zipser, D. (1988). Learning the hidden structure of speech. *Journal of the Acoustical Society of America, 83,* 1615-1626.

Ervin, S. (1964). Imitation and structural change in children's language. In E.H. Lenneberg (Ed.), *New Directions in the Study of*

Language. Cambridge, MA: MIT Press.

Fabbretti, D., Vicari, S., Pizzuto, E., & Volterra, V. (1993, July). *Language production abilities in Italian children with Down syndrome*. Paper presented at the Sixth International Congress for the Study of Child Language, Trieste, Italy.

Fahlman, S.E., & Lebiere, C. (1990). The cascade-correlation learning architecture. In D.S. Touretzky (Ed.), *Advances in Neural Information-Processing Systems II* (p. 524). San Mateo: Morgan Kaufman.

Fantz, R., Fagan, J., & Miranda, S. (1975). Early visual selectivity. In L. Cohen & P. Salapatek (Eds.), *Infant Perception: From sensation to cognition*. New York: Academic Press.

Farah, M.J., & McClelland, J. (1991). A computational model of semantic memory impairment: Modality specificity and emergent category specificity. *Journal of Experimental Psychology: General, 120,* 339-357.

Feldman, J., & Ballard, D. (1982). Connectionist models and their properties. *Cognitive Science, 6,* 205-254.

Fentress, J., & Kline, P. (Eds.). (in press). *The Hebb Legacy*. Cambridge University Press.

Fernald, A., & Kuhl, P. (1987). Acoustic determinants of infant preference for motherese speech. *Infant Behavior and Development, 10,* 279-93.

Fletcher, J.M. (1994). Afterword: Behavior-brain relationships in children. In S. Broman & J. Grafman (Eds.), *Atypical Cognitive Deficits in Developmental Disorders: Implications for Brain Function* (pp. 297-325). Hillsdale, NJ: Erlbaum.

Fletcher, P. (1990). Speech and language defects. *Nature, 346,* 226.

Fodor, J.A. (1983). *The Modularity of Mind: An Essay on Faculty Psychology*. Cambridge, MA: MIT Press.

Fodor, J.A. (1985). Multiple book review of *The Modularity of Mind. Behavioral and Brain Sciences, 8,* 1-42.

Fodor, J.A., & Pylyshyn, Z.W. (1988). Connectionism and cognitive architecture: A critical analysis. In S. Pinker & J. Mehler (Eds.), *Connections and Symbols* (pp. 3-71). (*Cognition* Special Issue). Cambridge, MA: MIT Press/Bradford Books.

Foldiak, P. (1991). Learning invariance in transformational

sequences. *Neural Computation, 3*, 194-200.

Fowler, A.E. (1993, July). *Phonological limits on reading and memory in young adults with Down syndrome.* Paper presented at the Sixth International Congress for the Study of Child Language, Trieste, Italy.

Friedlander, M.J., Martin, K.A.C., & Wassenhove-McCarthy, D. (1991). Effects of monocular visual deprivation on geniculocortical innervation of area 18 in cat. *The Journal of Neuroscience, 11*, 3268-3288.

Frost, D.O. (1982). Anomalous visual connections to somatosensory and auditory systems following brain lesions in early life. *Brain Research, 255(4)*, 627-635.

Frost, D.O. (1990). Sensory processing by novel, experimentally induced cross-modal circuits. *Annals of the New York Academy of Sciences, 608*, 92-109; discussion 109-12.

Galaburda, A.M. (1994). Language areas, lateralization and the innateness of language. *Discussions in Neuroscience, 10, 1/2*, 118-124.

Galaburda, A.M., & Livingstone, M. (1993). Evidence for a magnocellular defect in neurodevelopmental dyslexia. *Annals of the New York Academy of Sciences, 682*, 70-82.

Galaburda, A.M., Menard, M.T., & Rosen, G.D. (1994). Evidence for aberrant auditory anatomy in developmental dyslexia. *Proceedings of the National Academy of Sciences USA, 91*, 8010-8013.

Galaburda, A.M., & Pandya, D.N. (1983). The intrinsic architectonic and connectional organization of the superior temporal region of the rhesus monkey. *Journal of Comparative Neurology, 221(2)*, 169-84.

Gardner, H. (1983). *Frames of Mind.* New York: Basic Books.

Garnsey, S.M. (Ed.). (1993). Special issue: Event-related brain potentials in the study of language. *Language and Cognitive Processes, 8(4).*

Gasser, M., & Smith, L.B. (1991). The development of the notion of sameness: A connectionist model. *Proceedings of 13th Annual Conference of theCognitive Science Society,* 7197-23.

Gesell, A. (1929). *Infancy and Human Growth.* New York: Macmillan.

Gibson, E.J. (1969). *Principles of Perceptual Learning and Development.*

New York: Appleton-Century-Croft.

Giles, C.L., Griffin, R.D., & Maxwell, T. (1988). Encoding geometric invariances in higher-order neural networks. In D.Z. Anderson (Ed.), *Neural Information Processing Systems* (pp. 301-309). New York: American Institute of Physics.

Glushko, R.J. (1979). The organization and activation of orthographic knowledge in reading words aloud. *Journal of Experimental Psychology: Human Perception and Performance, 5,* 674-691.

Gold, E.M. (1967). Language identification in the limit. *Information and Control, 16,* 447-474.

Goldin-Meadow, S., & Mylander, C. (1984). Gestural communication in deaf children: The effects and non-effects of parental input on early language development. *Monographs of the Society for Research in Child Development, Serial No. 207,Vol 49, No. 3-4.*

Goldowitz, D. (1987). Cell partitioning and mixing in the formation of the CNS: Analysis of the cortical somatosensory barrels in chimeric mice. *Developmental Brain Research, 35,* 1-9.

Golinkoff, R.M., & Hirsh-Pasek, K. (1990). Let the mute speak: What infants can tell us about language acquisition. *Merrill-Palmer Quarterly, 36,* 67-92.

Goodglass, H. (1993). *Understanding Aphasia.* San Diego: Academic Press.

Gopnik, A., & Meltzoff, A. (1987). The development of categorization in the second year and its relation to other cognitive and linguistic developments. *Child Development, 58,* 1523-1531.

Gopnik M. (1990). Feature-blind grammar and dysphasia. *Nature, 344(6268),* 715.

Gopnik, M., & Crago, M.B. (1991). Familial aggregation of a developmental language disorder. *Cognition, 39,* 1-50.

Gopnik, M., Dalalakis, J., Fukuda, S.E., Fukuda, S., & Kehayia, E. (in press). Genetic language impairment: Unruly grammars. In J. Maynard-Smith (Ed.) *Proceedings of the British Academy.*

Gould, S.J. (1977). *Ontogeny and Phylogeny.* Cambridge, MA: Harvard University Press.

Gould, S.J., & Lewontin, R.C. (1979). The spandrels of San Marco and the Panglossian paradigm: A critique of the adaptationist programme. *Proceedings of the Royal Society of London, 205*, 281-288.

Greenfield, P.M. (1991). Language, tools and brain: The ontogeny and phylogeny of hierarchically organized sequential behavior. *Behavioral and Brain Sciences, 14*, 531-550.

Greenough, W.T., Black, J.E., & Wallace, C.S. (1993). Experience and brain development. In M. Johnson (Ed.), *Brain Development and Cognition: A Reader* (pp. 290-322). Oxford: Blackwell.

Greenough, W.T., McDonald, J.W., Parnisari, R.M., & Camel, J.E. (1986). Environmental conditions modulate degeneration and new dendrite growth in cerebellum of senescent rats. *Brain Research, 380*, 136-143.

Grossberg, S. (1976). Adaptive pattern classification and universal recoding: Part. I. Parallel development and coding of neural feature detectors. *Biological Cybernetics, 23*, 121-134.

Hagoort, P., Brown, C., & Groothusen, J. (1993). The syntactic positive shift (SPS) as an ERP measure of syntactic processing. *Language and Cognitive Processes, 8(4)*, 439-483.

Haith, M.M. (1980). *Rules that Newborn Babies Look By: The Organization of Newborn Visual Activity.* Hillsdale, NJ: Erlbaum.

Haith, M.M. (1990). The formation of visual expectations in early infancy: How infants anticipate the future. In C. Rovee-Collier (Ed.), *Abstracts of papers presented at the Seventh International Conference on Infant Studies. Special issue of Infant Behavior and Development, Vol. 13*, 11. Norwood, NJ: Ablex.

Hanson, S.J. (1990). Meiosis networks. In D.S. Touretzky (Ed.), *Advances in Neural Information-Processing Systems II* (p. 533). San Mateo: Morgan Kaufman.

Hare, M., & Elman, J.L. (1995). Learning and morphological change. *Cognition, 56*, 61-98.

Hebb, D.O. (1949). *The Organization of Behavior: A Neuropsychological Theory.* New York: Wiley.

Hecaen, H. (1983). Acquired aphasia in children: Revisited. *Neuropsychologia, 21*, 587.

Heilman, K.M., & Scholes, R.J. (1976). The nature of comprehension errors in Broca's, conduction and Wernicke's aphasics. *Cortex, 12*, 258-265.

Hellige, J.B. (1993). *Hemispheric Asymmetry: What's Right and What's*

Left. Cambridge, MA/London: Harvard University Press.

Herrnstein, R. J., & Murray, C. (1994). *The Bell Curve : Intelligence and Class Structure in American Life*. New York : Free Press.

Hertz, J.A., Krogh, A., & Palmer (Eds.) (1991). *Introduction to the theory of neural computation*. Lecture Notes Volume I, Santa Fe Institute, Studies in the Sciences of Complexity. Redwood City, CA: Addison-Wesley.

Hickey, T. (1993). Identifying formulas in first language acquisition. *Journal of Child Language, 20(1)*, 27-42.

Hinton, G.E., & Shallice, T. (1991). Lesioning a connectionist network: Investigations of acquired dyslexia. *Psychological Review, 98*, 74-95.

Hirsh-Pasek, K., Kemler Nelson, D., Jusczyk, P.W., Cassidy, K., Druss, B., & Kennedy, L. (1987). Clauses are perceptual units for young infants. *Cognition, 26*, 269-286.

Horn, G. (1985). *Memory, Imprinting, and the Brain*. Oxford: Clarendon Press.

Horn G. (1991). Learning, memory and the brain. *Indian Journal of Physiology and Pharmacology, 35(1)*, 3-9.

Hornik, K., Stinchcombe, M., & White, H. (1989). Multilayered feed-forward networks are universal approximators. *Neural Networks, 2(5)*, 359-366.

Hubel, D.H., & Wiesel, T.N. (1963). Receptive fields of cells in striate cortex of very young, visually inexperienced kittens. *Journal of Neurophysiology, 26*, 944-1002.

Hubel, D.H., & Wiesel, T.N. (1965). Receptive fields and functional architecture in two nonstriate visual areas (18 and 19) of the cat. *Journal of Neurophysiology, 28*, 229-289.

Hubel, D.H., & Wiesel, T.N. (1970). The period of susceptibility to the physiological effects of unilateral eye closure in kittens. *Journal of Physiology, 206(2)*, 419-436.

Hurst, J.A., Baraitser, M., Auger, E., Graham, F., & Norell, S. (1990). An extended family with a dominantly inherited speech disorder. *Developmental Medicine and Child Neurology, 32*, 347-355.

Hutchins, E. (1994). *Cognition in the World*. Cambridge, MA: MIT Press.

Hutchins, E., & Hazlehurst, B. (1991). Learning in the cultural process. In C. G. Langton, C. Taylor, J. D. Farmer, & S. Rasmussen

(Eds.), *Artificial* Life II (pp. 689-705). Redwood City, CA: Addison-Wesley.

Huttenlocher, P.R. (1979). Synaptic density in human frontal cortex: Developmental changes and effects of aging. *Brain Research, 163*, 195-205.

Huttenlocher, P.R. (1990). Morphometric study of human cerebral cortex development. *Neuropsychologia, 28:6*, 517-527.

Huttenlocher, P.R., & de Courten, C. (1987). The development of synapses in striate cortex of man. *Human Neurobiology, 6*, 1-9.

Huttenlocher, P.R., de Courten, C., Garey, L., & van der Loos, H. (1982). Synaptogenesis in human visual cortex synapse elimination during normal development. *Neuroscience Letters, 33*, 247-252.

Inhelder, B., & Piaget, J. (1958). *The Growth of Logical Thinking from Childhood to Adolescence.* New York: Basic Books.

Innocenti, G., & Clarke, S. (1984). Bilateral transitory projection to visual areas from auditory cortex in kittens. *Developmental Brain Research, 14*, 143-148.

Irle, E. (1990). An analysis of the correlation of lesion size, localization and behavioral effects in 283 published studies of cortical and subcortical lesions in old-world monkeys. *Brain Research Review, 15*, 181-213.

Isaacson, R.L. (1975). The myth of recovery from early brain damage. In N.G. Ellis (Ed.), *Aberrant Development in Infancy* (pp. 1-26). New York: Wiley.

Ivy, G., Akers, R., & Killackey, H.L. (1979). Differential distribution of callosal projection neurons in the neonatal and adult rat. *Brain Research, 173*, 532-537.

Jacobs, R.A. (1990). *Task decomposition through competition in a modular connectionist architecture.* Unpublished doctoral dissertation, Department of Computer & Information Science, University of Massachusetts, Amherst.

Jacobs, R.A., Jordan, M.I., & Barto, A.G. (1991). Task decomposition though competition in a modular connectionist architecture: The what and where vision tasks. *Cognitive Science, 15(2)*, 219-250.

Jacobs, R.A., Jordan, M.I., Nowlan, S.J., & Hinton, G.E. (1991). Adaptive mixtures of local experts. *Neural Computation, 3*, 79-

87.

Janowsky, J.S., & Finlay, B.L. (1986). The outcome of perinatal brain damage: The role of normal neuron loss and axon retraction. *Developmental Medicine, 28,* 375-389.

Janowsky, J.S., Shimamura, A., & Squire, L. (1989). Source memory impairment in patients with frontal lobe lesions. *Neuropsychologia, 27,* 1043-1056.

Jansson, G., & Johansson, G. (1973). Visual perception of bending motion. *Perception, 2(3),* 321-326.

Jernigan, T., & Bellugi, U. (1990). Anomalous brain morphology on magnetic resonance images in Williams Syndrome and Down Syndrome. *Archives of Neurology, 47,* 429-533.

Jernigan, T., & Bellugi, U. (1994). Neuroanatomical distinctions between Williams and Down Syndromes. In S. Broman & J. Grafman (Eds.)., *Atypical Cognitive Deficits in Developmental Disorders: Implications for Brain Function.* Hillsdale, NJ: Erlbaum.

Jernigan, T., Bellugi, U., & Hesselink, J. (1989). Structural differences on magnetic resonance imaging between Williams and Down Syndrome. Neurology, 39, (suppl 1): 277.

Jernigan, T. L., Bellugi, U., Sowell, E., Doherty, S., & Hesselink, J. R. (1993). Cerebral morphological distinctions between Williams and Down syndromes. *Archives of Neurology, 50,* 186-191.

Jernigan, T.L., Hesselink, J.R., Sowell, E., & Tallal, P.A. (1991). Cerebral structure on magnetic resonance imaging in language-impaired and learning-impaired children. *Archives of Neurology, 48(5),* 539-545.

Jernigan, T.L., Press, G.A., & Hesselink, J.R. (1990). Methods for measuring brain morphological features on magnetic resonance images: Validation and normal aging. *Archives of Neurology, 47,* 27-32.

Jernigan, T.L., & Tallal, P. (1990). Late childhood changes in brain morphology observable with MRI. *Developmental Medicine and Child Neurology, 32,* 379-385.

Jernigan, T.L., Trauner, D.A., Hesselink, J.R., & Tallal P.A. (1991). Maturation of human cerebrum observed in vivo during adolescence. *Brain, 114,* 2037-2049.

Johnson, J.S., & Newport, M. (1989). Critical period effects in second language learning: The influence of maturational state on the acquisition of English as second language. *Cognitive Psychology, 21,* 60-99.

Johnson, M.H. (1988). Parcellation and plasticity: Implications for ontogeny. *Behavioral and Brain Sciences, 11,* 547-549.

Johnson, M.H. (1990). Cortical maturation and the development of visual attention in early infancy. *Journal of Cognitive Neuroscience, 2,* 81-95.

Johnson, M.H. (in press). *The Cognitive Neuroscience of Development.* Oxford: Oxford University Press.

Johnson, M.H., & Bolhuis, J.J. (1991). Imprinting, predispositions and filial preference in the chick. In R.J. Andrew (Ed.), *Neural and Behavioural Plasticity* (pp. 133-156). Oxford: Oxford University Press.

Johnson, M.H., Bolhuis, J.J., & Horn, G. (1992). Predispositions and learning: Behavioural dissociations in the chick. *Animal Behaviour, 44,* 943-948.

Johnson, M.H., Davies, D.C., & Horn, G. (1989). A critical period for the development of filial preferences in dark-reared chicks. *Animal Behaviour, 37,* 1044-1046.

Johnson, M.H., & Horn, G. (1988). The development of filial preferences in the dark-reared chick. *Animal Behaviour, 36,* 675-683.

Johnson, M.H., & Karmiloff-Smith, A. (1992). Can neural selectionism be applied to cognitive development and its disorders? *New Ideas in Psychology, 10(1),* 35-46.

Johnson, M.H., & Morton, J. (1991). *Biology and Cognitive Development: The Case of Face Recognition.* Oxford: Blackwell.

Johnson, M.H., & Vecera, S. (1993). Cortical parcellation and the development of face processing. In B. de Boysson-Bardies, S. de Schonen, P. Jusczyk, P. MacNeilage, & J. Morton (Eds.), *Developmental Neurocognition: Speech and Face Processing in the First Year of Life* (pp. 135-148). Dordrecht: Kluwer Academic Press.

Johnston, J.G., & van der Kooy, D. (1989). Protooncogene expression identifies a transient columnar organization of the forebrain within the late embryonic ventricular zone. *Proceedings of the National Academy of Sciences USA, 86,* 1066-1070.

Johnston, J.R. (1994). Cognitive abilities of language-impaired children. In R. Watkins & M. Rice (Eds.), *Specific Language Impairments in Children: Current Directions in Research and Intervention.* Baltimore: Paul Brookes.

Johnston, J.R., & Kamhi, A.G. (1984). Syntactic and semantic aspects of the utterances of language-impaired children: The same can be less. *Merrill-Palmer Quarterly, 30,* 65-85.

Jordan, M.I. (1986). *Serial order: A parallel distributed processing approach* (ICS Tech. Rep. 8604). La Jolla: University of California, San Diego.

Judd, J.S. (1990). *Neural Network Design and the Complexity of Learning.* Cambridge, MA: MIT Press.

Jusczyk, P.W., & Bertoncini, J. (1988). Viewing the development of speech perception as an innately guided learning process. *Language and Speech, 31,* 217-238.

Jusczyk, P.W., Friederici, A.D., Wessels, J.M.I., Svenkerud, V., & Jusczyk, A.M. (1993). Infants' sensitivity to the sound pattern of native-language words. *Journal of Memory and Language, 32,* 402-420.

Kagan, J. (1981). *The Second Year: The Emergence of Self-Awareness.* Jerome Kagan, with Robin Mount et al. Cambridge, MA: Harvard University Press.

Kandel, E.R., Schwartz, J.H., & Jessell, T.M. (Eds.). (1995). *Essentials of Neural Science and Behavior.* Norwalk, CT : Appleton & Lange.

Karmiloff-Smith, A. (1979). *A functional approach to child language: A study of determiners and reference.* New York: Cambridge University Press.

Karmiloff-Smith, A. (1984). Children's problem solving. In M.E. Lamb, A.L. Brown, & B. Rogoff (Eds.), *Advances in Developmental Psychology* (Vol. 3). Hillsdale, NJ: Erlbaum.

Karmiloff-Smith, A. (1986). From metaprocesses to conscious access: Evidence from children's metalinguistic and repair data. *Cognition, 23,* 95-147.

Karmiloff-Smith, A. (1992a). *Beyond Modularity: A Developmental Perspective on Cognitive Science.* Cambridge, MA: MIT Press/ Bradford Books.

Karmiloff-Smith, A. (1992b). *Abnormal phenotypes and the challenges they pose to connectionist models of development.* Technical

Reports in Parallel Distributed Processing and Cognitive Neuroscience, TR.PDP.CNS.92.7, Carnegie Mellon University.

Karmiloff-Smith, A. (1992c). Nature, nurture and PDP: Preposterous Developmental Postulates? *Connection Science*, 4, 253-269.

Karmiloff-Smith, A., & Grant, J. (1993, March). *Within-domain dissociations in Williams syndrome: A window on the normal mind.* Poster presented at the 60th Annual Meeting of the Society for Research in Child Development, New Orleans, LA.

Karmiloff-Smith, A. & Inhelder, B. (1974). If you want to get ahead, get a theory. *Cognition, 3*, 195-212.

Karni, A., Meyer, G., Jezzard, P., Adams, M., Turner, R., & Ungerleider, L. (1995). Functional MRI evidence for adult motor cortex plasticity during motor skill learning. *Nature, 377,* 155-158.

Kegl, J., Senghas, A., & Coppola, M. (1995). Creation through contact: Sign language emergence and sign language change in Nicaragua. In M. DeGraf (Ed.), *Comparative Grammatical Change: The Interaction of Language Acquisition, Creole Genesis, and Diachronic Syntax.* Cambridge, MA: MIT Press.

Kendrick, K.M., & Baldwin, B.A. (1987). Cells in temporal cortex of conscious sheep can respond preferentially to the sight of faces. *Science, 236,* 448-50.

Kennedy, H., & Dehay, C. (1993). Cortical specification of mice and men. *Cerebral Cortex, 3(3),* 171-86.

Kerszberg, M., Dehaene, S., & Changeux, J.P. (1992). Stabilization of complex input output functions in neural clusters formed by synapse selection. Neural Networks, 5(3), 403-413.

Kilborn, K. (1991). Selective impairment of grammatical morphology due to induced stress in normal listeners: Implications for aphasia. *Brain and Language, 41,* 275-288.

Killackey, H.P. (1990). Neocortical expansion: An attempt toward relating phylogeny and ontogeny. *Journal of Cognitive Neuroscience, 2,* 1-17.

Killackey, H.P., Chiaia, N.L., Bennett-Clarke, C.A., Eck, M., & Rhoades, R. (1994). Peripheral influences on the size and organization of somatotopic representations in the fetal rat cortex. *Journal of Neuroscience, 14,* 1496-1506.

King, J.W., & Kutas, M. (1992). ERPs to sentences varying in syntactic complexity for good and poor comprehenders. *Psychophysi-*

ology, 29(4A), S44.

King, J.W., & Kutas, M. (1995). A brain potential whose latency indexes the length and frequency of words. *Cognitive Neuroscience Society Second Annual Meeting Poster Abstracts,* 68. Davis, CA: Cognitive Neuroscience Society.

Kinsbourne, M., & Hiscock, M. (1983). The normal and deviant development of functional lateralization of the brain. In M. Haith & J. Campos (Eds.), *Handbook of child psychology.* (Vol. II, 4th Edition, pp. 157-280). New York: Wiley.

Klima, E., S., Kritchevsky, M., & Hickok, G. (1993). *The neural substrate for sign language.* Symposium at the 31st Annual Meeting of the Academy of Aphasia, Tucson, AZ.

Kluender, R., & Kutas, M. (1993). Bridging the gap: Evidence from ERPs on the processing of unbounded dependencies. *Journal of Cognitive Neuroscience, 5(2),* 196-214.

Kluender, R., & Kutas, M. (in press). Interaction of lexical and syntactic effects in the processing of unbounded dependencies. *Language and Cognitive Processes.*

Kohonen, T. (1982). Clustering, taxonomy, and topological maps of patterns. In M. Lang (Ed.), *Proceedings of the Sixth International Conference on Pattern Recognition* (pp. 114-125). Silver Spring, MD: IEEE Computer Society Press.

Krashen, S.D. (1973). Lateralization, language learning, and the critical period. Some new evidence. *Language Learning, 23,* 63-74.

Kuczaj, S.A., II (1977). The acquisition of regular and irregular past tense forms. *Journal of Verbal Learning and Verbal Behavior, 16,* 589-600.

Kuhl, P.K. (1983). The perception of auditory equivalence classes for speech in early infancy. *Infant Behavior and Development, 6,* 263-285.

Kuhl, P.K. (1991) Perception, cognition, and the ontogenetic and phylogenetic emergence of human speech. In S. Brauth, W. Hall & R. Dooling (Eds.) *Plasticity of Development.* Cambridge, MA: MIT Press.

Kujala,T., Alho, K., Paavilainen, P., Summala, H., & Naatanen, R. (1992). Neural plasticity in processing of sound location by the early blind: An event-related potential study. *Electroencepha-*

lography and Clinical Neurophysiology, 84, 469-472.

Lachter, J., & Bever, T.G. (1988). The relation between linguistic structure and associative theories of language learning: a constructive critique of some connectionist learning models. In S. Pinker & J. Mehler (Eds.), *Connections and Symbols* (pp. 195-247). *(Cognition* Special Issue). Cambridge, MA: MIT Press/ Bradford Books.

Lane, H. (1976). *The Wild Boy of Aveyron.* Cambridge, MA: Harvard University Press.

Langdon, R.B., & Frost, D.O. (1991). Transient retinal axon collaterals to visual and somatosensory thalamus in neonatal hamsters. *Journal of Comparative Neurology, 310(2),* 200-214.

Lashley, K.S. (1950). In search of the engram. In *Symposia of the Society for Experimental Biology, No. 4, Physiological mechanisms and animal behaviour.* New York: Academic Press.

Lehky, S.R., & Sejnowski, T.J. (1988). Network model of shape-from-shading: Neural function arises from both receptive and projective fields. *Nature, 333,* 452-454.

Lenneberg, E.H. (1967). *Biological Foundations of Language.* New York:Wiley.

Leonard, L.B. (1992). The use of morphology by children with specific language impairment: Evidence from three languages. In R.S. Chapman, (Ed.), *Processes in Language Acquisition and Disorders.* St. Louis: Mosby Year Book.

Leslie, A.M. (1984). Infant perception of a manual pickup event. *British Journal of Developmental Psychology, 2,* 19-32.

Leslie, A.M. (1992). Leslie, A.M. (1992). Pretense, autism, and the "Theory of Mind" module. *Current Directions in Psychological Science, 1,* 18-21.

Leslie, A.M. (1994a). Pretending and believing - Issues in the theory of Tomm. *Cognition, 50(1-3),* 211-238.

Leslie, A.M. (1994b). ToMM, ToBy, and Agency: Core architecture and domain specificity. In L.A. Hirschfeld & S.A. Gelman (Eds.), *Mapping the Mind: Domain Specificity in Cognition and Culture* (pp. 119-148). Cambridge, UK: Cambridge University Press.

LeVay, S., Stryker, M.P., & Shatz, C.J. (1978). Ocular dominance columns and their development in layer IV of the cat's visual cortex: A quantitative study. *Journal of Comparative Neurology, 179*, 223-244.

Lightfoot, D. (1989). The child's trigger experience: Degree-0 learnability. *Behavioral and Brain Sciences, 12*, 321-275.

Linebarger, M., Schwartz, M., & Saffran, E. (1983). Sensitivity to grammatical structure in so-called agrammatic aphasics. *Cognition, 13*, 361-392.

Linsker, R. (1986). From basic network principles to neural architecture (series). *Proceedings of the National Academy of Sciences, USA 83*, 7508-7512, 8390-8394, 8779-8783.

Linsker, R. (1990). Perceptual neural organization: Some approaches based on network models and information theory. *Annual Review of Neuroscience, 13*, 257-281.

Locke, J.L. (1993). Learning to speak. *Journal of Phonetics, 21(2)*, 141-146.

Locke, J.L. (1994). Phases in the child's development of language: The learning that leads to human speech begins in the last trimester of pregnancy—long before they utter a word, infants are talking themselves into a language. *American Scientist, 85(5)*, 436-445.

Locke, J.L. (1995). More than words can say. *New Scientist, 145(1969)*, 30-33.

Lüders, H., Lesser, R., Hahn, J., Dinner, D., Morris, H., Resor, S., & Harrison, M. (1986). Basal temporal language area demonstrated by electrical stimulation. *Neurology, 36*, 505-509.

Lüders, H., Lesser, R., Dinner, D., Morris, H., Wyllie, E., & Godoy, J. (1991). Localization of cortical function: New information from extraoperative monitoring of patients with epilepsy. *Epilepsia, 29 (Suppl. 2)*, S56-S65.

Lüders, H., Lesser, R., Hahn, J., Dinner, D., Morris, H., Wyllie, E., & Godoy, J. (1991). Basal temporal language area. *Brain, 114*, 743-754.

MacWhinney, B. (1989). Competition and connectionism. In B. MacWhinney & E. Bates (Eds.), *The Crosslinguistic Study of Sentence Processing* (pp. 422-457). New York: Cambridge University Press.

MacWhinney, B. (1993). Connections and symbols: Closing the gap.

Cognition, 49(3), 291-296.

MacWhinney, B., Leinbach, J., Taraban, R., & McDonald, J. (1989). Language learning: Cues or rules? *Journal of Memory and Language, 28,* 255-277.

Mandler, J.M. (1988). How to build a baby: On the development of an accessible representational system. *Cognitive Development, 3,* 113-136.

Mandler, J.M. (1992). How to build a baby II: Conceptual primitives. *Psychological Review, 99(4),* 587-604.

Marchman, V. (1988). Rules and regularities in the acquisition of the English past tense. *Center for Research in Language Newsletter, 2,* April.

Marchman, V. (1993). Constraints on plasticity in a connectionist model of the English past tense. *Journal of Cognitive Neuroscience, 5(2),* 215-234.

Marchman, V., & Bates, E. (1994). Continuity in lexical and morphological development: A test of the critical mass hypothesis. *Journal of Child Language, 21(2),* 339-366.

Marchman, V., Wulfeck, B., & Weismer, S.E. (1995). *Productive use of English past-tense morphology in children with SLI and normal language* (Tech. Rep. No. CND-9514). La Jolla: University of California, San Diego, Center for Research in Language, Project in Cognitive and Neural Development.

Marcus, G., Ullman, M., Pinker, S., Hollander, M., Rosen, T.J., & Xu, F. (1992). Overregularization in language acquisition. *Monographs of the Society for Research in Child Development, 57.*

Mareschal, D., Plunkett, K., & Harris, P. (1995). Developing object permanence: A connectionist model. In J. D. Moore & J. F. Lehman (Eds.), *Proceedings of the Seventeenth Annual Conference of the Cognitive Science Society* (pp. 170-175). Mahwah, NJ: Erlbaum.

Marin-Padilla, M. (1970). Prenatal and early postnatal ontogenesis of the human motor cortex: A Golgi study. I. The sequential development of the cortical layers. *Brain Research, 23,* 167-183.

Marler, P., & Peters, S. (1988). Sensitive periods for song acquisition from tape recordings and live tutors in the swamp sparrow, *melospiza georgiana. Ethology, 77,* 76-84.

Marr, D. (1982). *Vision.* New York: Freeman.

Marshall, J.C. (1984). Multiple perspectives on modularity. *Cognition, 17,* 209-242.

Marslen-Wilson, W.D. (1993). Issues of process and representation. In G. Altmann & R. Shillcock (Eds.), *Cognitive Models of Speech Processing.* Hillsdale, NJ: Erlbaum.

Marslen-Wilson, W.D., & Welsh, A. (1979). Processing interactions during word recognition in continuous speech. *Cognitive Psychology, 10,* 29-63.

Martin, N., Dell, G.S., Saffran, E.M., & Schwartz, M.F. (1994). Origins of paraphasias in deep dysphasia: Testing the consequences of a decay impairment to an interactive spreading activation model of lexical retrieval. *Brain and Language, 47(1),* 52-88.

Matthews, J. (1994). Linguistic aspects of familial language impairment. *Special issue of the McGill Working Papers in Linguistics, 10(1&2).*

Mazoyer, B.M., Tzourio, N., Frak, V., Syrota, A., Murayama, N., Levrier, O., Salamon, G., Dehaene, S., Cohen, L., & Mehler, J. (1993). The cortical representation of speech. *Journal of Cognitive Neuroscience, 5(4),* 467-479.

McCabe, B.J., & Horn, G. (1994). Learning-related changes in Fos-like immunoreactivity in the chick forebrain after imprinting. *Proceedings of the National Academy of Sciences USA, 91,* 11417-11421.

McCarthy, D. (1954). Language development in children. In L. Carmichael (Ed.), *Manual of Child Psychology* . (2nd ed., pp.492-630). New York: John Wiley & Sons.

McClelland, J.L., (1989). Parallel distributed processing: Implications for cognition and development. *In R.G.M. Morris (Ed.), Parallel Distributed Processing: Implications for Psychology and Neurobiology* (pp. 9-45). Oxford: Clarendon Press.

McClelland, J.L., & Elman, J.L. (1986). The TRACE model of speech perception. *Cognitive Psychology, 18,* 1-86.

McClelland, J.L., & Jenkins (1991). Nature, nurture and connectionism: Implications for connectionist models of development. In K. van Lehn (Ed.), *Architectures for Intelligence — the Twenty-*

second (1988) Carnegie Symposium on Cognition. Hillsdale NJ: Erlbaum.

McClelland, J.L., & Rumelhart, D.E. (1981). An interactive activation model of context effects in letter perception: Part 1. An account of basic findings. *Psychological Review, 86,* 287-330.

McKinney, M.L., & McNamara, K.J. (1991). *Heterochrony: The Evolution of Ontogeny.* New York and London: Plenum Press.

McNaughton, B.L., & Nadel, L. (1990). Hebb-Marr networks and the neurobiological representation of action in space. In M.A. Gluck & D.E. Rumelhart (Eds.), *Neuroscience and Connectionist Theory.* Hillsdale, NJ: Erlbaum.

McShane, J. (1979). The development of naming. *Linguistics, 17,* 879-905.

Mecklinger, A., Schriefers, H. Steinhauer, K., & Friederici, A.D. (1995). Processing relative clauses varying on syntactic and semantic dimensions: An analysis with event-related potentials. *Memory & Cognition, 23(4),* 477-494.

Mehler, J., & Christophe, A. (1994). Maturation and learning of language in the first year of life. In M.S. Gazzaniga (Ed.), *The Cognitive Neurosciences: A Handbook for the Field* (pp. 943-954). Cambridge MA: MIT Press.

Mehler, J., & Fox, R., (Eds.). (1985). *Neonate Cognition: Beyond the Blooming Buzzing Confusion.* Hillsdale, NJ: Erlbaum.

Mehler, J., Lambertz, G., Jusczyk, P., & Amiel-Tison, C. (1986). Discrimination de la langue maternelle par le nouveau-né. *C.R. Academie des Sciences, 303, Serie III,* 637-640.

Mel, B.W. (1990). *The sigma-pi column: A model for associative learning in cerebral neocortex* (CNS Memo #6). Pasadena, CA: California Institute of Technology, Computational and Neural Systems Program.

Meltzoff, A.N. (1988). Infant imitation and memory: Nine-month-olds in immediate and deferred tests. *Child Development, 59,* 217-225.

Menn, L. (1971). Phonotactic rules in beginning speech. *Lingua, 26,* 225-251.

Menn, L., & Obler, L.K. (Eds.). (1990). *Agrammatic Aphasia: Cross-language Narrative Sourcebook.* Amsterdam/Philadelphia: John

Benjamins.

Merzenich, M.M. (in press). *Cortical plasticity: Shaped, distributed representations of learned behaviors.* In B. Julesz & I. Kovacs (Eds.), *Maturational windows and cortical plasticity in human development: Is there a reason for an optimistic view?* Reading, MA: Addison Wesley.

Merzenich, M.M., Kaas, J.H., Wall, J.T., Sur, M., Nelson, R.J., & Felleman, D.J. (1983). Progression of change following median nerve section in the cortical representation of the hand in areas 3b and 1 in adult owl and squirrel monkeys. *Neuroscience, 10(3),* 639-65.

Merzenich, M.M., Recanzone, G., Jenkins, W.M., Allard, T.T., & Nudo, R.J. (1988). Cortical representational plasticity. In P. Rakic & W. Singer (Eds.), *Neurobiology of Neocortex* (pp. 41-67). New York: John Wiley & Sons.

Miceli, C., Mazzucchi, A., Menn, L., & Goodglass, H. (1983). Contrasting cases of Italian agrammatic aphasia without comprehension disorder. *Brain and Language, 19,* 65-97.

Miller, K.D., Keller, J.B., & Stryker, M.P. (1989). Ocular dominance column development: Analysis and simulation. *Science, 245,* 605-615.

Miller, K.D., Stryker, M.P., & Keller, J.B. (1988). Network model of ocular dominance column formation: Computational results. *Neural Networks, 1,* S267.

Milner, B., Petrides, M., & Smith, M. (1985). Frontal lobes and the temporal organization of behavior. *Human Neurobiology, 4,* 137-142.

Minsky, M., & Papert, S. (1969). *Perceptrons.* Cambridge, MA: MIT Press.

Mishkin, M., Ungerleider, L.G., & Macko, K.A. (1983). Object vision and spatial vision: Two cortical pathways. *Trends in Neurosciences, 6,* 414-417.

Miyake, A., Carpenter, P.A., & Just, M.A. (1994). A capacity approach to syntactic comprehension disorders: Making normal adults perform like aphasic patients. *Cognitive Neuropsychology, 11(6),* 671-717.

Molfese, D. (1989). Electrophysiological correlates of word meanings in 14-month-old human infants. *Developmental Neuropsy-*

chology, 5, 70-103.

Molfese, D. (1990). Auditory evoked responses recorded from 16-month-old human infants to words they did and did not know. *Brain and Language, 38,* 345-363.

Molfese, D.L., & Segalowitz, S.J. (1988). *Brain Lateralization in Children: Developmental Implications.* New York: Guilford Press.

Molnar, Z., & Blakemore, C. (1991). Lack of regional specificity for connections formed between thalamus and cortex in coculture. *Nature, 351 (6326),* 475-7.

Morrison, J.H., & Magistretti, P.J. (1983). *Trends in Neurosciences, 6,* 146.

Munakata, Y., McClelland, J.L., Johnson, M.H., & Siegler, R.S. (in press). Rethinking infant knowledge: Toward an adaptive process account of successes and failures in object permanence. *Psychological Review.*

Münte, T.F., Heinze, H., & Mangun, G.R. (1993). Dissociation of brain activity related to syntactic and semantic aspects of language. *Journal of Cognitive Neuroscience, 5(3),* 335-344.

Nasrabadi, N.M., & Feng, Y. (1988). Vector quantization of images based upon the Kohonen Self-Organizing Feature Maps. In *IEEE International Conference on Neural Networks, 1,* 101-108. New York: IEEE.

Nelson, K. (1973). Structure and strategy in learning to talk. *Monographs of the Society for Research in Child Development, 38,* (1-2, Serial No. 149).

Neville H.J. (1990). Intermodal competition and compensation in development. Evidence from studies of the visual system in congenitally deaf adults. *Annals of the New York Academy of Sciences, 608,* 71-87; discussion 87-91.

Neville, H.J. (1991). Neurobiology of cognitive and language processing: Effects of early experience. In K. Gibson & A. C. Petersen, (Eds.), *Brain Maturation and Cognitive Development: Comparative and Cross-cultural Perspectives.* Hawthorne, NY: Aldine de Gruyter Press. (V, 62)

Neville, H.J., & Lawson, D. (1987). Attention to central and peripheral visual space in a movement detection task: An event-related potential and behavioral study. II. Congenitally deaf adults. *Brain Research, 45,* 268-2283.

Neville, H.J., Mills, D., & Bellugi, U. (1994). Effects of altered auditory sensitivity and age of language acquisition on the development of language-relevant neural systems: Preliminary studies of Williams Syndrome. In S. Broman & J. Grafman (Eds.)., *Atypical Cognitive Deficits in Developmental Disorders: Implications for Brain Function* (pp. 67-83). Hillsdale, NJ: Erlbaum.

Neville, H.J., Nicol, J., Barss, A., Forster, K., & Garrett, M. (1991). Syntactically based sentence-processing classes: Evidence from event-related brain potentials. *Journal of Cognitive Neuroscience, 3,* 155-170.

Newport, E.L. (1981). Constraints on structure: Evidence from American Sign Language and language learning. In W.A.Collins (Ed.), *Aspects of the Development of Competence: Minnesota Symposia on Child Psychology. Vol.14.* Hillsdale, NJ: Erlbaum.

Newport, E.L. (1988). Constraints on learning and their role in language acquisition: Studies of the acquisition of American Sign Language. *Language Sciences, 10,* 147-172.

Newport, E.L. (1990). Maturational constraints on language learning. *Cognitive Science, 14,* 11-28.

Nichelli, P., Grafman, J., Pietrini, P., Alway, D., Carton, J.C., & Miletich, R. (1994). Brain activity in chess playing. *Nature, 369 (6477),* 191.

Niederer, J., Maimon, G., & Finlay, B. (1995). Failure to reroute or compress thalamocortical projections after prenatal posterior cortical lesions. *Society of Neuroscience Abstracts, 21.*

Norman, D.A., & Shallice, T. (1983). Attention to action: Willed and automatic control of behavior. In R.J. Davidson, G.E. Schwartz, & D. Shapiro (Eds.), *Consciousness and Self Regulation: Advances in Research, Vol. IV.* New York: Plenum Press.

Nowakowski, R. S. (1993). Basic concepts of CNS development. In M. Johnson (Ed.), *Brain Development and Cognition: A Reader* (pp. 54-92). [Reprinted from *Child Development, 58,* 568-595, 1987]. Oxford: Blackwell.

Ojemann, G.A. (1991). Cortical organization of language. *Journal of Neuroscience, 11(8),* 2281-2287.

O'Leary, D.D. (1993). Do cortical areas emerge from a protocortex?

In M. Johnson (Ed.), *Brain Development and Cognition: A Reader* (pp.323-337). Oxford: Blackwell Publishers.

O'Leary, D.D., & Stanfield, B.B. (1985). Occipital cortical neurons with transient pyramidal tract axons extend and maintain collaterals to subcortical but not intracortical targets. *Brain Research, 336(2),* 326-333.

O'Leary, D.D., & Stanfield, B.B. (1989). Selective elimination of extended by developing cortical neurons is dependent on regional locale: Experiments utilizing fetal cortical transplants. *Journal of Neuroscience, 9(7),* 2230-2246.

Oppenheim, R. (1981). Neuronal death and some related regressive phenomena during neurogenesis: A selective historical review and progress report. In W. M. Cowan (Ed.), *Studies of Developmental Neurobiology.* New York: Oxford University Press.

Oppenheim, J.S., Skerry, J.E., Tramo, M.J., & Gazzaniga, M.S. (1989). Magnetic resonance imaging morphology of the corpus callosum in monozygotic twins. *Annals of Neurology, 26,* 100-104.

Orban, G.A., Dupont, P., DeBruyn, B., Vogels, R., Vandenberghe, R., Bormans, G., & Mortelmans, L. (1995). *Functional specialisation and task dependency of human visual cortex.* Manuscript submitted for publication.

O'Reilly, R.C. (1992). *The self-organization of spatially invariant representations.* Parallel Distributed Processing and Cognitive Neuroscience., PDP.CNS.92.5, Carnegie Mellon University, Department of Psychology.

O'Reilly, R.C., & Johnson, M. (1994). Object recognition and sensitive periods: A computational analysis of visual imprinting. *Neural Computation, 6(3),* 357-389.

O'Reilly, R.C., & McClelland, J.L. (1992). *The self-organization of spatially invariant representations* (PDP.CNS.92.5). Pittsburgh, PA: Carnegie Mellon University, Department of Psychology.

Osterhout, L., & Holcomb, P.J. (1993). Event-related potentials and syntactic anomaly: Evidence of anomaly detection during the perception of continuous speech. *Language and Cognitive Processes, 8(4),* 413-437.

Oyama, S. (1985). *The Ontongeny of Information: Developmental Systems and Evolution.* Cambridge: Cambridge University Press.

Oyama, S. (1992). The problem of change. In M. Johnson (Ed.), *Brain*

Development and Cognition: A Reader (pp.19-30). Oxford: Blackwell Publishers.

Pallas, S.L., & Sur, M. (1993). Visual projections induced into the auditory pathway of ferrets: II. Corticocortical connections of primary auditory cortex. *Journal of Comparative Neurology, 337(2)*, 317-33.

Pandya, D.N., & Yeterian, E.H. (1985). Architecture and connections of cortical association areas. In A. Peters & E.G. Jones (Eds.), *Cerebral Cortex: Vol. 4. Association and Auditory Cortices* (pp. 3-61). New York: Plenum.

Pandya, D.N., & Yeterian, E.H. (1990). Architecture and connections of cerebral cortex: Implications for brain evolution and function. In A.B. Scheibel & A.F. Wechsler (Eds.), *Neurobiology of Higher Cognitive Function.* New York: Guilford Press.

Parmelee, A.H., & Sigman, M.D. (1983). Perinatal brain development and behavior. In M.M. Haith & J. Campos (Eds.), *Infancy and the Biology of Development: Vol. 2. Handbook of Child Psychology.* New York: Wiley.

Pearlmutter, B.A. (1989). Learning state space trajectories in recurrent neural networks. *Neural Computation, 1*, 263-269.

Pennington, B.F. (1991). Genetic and neurological influences on reading disability: An overview. *Reading and Writing, 3(3)*, 191-201.

Perrett, D.I., Rolls, E.T., & Caan, W (1982). Visual neurones responsive to faces in the monkey temporal cortex. *Experimental Brain Research, 47(3)*, 329-342.

Peters, A. (1983). *The Units of Language Acquisition.* Cambridge: Cambridge University Press.

Petersen, S.E., Fiez, J.A., & Corbetta, M. (1991). Neuroimaging. *Current Opinion in Neurobiology, 2*, 217-222.

Petersen, S.E., Fox, P., Posner, M., Mintun, M., & Raichle, M. (1988). Positron emission tomographic studies of the cortical anatomy of single-word processing. *Nature, 331*, 585-589.

Petitto, L. (1987). On the autonomy of language and gesture: Evidence from the acquisition of personal pronouns in American Sign Language. *Cognition, 27*, 1-52.

Piaget, J. (1952). *The Origins of Intelligence in Children.* New York:

International University Press.

Piaget, J. (1955). Les stades du développement intellectuel de l'enfant et de l'adolescent. In P. Osterrieth et al. (Eds.), *Le problème des stades en psychologie de l'enfant.* Paris: Presses Universitaires France

Piatelli-Palmarini, M. (1989). Evolution, selection and cognition: From "learning" to parameter setting in biology and the study of language. *Cognition, 31,* 1-44.

Pineda, F.J. (1989). Recurrent back-propagation and the dynamical approach to adaptive neural computation. *Neural Computation, 1,* 161-172.

Pinker, S. (1979). Formal models of language learning. *Cognition, 7,* 217-283.

Pinker, S. (1991). Rules of language. *Science, 253,* 530-535.

Pinker, S. (1994a). On language. *Journal of Cognitive Neuroscience, 6(1),* 92-97.

Pinker, S. (1994b). *The Language Instinct: How the Mind Creates Language.* New York: William Morrow.

Pinker, S., & Bloom, P. (1990). Natural language and natural selection. *Behavioral and Brain Sciences, 13,* 707-784.

Pinker, S., & Prince, A. (1988). On language and connectionism: Analysis of a parallel distributed processing model of language acquisition. *Cognition, 28,* 73-193.

Plante, E. (1992, November). *The biology of developmental language disorders.* Panel presentation at the Annual Meeting of the American Speech-Language-Hearing Association, San Antonio, TX.

Plaut, D.C. (1995). Double dissociation without modularity: Evidence from connectionist neuropsychology. *Journal of Clinical and Experimental Neuropsychology, 17(2),* 291-321.

Plaut, D.C., McClelland, J.L., Seidenberg, M.S., & Patterson, K.E. (1994). *Parallel distributed processing and cognitive neuroscience* (Tech. Rep. PDP.CNS.94.5). Pittsburgh PA: Carnegie Mellon University, Department of Psychology.

Plaut, D.C., & Shallice, T. (1993). Deep dyslexia: A case study of connectionist neuropsychology. *Cognitive Neuropsychology, 10(5),* 377-500.

Plunkett, K. (1986). Learning strategies in two Danish children's language development. *Scandinavian Journal of Psychology, 27,* 64-73.

Plunkett, K. (1993). Lexical segmentation and vocabulary growth in early language acquisition. *Journal of Child Language, 20,* 43-60.

Plunkett, K., & Marchman, V. (1991). U-shaped learning and frequency effects in a multi-layered perceptron: Implications for child language acquisition. *Cognition, 38,* 43-102.

Plunkett, K., & Marchman, V. (1993). From rote learning to system building: Acquiring verb morphology in children and connectionist nets. *Cognition , 48,* 21-69.

Plunkett, K., & Sinha, C. (1992). Connectionism and developmental theory. *British Journal of Developmental Psychology, 10,* 209-254.

Plunkett, K., Sinha, C., Møller, M.F., & Strandsby, O. (1992). Symbol grounding of the emergence of symbols? Vocabulary growth in children and a connectionist net. *Connection Science, 4(3-4),* 293-312.

Poeppel, D. (in press). A critical review of PET studies of phonological processing. *Brain and Language.*

Poggio, T., & Girosi, F. (1990). Regularization algorithms for learning that are equivalent to multilayer networks. *Science 247,* 978-982.

Poizner, H., Klima, E., & Bellugi, U. (1987). *What the Hands Reveal About the Brain.* Cambridge, MA: Bradford Books, MIT Press.

Pons, T.P., Garraghty, P.E., & Mishkin, M. (1992). Serial and parallel processing of tactual information in somatosensory cortex of rhesus monkeys. *Journal of Neurophysiology, 68(2),* 518-27.

Pons, T.P., Garraghty, P.E., Ommaya, A.K., Kaas, J.H. Taub, E., & Mishkin M. (1991). Massive cortical reorganization after sensory deafferentation in adult macaques [see comments]. *Science, 252(5014),* 1857-60.

Port, R., & van Gelder, T. (1995). *Mind as Motion: Dynamical Perspectives on Behavior and Cognition.* Cambridge, MA: MIT Press.

Posner, M., & Keele, S. (1968). On the genesis of abstract ideas. *Journal of Experimental Psychology, 77,* 353-363.

Posner, M.I., & Petersen, S.E. (1990). The attention system of the human brain. *Annual Review of Neuroscience, 13,* 25-42.

Posner, M., & Snyder, C. (1975). Attention and cognitive control. In R. Solso (Ed.), *Information Processing and Cognition.* Hillsdale,

NJ: Erlbaum.

Premack, D. (1990). The infant's theory of self-propelled objects. *Cognition, 36*, 1-16.

Rakic, P. (1975). Timing of major ontogenetic events in the visual cortex of the rhesus monkey. In N. Buchwald & M. Brazier (Eds.), *Brain Mechanisms in Mental Retardation*. New York: Academic Press.

Rakic, P. (1988). Specification of cerebral cortical areas. *Science, 241*, 170-176.

Rakic, P., Bourgeois, J.P., Eckenhoff, M.F., Zecevic, N., & Goldman-Rakic, P.S. (1986). Concurrent overproduction of synapses in diverse regions of the primate cerebral cortex. *Science, 232*, 232-235.

Ramachandran, V.S. (1993). Behavioral and magnetoencephalographic correlates of plasticity in the adult human brain. *Proceedings of the National Academy of Sciences, 90*, 10413-10420.

Ramachandran, V.S. , Rogers-Ramachandran, D., & Stewart, M. (1992). Perceptual correlates of massive cortical reorganization. *Science, 258*, 1159-1160.

Rasmussen, T., & Milner, B. (1977). The role of early left brain injury in determining lateralization of cerebral speech functions. *Annals of the New York Academy of Sciences, 229*, 355-369.

Reilly, J. (1994, June). *Affective expression in infants with early focal brain damage*. Workshop at the 9th International Conference on Infant Studies, Paris.

Reilly, J., Bates, E., & Marchman, V. (in press). Narrative discourse in children with early focal brain injury. In M. Dennis (Ed.), *Special Issue, Discourse in Children with Anomalous Brain Development or Acquired Brain Injury, Brain and Language*.

Reznick, J.S., & Goldfield, B.A. (1992). Rapid change in lexical development in comprehension and production. *Developmental Psychology, 28(3)*, 406-413.

Rice, M. (Ed.). (in press). *Towards a Genetics of Language*. Hillsdale, NJ: Erlbaum.

Riva, D., & Cazzaniga, L. (1986). Late effects of unilateral brain lesions before and after the first year of life. *Neuropsychologia, 24*, 423-428.

Riva, D., Cazzaniga, L., Pantaleoni, C., Milani, N., & Fedrizzi, E.

(1986). *Journal of Pediatric Neurosciences, 2,* 239-250.

Rochat, Ph. (1984). *Vision et toucher chez l'enfant: La construction de paramètres spatiaux de l'objet.* New York: Peter Lang.

Rodier, P. (1980). Chronology of neuron development. *Developmental Medicine and Child Neurology, 22,* 525-545.

Roe, A.W., Pallas, S.L., Hahm, J.O., & Sur, M. (1990). A map of visual space induced in primary auditory cortex. *Science, 250 (4982),* 818-20.

Roeper, T. (1988). Grammatical principles of first language acquisition: Theory and evidence. In F.J. Newmeyer (Ed.), *Linguistics: The Cambridge Survey: Vol. II. Linguistic Theory: Extensions and Implications.* Cambridge: Cambridge University Press.

Rolls, E.T. (1989). Parallel distributed processing in the brain: Implications of the functional architecture of neuronal networks in the hippocampus. In R.G.M. Morris (Ed.), *Parellel Distributed Processing: Implications for Psychology and Neuroscience* (pp. 286-308). Oxford: Oxford University Press.

Rosenblatt, F. (1958). The perceptron: A probabilistic model for information storage and organization in the brain. *Psychological Review, 65,* 368-408. Reprinted in Anderson & Rosenfeld (1988), pp. 92-114.

Rosenblatt, F. (1959). Two theorems of statistical separability in the perceptron. In *Mechanisation of thought processes: Proceedings of a symposium held at the National Physical Laboratory, November 1958. Vol 1* (pp.421-456). London: HM Stationery Office.

Rosenblatt, F. (1962). *Principles of neurodynamics.* New York: Spartan.

Rumelhart, D.E., Hinton, G., & McClelland, J.L. (1986). A general framework for parallel distributed processing. In D.E. Rumelhart & J.L. McClelland (Eds.), *Parallel Distributed Processing: Explorations in the Microstructure of Cognition: Vol. 1. Foundations* (pp. 45-76). Cambridge, MA: MIT Press.

Rumelhart, D.E., Hinton, G., & Williams, R. (1986). Learning internal representations by error propagation. In D.E. Rumelhart & J.L. McClelland (Eds.), *Parallel Distributed Processing: Explorations in the Microstructure of Cognition: Vol. 1. Foundations* (pp. 318-362). Cambridge, MA: MIT Press.

Rumelhart D.E., & McClelland, J.L. (1981). Interactive processing through spreading activation. In A.M. Lesgold & C.A. Perfetti

(Eds.), *Interactive Processes in Reading*. Hillsdale, NJ: Erlbaum.

Rumelhart D.E., & McClelland, J.L. (1982). An interactive activation model of context effects in letter perception: Part 2. The contextual enhancement effect and some tests and extensions of the model. *Psychological Review, 89*, 60-94.

Rumelhart, D.E., & McClelland, J.L. (1986). On learning the past tenses of English verbs. In D.E. Rumelhart & J.L. McClelland (Eds.), *Parallel distributed processing: Explorations in the microstructure of cognition. Volume 2. Psychological and biological models* (pp. 216-271). Cambridge, MA: MIT Press.

Rumelhart, D.E., & Zipser, D. (1986). Feature discovery by competitive learning. In D.E. Rumelhart & J.L. McClelland (Eds.), *Parallel Distributed Processing: Explorations in the Microstructure of Cognition: Vol. 1. Foundations* (pp. 318-362). Cambridge, MA: MIT Press.

Russell, J. (1994). *Object permanence, Piagetian theory and connectionism*. Manuscript submitted for publication.

Sachs, J., Bard, B., & Johnson, M.L. (1981). Language learning with restricted input: Case studies of two hearing children of deaf parents. *Applied Psycholinguistics, 2(1)*, 33-54.

Sanides, F. (1972). Representation in the cerebral cortex and its areal lamination patterns. In G.H. Bourne (Ed.), *Structure and Function of Nervous Tissue* (Vol. 5, pp. 329-453). New York: Raven.

Sankoff, G. (1980). *The Social Life of Language*. Philadelphia: University of Pennsylvania Press.

Sargent, P.L., Nelson, C.A., & Carver, L.J. (1992). Cross species recognition in infants and adult humans: ERP and behavioral measures. *ICIS*, Miami.

Satz, P., Strauss, E., & Whitaker, H. (1990). The ontogeny of hemispheric specialization: Some old hypotheses revisited. *Brain and Language, 38:4*, 596-614.

Schlaggar, B.L., Fox, K., & O'Leary, D.D. (1993). Postsynaptic control of plasticity in developing somatosensory cortex [see comments]. *Nature, 1, 364*, 623-6.

Schlaggar, B.L., & O'Leary, D.D. (1991). Potential of visual cortex to develop an array of functional units unique to somatosensory cortex. *Science, 252*, 1556-1560.

Schneider, W., Noll, D.C., & Cohen, J.D. (1993). Functional topographic mapping of the cortical ribbon in human vision with conventional MRI scanners. *Nature, 365,* 150-153.

Schneider, W., & Shiffrin, R. (1977). Controlled and automatic human information processing: 1. Detection, search and attention. *Psychological Review, 84,* 321-330.

Seidenberg, M.S., & McClelland, J.L. (1989). A distributed developmental model of visual word recognition and naming. *Psychological Review, 96,* 523-568.

Sejnowski, T.J., & Rosenberg, C.R. (1987). *NETtalk: A parallel network that learns to read aloud* (Electrical Engineering and Computer Science Tech. Rep. JHU/EECS-86/01). Baltimore, MD: The Johns Hopkins University. Reprinted in Anderson & Rosenfeld (1988), pp. 663-672.

Sereno, M.I. (1990). Language and the primate brain. *Center for Research in Language Newsletter Vol. 4* No. 4. University of California, San Diego.

Sereno, M.I., & Sereno, M.E. (1991). Learning to see rotation and dilation with a Hebb rule. In R.P. Lippman, J. Moody, & D.S. Touretzky (Eds.), *Advances in Neural Information- Processing Systems 3* (pp. 320-326). San Mateo, CA: Morgan Kaufman.

Shallice, T. (1988). *From Neuropsychology to Mental Structure.* New York: Cambridge University Press.

Shankweiler, D., Crain, S., Gorrell, P., & Tuller, B. (1989). Reception of language in Broca's aphasia. *Language and Cognitive Processes, 4(1),* 1-33.

Shatz, C.J. (1992a). How are specific connections formed between thalamus and cortex? *Current Opinion in Neurobiology, 2(1),* 78-82.

Shatz, C.J. (1992b). Dividing up the neocortex [see comments]. *Science, 258 (5080),* 237-8.

Shatz, C.J. (1992c). The developing brain. *Scientific American, 267(3),* 60-7.

Shiffrin, R. M., & Schneider, W. (1977). Controlled and automatic human information processing: II. Perceptual learning, automatic attending and a general theory. *Psychological Review, 84,* 127-190.

Shimamura, A., Janowsky, J., & Squire, L. (1990). Memory for the temporal order of events in patients with frontal lobe lesions and amnesic patients. *Neuropsychologia, 28(8)*, 803-813.

Shrager, J., & Johnson, M.H. (in press). Modeling the development of cortical function. In B. Julesz & I. Kovacs (Eds.), *Maturational windows and cortical plasticity in human development: Is there a reason for an optimistic view?* Reading, MA: Addison Wesley.

Shultz, T.R., Schmidt, W.C., Buckingham, D., & Mareschal, D. (1995). Modeling cognitive development with a generative connectionist algorithm. In G. Halford & T. Simon (Eds.), *Developing Cognitive Competence: New Approaches to Process Modelling.* Hillsdale, NJ: Erlbaum.

Siegler, R. (1981). Developmental sequences within and between concepts. *Monographs of the Society for Research in Child Development, 46*, Whole No. 2.

Siegler, R.S., & Munakata, Y. (1993). Beyond the immaculate transition: Advances in the understanding of change. *SRCD Newsletter*, Winter 1993, 3-13.

Simonds, R.J., & Scheibel, A.B. (1989). The postnatal development of the motor speech area: A preliminary study. *Brain and Language, 37*, 42-58.

Slater, A. (1992). The visual constancies in early infancy. *The Irish Journal of Psychology, 13(4)*, 411-424.

Slater, A., & Morison, V. (1991). Visual attention and memory at birth. In M.J. Weiss & P. Velazo (Eds.), *Newborn Attention.* Norwood, NJ: Ablex.

Slater, A., Morison, V., & Rose, D. (1983). Perception of shape by the newborn baby. *British Journal of Developmental Psychology, 1*, 135-142.

Smith, L.B. (1989). A model of perceptual classification in children and adults. *Psychological Review, 96*, 125-144.

Smith, L.B. (1996). Paper presented at the Conference on Human Development, University of California, San Diego. January, 1996.

Smith, L.B., & Thelen, E. (Eds.). (1993). *A Dynamic Systems Approach to Development: Applications.* Cambridge, MA: MIT Press.

Smith, N., & Tsimpli, I.-M. (1995). *The Mind of a Savant: Language Learning and Modularity.* Oxford, UK : Blackwell.

Spelke, E.S. (1991). Physical knowledge in infancy: reflections on

Piaget's theory. In S. Carey & R. Gelman, (Eds.), *Epigenesis of the Mind: Essays in biology and knowledge*. Hillsdale, NJ: Erlbaum.

Spelke, E.S. (1994). Initial knowledge: Six suggestions. *Cognition, 50,* 431-445.

Spelke, E.S., Breinlinger, K., Macomber, J., & Jacobson, K. (1992). Origins of knowledge. *Psychological Review, 99(4),* 605-632.

Stanfield, B.B., & O'Leary, D.D. (1985). Fetal occipital cortical neurones transplanted to the rostral cortex can extend and maintain a pyramidal tract axon. *Nature, 313(5998),* 135-137.

Starkey, P., Spelke, E.S., & Gelman, R. (1990). Numerical abstraction by human infants. *Cognition, 36,* 97-127.

Stiles, J. (1994, June). *Spatial cognitive development in children with focal brain injury.* Workshop at the 9th International Conference on Infant Studies, Paris.

Stiles, J. (in press). Plasticity and development: Evidence from children with early focal brain injury. In B. Julesz & I. Kovacs (Eds.), *Maturational Windows and Cortical Plasticity in Human Development: Is There Reason for an Optimistic View?* Addison-Wesley .

Stiles, J., & Thal, D. (1993). Linguistic and spatial cognitive development following early focal brain injury: Patterns of deficit and recovery. In M. Johnson (Ed.), *Brain Development and Cognition: A Reader* (pp.643-664). Oxford: Blackwell Publishers.

Stornetta, W.S., Hogg, T., & Huberman, B.A. (1988). A dynamical approach to temporal pattern processing (pp. 750-759). In D.Z. Anderson (Ed.), *Neural Information Processing Systems.* (Denver 1987). New York: American Institute of Physics.

Strauss, M.S., & Curtis, L.E. (1981). Infants' perception of numerosity. *Child Development, 52,* 1146-1152.

Sur, M., Garraghty, P.E., & Roe, A.W. (1988). Experimentally induced visual projections into auditory thalamus and cortex. *Science, 242,* 1437-1441.

Sur, M., Pallas, S.L., & Roe, A.W. (1990). Cross-modal plasticity in cortical development: differentiation and specification of sensory neocortex. *Trends in Neuroscience, 13,* 227-233.

Sutherland, S. (1993, March 7). Evolution between the ears. *New*

York Times.

Sutton, R.S. (1984). *Temporal credit assignment in reinforcement learning.* Ph.D. Thesis, University of Massachusetts, Amherst.

Szathmary, E., & Smith, E.M. (1995). The major evolutionary transitions. *Nature, 374(6519)*, 227-232.

Tallal, P. (1988). Developmental language disorders. In J.F. Kavanagh & T.J. Truss, Jr. (Eds.), *Learning Disabilities: Proceedings of the National Conference*, 181-272. Parkton, MD: York Press.

Tallal, P., Miller, G., Bedi, G., Jenkins, W.M., Wang, X., Nagarajan, S.S., & Merzenich, M.M. (1995). Training with temporally modified speech results in dramatic improvements in speech perception. *Society for Neuroscience Abstracts, 21(1)*, 173.

Tallal, P., Ross, R., & Curtiss, S. (1989). Familial aggregation in specific language impairment. *Journal of Speech and Hearing Disorders, 54*, 157-173.

Tallal, P., Sainburg, R.L., & Jernigan, T. (1991). The neuropathology of developmental dysphasia: Behavioral, morphological, and physiological evidence for a pervasive temporal processing disorder. *Reading and Writing, 3*, 363-377.

Tallal, P., Stark, R., & Mellits, D. (1985). Identification of language-impaired children on the basis of rapid perception and production skills. *Brain and Language, 25*, 314-322.

Tallal, P., Townsend, J., Curtiss, S., & Wulfeck, B. (1991). Phenotypic profiles of language-impaired children based on genetic/family history. *Brain and Language, 41*, 81-95.

Tank, D.W., & Hopfield, J.J. (1987). Neural computation by time compression. *Proceedings of the National Academy of Sciences USA, 84*, 1896-1900.

Templin, M.C. (1957). *Certain Language Skills in Children: Their Development and Interrelationships.* Minneapolis, MN: University of Minnesota Press.

Thal, D., Marchman, V., Stiles, J., Aram, D., Trauner, D., Nass, R., & Bates, E. (1991). Early lexical development in children with focal brain injury. *Brain and Language, 40*, 491-527.

Thal, D., Tobias, S., & Morrison, D. (1991). Language and gesture in late talkers: A one-year follow-up. *Journal of Speech and Hearing*

Research, 34(3), 604-612.

Thal, D., Wulfeck, B., & Reilly, J. (1993, November). *Brain and language: A cross-population perspective.* Symposium presented at the Annual Meeting of the American Speech-Language-Hearing Association, Anaheim, CA.

Thatcher, R.W. (1992). Cyclic cortical reorganization during early childhood. *Brain and Cognition, 20,* 24-50.

Thelen, E. (1985). Developmental origins of motor coordination: Leg movements in human infants. *Developmental Psychobiology, 18,* 323-333.

Thelen, E. (1994). Three-month old infants can learn task-specific patterns of inter-limb coordination. *Psychological Science, 5,* 280-285.

Thelen, E., & Smith, L.B. (1994). *A dynamic Systems Approach to the Development of Cognition and Action.* Cambridge, MA: MIT Press.

Thomas, C.E., Tramo, M.J., Loftus, W.C., Newton, C.H., & Gazzaniga, M.S. (1990). Gross morphometry of frontal, parietal, and temporal cortex in monozygotic twins. *Society for Neuroscience Abstracts, 16,* 1151.

Thompson, D.W. (1917/1968). *On Growth and Form* (2d ed., reprinted). Cambridge [Eng.] University Press. (Original work published 1917.)

Townsend, J, Wulfeck, B., Nichols, S., & Koch, L. (1995). *Attentional deficits in children with developmental language disorder* (Tech. Rep. No. CND-9503). La Jolla: University of California, San Diego, Center for Research in Language, Project in Cognitive and Neural Development.

Tramo, M.J., Loftus, W.C., Thomas, C.E., Green, R.L., Mott, L.A., & Gazzaniga, M.S. (1994) *The surface area of human cerebral cortex and its gross morphological subdivisions.* Manuscript, University of California, Davis.

Trauner D., Chase, C., Walker, P., & Wulfeck, B. (1993). Neurologic profiles of infants and children after perinatal stroke. *Pediatric Neurology, 9(5),* 383-386.

Trauner, D., Wulfeck, B., Tallal, P., & Hesselink, J. (1995). *Neurologic and MRI profiles of language-impaired children* (Tech. Rep. No. CND-9513). La Jolla: University of California, San Diego, Center for Research in Language, Project in Cognitive and Neural Development.

Turing, A.M. (1952). The chemical basis of morphogenesis. *Philosophical Transactions of the Royal Society, London, B 237*, 37.

Tyler, L.K. (1992). *Spoken Language Comprehension: An Experimental Approach to Normal and Disordered Processing.* Cambridge, MA: MIT Press.

Ungerleider, L.G., & Mishkin, M. (1982). Two cortical visual systems. In D.J. Ingle, M.A. Goodale, & J.W. Mansfield (Eds.), *Analysis of Visual Behavior.* Cambridge, MA: MIT Press.

Van der Lely, H.K.J. (1994). Canonical linking rules: Forward versus reverse linking in normally developing and specifically language-impaired children. *Cognition, 51(1)*, 29-72.

Van Geert, P. (1991). A dynamic systems model of cognitive and language growth. *Psychological Review, 98*, 3-53.

Van Geert, P. (1994). *Dynamic Systems of Development: Change Between Complexity and Chaos.* New York/London: Harvester Wheatsheaf.

Van Gelder, T. (1990). Compositionality: A connectionist variation on a classical theme. *Cognitive Science, 14*, 355-384.

Vargha-Khadem, F., Isaacs, E.B., Papaleloudi, H., Polkey, C.E., & Wilson, J. (1991). Development of language in 6 hemispherectomized patients. *Brain, 114*, 473-495.

Vargha-Khadem, F., Isaacs, E.B., van der Werf, S., Robb, S., & Wilson, J. (1992). Development of intelligence and memory in children with hemiplegic cerebral palsy: The deleterious consequences of early seizures. *Brain, 115*, 315-329.

Vargha-Khadem, F., O'Gorman, A., & Watters, G. (1985). Aphasia and handedness in relation to hemispheric side, age at injury and severity of cerebral lesion during childhood. *Brain, 108*, 677-696.

Vargha-Khadem, F., & Passingham, R. (1990). Speech and language defects. *Nature, 346*, 226.

Vargha-Khadem, F., & Polkey, C.E. (1992). A review of cognitive outcome after hemidecortication in humans. In F.D. Rose & D.A. Johnson (Eds.), *Recovery from Brain Damage: Advances in Experimental Medicine and Biology*: Vol. 325 (pp. 137-151).

Reflections and Directions. New York: Plenum Press.

Vargha-Khadem, F., Watkins, K., Alcock, K., Fletcher, P., & Passing-ham (1995). Praxic and nonverbal cognitive deficits in a large family with a genetically transmitted speech and language disorder. *Proceedings of the National Academy of Sciences USA,* 92, 930-933.

Veraart, C., DeVolder, A., Wanet-Defalque, M., Bol, A., Michel, C., & Goffinet, A. (1990). Glucose utilization in human visual cortex is abnormally elevated in blindness of early onset but decreased in blindness of late onset. *Brain Research, 510,* 115-121.

Vinter, A. (1986). The role of movement in eliciting early imitation. *Child Development, 57,* 66-71.

Volpe, J.J. (1987). *Neurology of the Newborn* (2nd ed.). Philadelphia: Saunders.

Volterra, V., & Erting, C. (Eds.). (1990). *From Gesture to Language in Hearing and Deaf Children.* New York: Springer-Verlag.

Volterra, V., Sabbadini, L., Capirci, O., Pezzini, G., & Osella, T. (1995). Language development in Italian children with Will-iams Syndrome. *Journal of Genetic Counseling, 6(1),* 137-138.

Von Hofsten, C (1989). Transition mechanisms in sensori-motor development. In A. de Ribaupierre (Ed.), *Transition Mecha-nisms in Child Development: The Longitudinal Perspective* (pp. 223-259). Cambridge, UK: Cambridge University Press.

Waddington, C. H. (1975). *The Evolution of an Evolutionist.* Ithaca, NY: Cornell University Press.

Wanet-Defalque, M., Veraart, C., DeVolder, A., Metz, R., Michel, C., Dooms, G., & Goffinet, A. (1988). High metabolic activity in the visual cortex of early blind human subjects. *Brain Research,* 446 (2), 369-373.

Wang, P.P., & Bellugi, U. (1994). Evidence from two genetic syn-dromes for a dissociation between verbal and visual-spatial short-term memory. *Journal of Clinical and Experimental Neurop-sychology, 16(2),* 317-322.

Webster, M.J., Bachevalier, J., & Ungerleider, L.G. (in press). Devel-opment and plasticity of visual memory circuits. In B. Julesz & I. Kovacs (Eds.), *Maturational Windows and Adult Cortical Plas-ticity in Human Development: Is There Reason for an Optimistic*

View? Reading, MA: Addison-Wesley.

Welsh, M.C., & Pennington, B.F. (1988). Assessing frontal lobe functioning in children: Views from developmental psychology. *Developmental Neuropsychology, 4,* 199-230.

Werker, J., & Tees, R. (1984). Cross-language speech perception: Evidence for perceptual reorganization during the first year of life. *Infant Behavior and Development, 7,* 49-63.

Werner, H. (1948). *Comparative Psychology of Mental Development,* with a foreword by Gordon W. Allport (revised edition.). New York: International Universities Press.

Wexler, K., & Culicover, P.W. (1980). *Formal Principles of Language Acquisition.* Cambridge, MA: MIT Press.

Widrow, G., & Hoff, M.E. (1960). Adaptive switching circuits. *Institute of Radio Engineers, Western Electronic Show and Convention, Convention Record, Part 4,* 96-104.

Williams, R.J., & Zipser, D. (1989). A learning algorithm for continually running fully recurrent neural networks. *Neural Computation, 1,* 270-280.

Woods, B.T., & Teuber, H.L. (1978). Changing patterns of childhood aphasia. *Annals of Neurology, 3,* 272-280.

Wulfeck, B. (1988). Grammaticality judgments and sentence comprehension in agrammatic aphasia. *Journal of Speech and Hearing Research, 31,* 72-81.

Wulfeck, B., Trauner, D., & Tallal, P. (1991). Neurologic, cognitive and linguistic features of infants after early stroke. *Pediatric Neurology, 7,* 266-269.

Wynn, K. (1992). Addition and subtraction by human infants. *Nature, 358,* 749-750.

Yakovlev, P., & Lecours, A. (1967). The myelino-genetic cycle of regional maturation of the brain. In A. Minkowski (Ed.), *Regional Development of the Brain in Early Life.* Philadelphia: Davis Co.

Younger, B.A., & Cohen, L.B. (1982, March). *Infant perception of correlated attributes.* Paper presented at the International Conference on Infant Studies, Austin.

Zaidel, E. (1990). Language functions in the two hemispheres following complete cerebral commisurotomy and hemispherectomy. In F. Boller & J. Grafman (Eds.), *Handbook of*

Neuropsychology (Vol. 4). Amsterdam: Elsevier Science Publishers, N.V.

Zelazo, P.D., & Reznick, J.S. (1991). Age-related asynchrony of knowledge and action. *Child Development, 62,* 719-735.

Zemel, R.S., & Hinton, G.E. (1995). Learning population codes by minimizing description length. *Neural Computation, 7(3),* 549-564).

Zipser, D., & Andersen, R.A. (1988). A back-propagation programmed network that simulates response properties of a subset of posterior parietal neurons. *Nature, 331,* 679-684.

Zurif, E., & Caramazza, A. (1976). Psycholinguistic structures in aphasia: Studies in syntax and semantics. In H. & H.A. Whitaker (Eds.), *Studies in Neurolinguistics* (Vol. I). New York: Academic Press.

Zurif, E., Swinney, D., Prather, P., Solomon, J., & Bushell, C. (1993). An on-line analysis of syntactic processing in Broca's and Wernicke's aphasia. *Brain and Language, 45,* 448-464.

Subject index

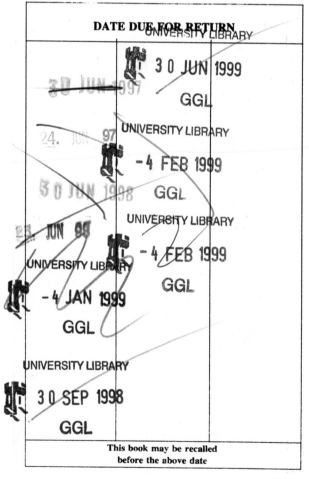